Advances in Experimental Medicine and Biology

Volume 1017

More information about this series at http://www.springer.com/series/5584

Guang-Hui Dong
Editor

Ambient Air Pollution and Health Impact in China

Springer

Editor
Guang-Hui Dong
Guangzhou Key Laboratory of Environmental Pollution and Health Risk Assessment
Department of Preventive Medicine
School of Public Health
Sun Yat-sen University
Guangzhou, Guangdong, China

ISSN 0065-2598 ISSN 2214-8019 (electronic)
Advances in Experimental Medicine and Biology
ISBN 978-981-10-5656-7 ISBN 978-981-10-5657-4 (eBook)
DOI 10.1007/978-981-10-5657-4

Library of Congress Control Number: 2017955384

Printed on acid-free paper

This Springer imprint is published by Springer Nature
The registered company is Springer Nature Singapore Pte Ltd.
The registered company address is: 152 Beach Road, #21-01/04 Gateway East, Singapore 189721, Singapore

Contents

Contributors

Zhipeng Bai State Key Laboratory of Environmental Criteria and Risk Assessment, Chinese Research Academy of Environmental Sciences, Beijing, China

Atmospheric Chemistry & Aerosol Division, Chinese Research Academy of Environmental Sciences, Beijing, China

Zhongqiang Cao Wuhan Children's Hospital (Wuhan Maternal and Child Healthcare Hospital), Tongji Medical College, Huazhong University of Science & Technology, Wuhan, China

Renjie Chen Department of Environmental Health, School of Public Health, Fudan University, Shanghai, China

Wen Chen Guangzhou Key Laboratory of Environmental Pollution and Health Risk Assessment, Department of Preventive Medicine, School of Public Health, Sun Yat-sen University, Guangzhou, Guangdong, China

Yawen Chen Wuhan Children's Hospital (Wuhan Maternal and Child Healthcare Hospital), Tongji Medical College, Huazhong University of Science & Technology, Wuhan, China

Zan-Xiong Chen Maternal and Children's Health Hospital of Maoming City, Maoming, Guangdong Province, China

Furong Deng Department of Occupational and Environmental Health Sciences, School of Public Health, Peking University, Beijing, China

Guang-Hui Dong Guangzhou Key Laboratory of Environmental Pollution and Health Risk Assessment, Department of Preventive Medicine, School of Public Health, Sun Yat-sen University, Guangzhou, Guangdong, China

Xinbiao Guo Department of Occupational and Environmental Health Sciences, School of Public Health, Peking University, Beijing, China

Bin Han State Key Laboratory of Environmental Criteria and Risk Assessment, Chinese Research Academy of Environmental Sciences, Beijing, China

Atmospheric Chemistry & Aerosol Division, Chinese Research Academy of Environmental Sciences, Beijing, China

Yiling He Department of Health Policy and Management, School of Public Health, Sun Yat-sen University, Guangzhou, China

Li-Wen Hu Guangzhou Key Laboratory of Environmental Pollution and Health Risk Assessment, Department of Preventive Medicine, School of Public Health, Sun Yat-sen University, Guangzhou, Guangdong, China

Cunrui Huang Department of Health Policy and Management, School of Public Health, Sun Yat-sen University, Guangzhou, China

Haidong Kan Department of Environmental Health, School of Public Health, Fudan University, Shanghai, China

Min-Li Kong Maternal and Children's Health Hospital of Maoming City, Maoming, Guangdong Province, China

Wayne R. Lawrence Department of Epidemiology and Biostatistics, School of Public Health, State University of New York, Albany, NY, USA

Xing Li Guangdong Provincial Institute of Public Health, Guangdong Provincial Center for Disease Control and Prevention, Guangzhou, China

Hualiang Lin Guangdong Provincial Institute of Public Health, Guangdong Provincial Center for Disease Control and Prevention, Guangzhou, China

Wei-Wei Lin Guangzhou Key Laboratory of Environmental Pollution and Health Risk Assessment, Department of Preventive Medicine, School of Public Health, Sun Yat-sen University, Guangzhou, Guangdong, China

Tao Liu Guangdong Provincial Institute of Public Health, Guangdong Provincial Center for Disease Control and Prevention, Guangzhou, China

Yimin Liu Laboratory of Occupational Environment and Health Effects, Guangzhou Key Medical Discipline of Occupational Health Guardianship, Guangzhou Prevention and Treatment Center for Occupational Diseases, Guangzhou Twelfth People's Hospital, Guangzhou, China

Rui Ma Department of Health Policy and Management, School of Public Health, Sun Yat-sen University, Guangzhou, China

Wenjun Ma Guangdong Provincial Institute of Public Health, Guangdong Provincial Center for Disease Control and Prevention, Guangzhou, China

Yue Niu Department of Environmental Health, School of Public Health, Fudan University, Shanghai, China

Lihong Ren State Key Laboratory of Environmental Criteria and Risk Assessment, Chinese Research Academy of Environmental Sciences, Beijing, China

Atmospheric Chemistry & Aerosol Division, Chinese Research Academy of Environmental Sciences, Beijing, China

Meng Ren Department of Health Policy and Management, School of Public Health, Sun Yat-sen University, Guangzhou, China

Shengzhi Sun School of Public Health, Li Ka Shing Faculty of Medicine, The University of Hong Kong, Hong Kong, China

Yafei Tan Wuhan Children's Hospital (Wuhan Maternal and Child Healthcare Hospital), Tongji Medical College, Huazhong University of Science & Technology, Wuhan, China

Linwei Tian School of Public Health, Li Ka Shing Faculty of Medicine, The University of Hong Kong, Hong Kong, China

Qiong Wang Department of Health Policy and Management, School of Public Health, Sun Yat-sen University, Guangzhou, China

Suhan Wang Department of Health Policy and Management, School of Public Health, Sun Yat-sen University, Guangzhou, China

Xiaojie Wang School of Public Health, Guangdong Pharmaceutical University, Guangzhou, China

Jianpeng Xiao Guangdong Provincial Institute of Public Health, Guangdong Provincial Center for Disease Control and Prevention, Guangzhou, China

Yan-Qi Xie Maternal and Children's Health Hospital of Maoming City, Maoming, Guangdong Province, China

Bo-Yi Yang Guangzhou Key Laboratory of Environmental Pollution and Health Risk Assessment, Department of Preventive Medicine, School of Public Health, Sun Yat-sen University, Guangzhou, Guangdong, China

Di Yang Department of Occupational and Environmental Health Sciences, School of Public Health, Peking University, Beijing, China

Rong Yang Wuhan Children's Hospital (Wuhan Maternal and Child Healthcare Hospital), Tongji Medical College, Huazhong University of Science & Technology, Wuhan, China

Wen Yang State Key Laboratory of Environmental Criteria and Risk Assessment, Chinese Research Academy of Environmental Sciences, Beijing, China

Atmospheric Chemistry & Aerosol Division, Chinese Research Academy of Environmental Sciences, Beijing, China

Xuan Yang Department of Occupational and Environmental Health Sciences, School of Public Health, Peking University, Beijing, China

Weilin Zeng Guangdong Provincial Institute of Public Health, Guangdong Provincial Center for Disease Control and Prevention, Guangzhou, China

Xiao-Wen Zeng Guangzhou Key Laboratory of Environmental Pollution and Health Risk Assessment, Department of Preventive Medicine, School of Public Health, Sun Yat-sen University, Guangzhou, Guangdong, China

Bin Zhang Wuhan Children's Hospital (Wuhan Maternal and Child Healthcare Hospital), Tongji Medical College, Huazhong University of Science & Technology, Wuhan, China

Jinzhu Zhao Wuhan Children's Hospital (Wuhan Maternal and Child Healthcare Hospital), Tongji Medical College, Huazhong University of Science & Technology, Wuhan, China

Chapter 1
Urgency to Assess the Health Impact of Ambient Air Pollution in China

Bo-Yi Yang, Yimin Liu, Li-Wen Hu, Xiao-Wen Zeng, and Guang-Hui Dong

Abstract As the world's second-largest economy, China is going on suffering from environmental pollution, especially for ambient air pollution, which has become a major threat to public health; public awareness of the detrimental effects of air pollution on health is increasing—particularly in relation to haze days. Considering the nonlinear relationship of ambient air pollution exposure and health impacts, and the differences in specific sources of air pollution with those in North America and Europe, conducting health impact assessments of ambient air pollution in China has thus become an urgent task for public health practitioners. Systematic review of the health effects of exposure to ambient air pollution from quantitative studies conducted in Chinese could provide vital information for epidemiology-based health impact assessments and the implementation of a national environmental protection policy.

Keywords Ambient air pollution • Health impact • China

With the rapid economic development, China has become the world's second-largest economy. However, behind this beautiful scene, environmental pollution induced by urbanization and industrialization has led to devastating health impacts in Chinese population [5]. According to the reports of the World Health Organization, the number of patients and deaths because of lungs, stomach, liver, and esophagus diseases accounted for 30%, 40%, 50%, and 50% of the global total, respectively

---

B.-Y. Yang • L.-W. Hu • X.-W. Zeng • G.-H. Dong (✉)
Guangzhou Key Laboratory of Environmental Pollution and Health Risk Assessment, Department of Preventive Medicine, School of Public Health, Sun Yat-sen University, 74 Zhongshan 2nd Road, Yuexiu District, Guangzhou 510080, Guangdong, China
e-mail: donggh5@mail.sysu.edu.cn; donggh512@hotmail.com

Y. Liu
Laboratory of Occupational Environment and Health Effects, Guangzhou Key Medical Discipline of Occupational Health Guardianship, Guangzhou Prevention and Treatment Center for Occupational Diseases, Guangzhou Twelfth People's Hospital, Guangzhou 510620, China

G.-H. Dong (ed.), *Ambient Air Pollution and Health Impact in China*, Advances in Experimental Medicine and Biology 1017, DOI 10.1007/978-981-10-5657-4_1

[22]. One culprit behind these figures is the deteriorating environmental pollution. According to an authoritative estimate made by the World Bank, the annual losses caused by environmental pollution amounted to about 10% of total gross domestic product (GDP) [17, 21].

Compared with the pollution condition of other environmental media, air pollution in China is becoming more and more serious and has aroused great attention for the frequent haze events in recent years [18, 19]. For example, results of the Global Burden of Diseases, Injuries, and Risk Factors Study 2010 [9] indicated that ambient air pollution has ranked as the fourth in terms of the age-standardized disability-adjusted life years (DALYs) rate in 2010 [18]. In 2014–2015, the total DALYs simply caused by particles with an aerodynamic diameter of ≤ 2.5 μm ($PM_{2.5}$) and ≤ 10 μm (PM_{10}) in China was 7.2 and 20.66 million, and mortality and chronic bronchitis shared about 93% of the total DALYs for PM_{10} [16].

How dangerous is the ambient air pollution in China? It may be more vivid and meaningful to compare air pollution to cigarette smoking than just citing the numbers of yearly deaths. As shown in Fig. 1.1, we may have an idea of air pollution in terms of cigarettes equivalent based on the satellite monitoring data by applying

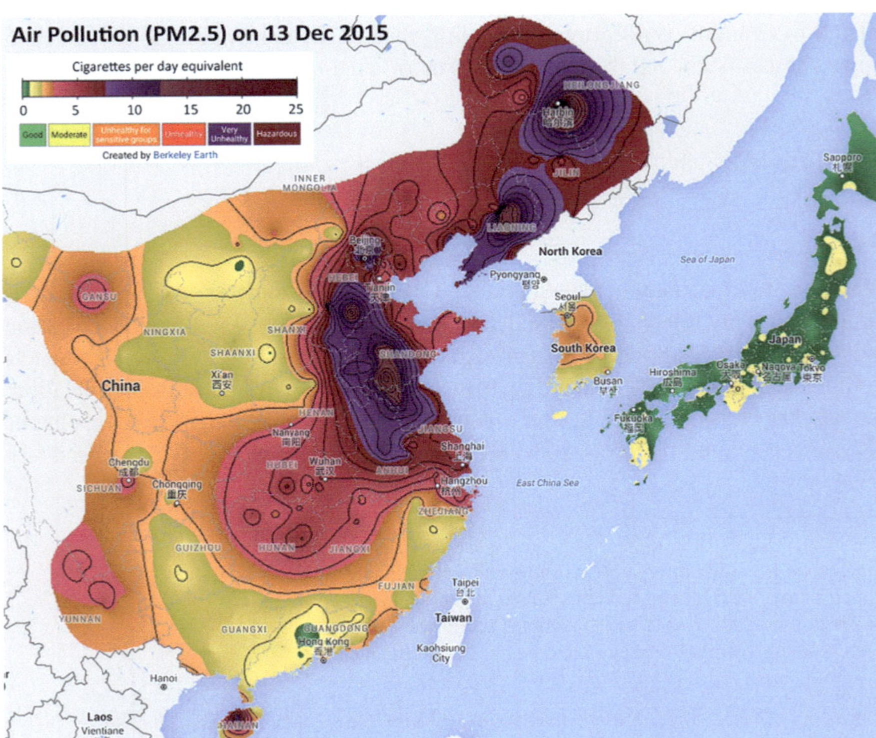

Fig. 1.1 Map of air pollution and cigarette equivalence in East China (Cited from Berkeley Earth: http://berkeleyearth.org/air-pollution-and-cigarette-equivalence/)

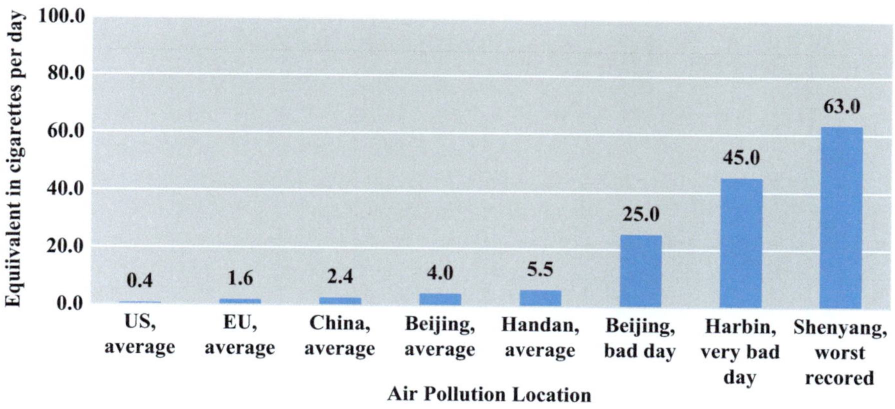

Fig. 1.2 Air pollution approximated as cigarettes equivalent in China with the United States and Europe (Cited from Berkeley Earth: http://berkeleyearth.org/air-pollution-and-cigarette-equivalence/)

kriging interpolation to derive pollution maps for China [15]. For example, in 2015, the average $PM_{2.5}$ in Beijing over the year was about 85 μg/m^3, equivalent to about four cigarettes per day. The average value in the industrial city of Handan, about 200 km south of Beijing, was about 120 μg/m^3, equivalent to 5.5 cigarettes/day. When in Beijing the level rose to 550 μg/m^3, it was equivalent to 25 cigarettes per day. In Harbin, the air pollution had reached the limit of the scale, 999 μg/m^3. That would be equivalent to 45 cigarettes per day.

According to the rule of thumb, one cigarette per day is the rough equivalent of a $PM_{2.5}$ level of 22 μg/m^3; the EPA estimates that the average air pollution in the United States in 2013 was 9.0 μg/m^3 indicating equivalent to 0.41 cigarettes per day for every person in the United States (Fig. 1.2). For Europe, air pollution is equivalent in detrimental health effects to smoking 1.6 cigarettes per day. In China, the numbers are far worse; on bad days, the health effects of air pollution are comparable to the harm done smoking two packs per day (40 cigarettes). Moreover, a recent peak reported in the city of Shenyang set a new record of 1,400 μg/m^3, equivalent to over three packs of cigarettes per day for every man, woman, and child living there.

It is well known that the relationship of air pollution with health effect is nonlinear [4, 13, 14]. As shown in Fig. 1.3, at low pollution levels, we can find a greater rake ratio for the relationship of air pollution with health effects; however, at a high pollution level, the slope is becoming a little flat. For example, based on the data from a global analysis, it showed that dose–response relationships between air pollution and chronic obstructive pulmonary disease (COPD) mortality are nonlinear, with mortality risks increasing rapidly at low $PM_{2.5}$ levels (<100 μg/m^3) and reaching a plateau at higher levels (>300 μg/m^3) [4]. This plateau level can be applied to the Chinese population with regard to air pollutant levels throughout China. So, it is not surprising that the results from the low air pollution levels are inconsistent with

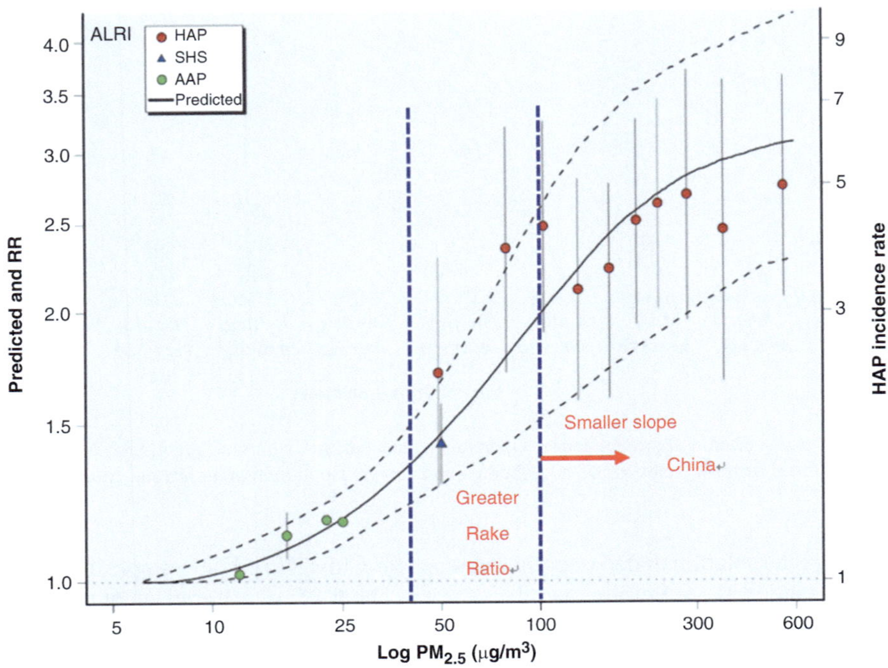

Fig. 1.3 Nonlinear relationship of air pollution with health effects (Cited from Burnett et al. [4])

the results from the high pollution levels. Although, many studies have assessed the effects of ambient air pollution on human health, most of these studies were from the developed countries where ambient air pollution levels were usually below 40 μg/m^3. Furthermore, the nonlinear relationship indicates that the associations from the developed countries may be not suitable for Chinese population.

With the transformation of traditional ideas that ambient air pollution and the associated health effects have typically been regarded as local or regional problems, with local or regional solutions, it is increasingly recognized that air quality in a given location can be substantially affected by atmospheric transport from distant sources, including sources from other continents [1, 6, 7, 10–12, 20], and this transport of pollution indicates that health impacts of air pollution are not only a local issue [2, 3, 8, 23]. For example, recently, Zhang et al. [24] estimated premature mortality caused by $PM_{2.5}$ pollution as a result of atmospheric transport and the production and consumption of goods and services in different world regions. They found that, of the 3.45 million premature deaths related to $PM_{2.5}$ pollution in 2007 worldwide, about 12% (411,100 deaths) were related to air pollutants emitted in a region of the world other than that in which the death occurred, and about 22% (762,400 deaths) were associated with goods and services produced in one region for consumption in another. For example, $PM_{2.5}$ pollution produced in China in 2007 is linked to more than 64,800 premature deaths in regions other than China, including more than 3,100 premature deaths in Western Europe and the United States. On the other hand, consumption in Western Europe and the United States is linked to more than 108,600 premature deaths in China.

Considering the above contents, and the fact that air pollution in China is mainly different from that in developed countries in terms of its magnitude, it will be very meaningful to show the health impact of air pollution in Chinese population. So, the aim of this book is to provide up-to-date review of the magnitude of adverse health effects of ambient air pollution in Chinese population. To the end, epidemiological evidence dating back to the 1990s, including data on mortality, morbidity, and hospital utilization for all non-accidental causes, cardiovascular diseases, stroke, respiratory diseases, mental health, and birth deficit caused by ambient PM_{10}, $PM_{2.5}$, sulfur dioxide, nitrogen dioxide, ozone, and carbon monoxide pollution in all age groups, was systematically assessed.

Acknowledgments This study was funded by grants from the National Key Research and Development Program of China (2016YFC0207000), the National Natural Science Foundation of China (No. 81472936, 81673127, and 81673128), the Guangdong Province Natural Science Foundation (2014A050503027), and the Fundamental Research Funds for the Central Universities (No. 16ykzd02, 17ykpy16).

References

1. Akimoto H. Global air quality and pollution. Science. 2003;302:1716–9.
2. Anenberg SC, West JJ, Yu H, Chin M, Schulz M, Bergmann D, Bey I, Bian H, Diehl T, Fiore A, Hess P, Marmer E, Montanaro V, Park R, Shindell D, Takemura T, Dentener F. Impacts of intercontinental transport of anthropogenic fine particulate matter on human mortality. Air QualAtmos Health. 2014;7(3):369–79.
3. Anenberg SC, West IJ, Fiore AM, Jaffe DA, Prather MJ, Bergmann D, Cuvelier K, Dentener FJ, Duncan BN, Gauss M, Hess P, Jonson JE, Lupu A, Mackenzie IA, Marmer E, Park RJ, Sanderson MG, Schultz M, Shindell DT, Szopa S, Vivanco MG, Wild O, Zeng G. Intercontinental impacts of ozone pollution on human mortality. Environ SciTechnol. 2009;43(17):6482–7.
4. Burnett RT, Pope CA 3rd, Ezzati M, Olives C, Lim SS, Mehta S, Shin HH, Singh G, Hubbell B, Brauer M, Anderson HR, Smith KR, Balmes JR, Bruce NG, Kan H, Laden F, Prüss-Ustün A, Turner MC, Gapstur SM, Diver WR, Cohen A. An integrated risk function for estimating the global burden of disease attributable to ambient fine particulate matter exposure. Environ Health Perspect. 2014;122(4):397–403.
5. Chen RJ. The health effects of complex air pollution in 17 Chinese cities. Doctoral dissertation 2013.
6. Cooper OR, Forster C, Parrish D, Trainer M, Dunlea E, Ryerson T, Hübler G, Fehsenfeld F, Nicks D, Holloway J, de Gouw J, Warneke C, Roberts JM, Flocke F, Moody J. A case study of transpacific warm conveyor belt transport: influence of merging airstreams on trace gas import to North America. J Geophys Res Atmos. 2004;109(D23):D23S08.
7. Dentener F, Keating T, Akimoto H. Hemispheric transport of air pollution 2010. Part A: ozone and particulate matter. Report No. ECE/EN.AIR/100, air pollution studies, UNECE 2010; No. 17, 135–215.
8. Duncan BN, West JJ, Yoshida Y, Fiore AM, Ziemke JR. The influence of European pollution on ozone in the near east and northern Africa. AtmosChem Phys. 2008;8:2267–83.

9. GBD 2015 Risk Factors Collaborators. Global, regional, and national comparative risk assessment of 79 behavioural, environmental and occupational, and metabolic risks or clusters of risks, 1990–2015: a systematic analysis for the global burden of disease study. Lancet. 2016;388:1659–724.
10. Jaffe D, Anderson T, Covert D, Kotchenruther R, Trost B, Danielson J, Simpson W, Berntsen T, Karlsdottir S, Blake D, Harris J, Carmichael G, Uno I. Transport of Asian air pollution to North America. Geophys Res Lett. 1999;26:711–4.
11. Lin J, Pan D, Davis SJ, Zhang Q, He K, Wang C, Streets DG, Wuebbles DJ, Guan D. China's international trade and air pollution in the United States. Proc Natl AcadSci USA. 2014;111(5):1736–41.
12. Liu J, Mauzerall DL, Horowitz LW. Evaluating inter-continental transport of fine aerosols: (2) global health impact. Atmos Environ 2009. 2009;43:4339–47.
13. Pope CA, Burnett RT, Krewski D, Jerrett M, Shi Y, Calle EE, Thun MJ. Cardiovascular mortality and exposure to airborne fine particulate matter and cigarette smoke: shape of the exposure-response relationship. Circulation. 2009;120:941–8.
14. Pope CA, Burnett RT, Turner MC, Cohen A, Krewski D, Jerrett M, Gapstur SM, Thun MJ. Lung cancer and cardiovascular disease mortality associated with ambient air pollution and cigarette smoke: shape of the exposure-response relationships. Environ Health Perspect. 2011;119:1616–21.
15. Rohde RA, Muller RA. Air pollution in China: mapping of concentrations and sources. PLoS One. 2015;10(8):e0135749.
16. Maji KJ, Arora M, Dikshit AK. Burden of disease attributed to ambient $PM_{2.5}$ and PM_{10} exposure in 190 cities in China. Environ Sci Pollut Res Int. 2017;24(12):11559–72.
17. The World Bank and the Chinese State Environment Protection Agency. Cost of pollution in China: economic estimates of physical damages//Call for a Green China. Beijing, China. 2007.
18. Yang G, Wang Y, Zeng Y, Gao GF, Liang X, Zhou M, Wan X, Yu S, Jiang Y, Naghavi M, Vos T, Wang H, Lopez AD, Murray CJ. Rapid health transition in china, 1990–2010: findings from the global burden of disease study 2010. Lancet. 2013;381:1987–2015.
19. Yang J, Zhou M, Yin P, Li M, Ou CQ, Gu S, Liu Q. Mortality as a function of dust-haze in China: a multi-city time-series study. Lancet. 2016;388(Suppl 1):S19.
20. Verstraeten WW, Neu JL, Williams JE, Bowman KW, Worden JR, Boersma KF. Rapid increases in tropospheric ozone production and export from China. Nat Geosci. 2015;8:690–5.
21. Wang XP, Mauzerall DL. Evaluating impacts of air pollution in China on public health: implications for future air pollution and energy policies. Atmos Environ. 2006;40:1706–21.
22. WB. Consumer price index (2010 = 100). The World Bank. 2016.
23. West JJ, Naik V, Horowitz LW, Fiore AM. Effect of regional precursor emission controls on long-range ozone transport – part 2: steady-state changes in ozone air quality and impacts on human mortality. AtmosChem Phys. 2009;9:6095–107.
24. Zhang Q, Jiang X, Tong D, Davis SJ, Zhao H, Geng G, Feng T, Zheng B, Lu Z, Streets DG, Brauer M, van Donkelaar A, Martin RV, Huo H, Liu Z, Pan D, Kan H, Yan Y, Lin J, He K, Guan D. Transboundary health impacts of transported global air pollution and international trade. Nature. 2017;543:705–9.

Chapter 2
Characteristics of Major Air Pollutants in China

Lihong Ren, Wen Yang, and Zhipeng Bai

Abstract Following the rapid development of China's economy, air pollution has become more and more serious. Air pollution in China presents complex pollution characterized by high $PM_{2.5}$ and O_3 concentration. This study presents an overview of the status of air quality and emission in China and discusses the temporal and spatial distribution of major pollutants (PM_{10}, $PM_{2.5}$, SO_2, NO_X, and O_3). The results show that the reduced emissions have improved the air quality in China. However, the Chinese National Ambient Air Quality Standard (CNAAQS) for PM_{10} and $PM_{2.5}$ still be exceeded in many cities of China in 2015. A total of 77.5% (for $PM_{2.5}$) and 65.4% (for PM_{10}) of the monitoring cities were found to be exceeded CNAAQS. The average annual O_3 concentration was increasing during 2013–2015, and 16% of the total cities in 2015 did not meet the CNAAQS, indicating that O_3 pollution should be paid more attention. For NO_2 and SO_2, the exceedances of CNAAQS are rare. $PM_{2.5}$, PM_{10}, and SO_2 concentrations are higher in northern than in southern regions. High NO_2 occurred in Beijing-Tianjin-Hebei and Yangtze River delta region. Secondary particles formation and motor vehicle exhaust were the main sources of $PM_{2.5}$ in megacities. Dust was the main source for PM_{10}. The formation of O_3 is VOC-limited in urban areas of China and NO_X-limited in nonurban areas.

Keywords Air pollutants • Pollution characteristics • Emission status • China

L. Ren • W. Yang • Z. Bai (✉)
State Key Laboratory of Environmental Criteria and Risk Assessment, Chinese Research Academy of Environmental Sciences, Beijing 100012, China

Atmospheric Chemistry & Aerosol Division, Chinese Research Academy of Environmental Sciences, Beijing 100012, China
e-mail: baizp@craes.org.cn

G.-H. Dong (ed.), *Ambient Air Pollution and Health Impact in China*, Advances in Experimental Medicine and Biology 1017, DOI 10.1007/978-981-10-5657-4_2

2.1 Introduction

In recent years, China's economy has developed rapidly, the process of urbanization and industrialization has been speeded up, and energy consumption has increased. The statistics shows that China's gross domestic product (GDP) annual growth rate was 6.9–9.5% during 2011–2015 [17]. According to China Vehicle Environmental Management Annual Report, the vehicle population in 2015 was 279 million [13–15]. Since the late 1970s, the total energy consumption has greatly increased from 571 million tonnes of coal equivalence (Mtce) in 1978 to 4300 Mtce in 2015 [18]. Coal is the major fraction of energy consumption, accounting for 70% of China's energy consumption. Coal burning is the major source of ambient sulfur dioxide (SO_2), nitrogen oxides (NO_X), and soot.

With the development of economy, regional complex air pollution (characterized as a complex status of ozone (O_3) and fine particle ($PM_{2.5}$)) is one of the major environmental problems. Since 2013, heavy pollution events occurred frequently in China with 75% of cities and eight million people suffering from haze pollution, which constrains sustainable development of society and economy and threatens human health [8]. Air quality in China ranked the second to last in 180 countries, only better than Bangladesh [6].

Air pollutants can cause a variety of health problems. Exposure to high concentration of particulate matters can increase mortality or morbidity; excessive O_3 can cause breathing problems, trigger asthma, reduce lung function, and cause lung diseases; long-term and peak exposures to high NO_X can increase symptoms of bronchitis in asthmatic children; and SO_2 can affect the respiratory system and the functions of the lungs and causes irritation of the eyes [27]. The 1948 Donora smog caused by SO_2 and its oxides killed 20 people and sickened 5911 people [4]. The 1952 Los Angeles photochemical smog episode killed 400 people and a great many of people with red eyes, swollen throat, inflammation, and other respiratory diseases [23]. Therefore, it is both important and valuable to study the atmospheric pollution status and the variation characteristics of air pollution and its influential factors.

In this study, based on the data of air pollutants (PM_{10}, $PM_{2.5}$, SO_2, NO_X, and O_3) obtained from the national air pollution monitoring network in China, we will present an overview and analysis of air quality in China, analyze emission and pollution characteristics of major atmospheric pollutants, and then discuss temporal and spatial distributions of these pollutants. The study results will provide basic information for studying the health effects of air pollutants.

2.2 Chinese National Ambient Air Quality Standard

The Chinese National Ambient Air Quality Standard was issued firstly in 1982, when concentration limits for total suspended particulates (TSP), SO_2, NO_2, lead, and BaP were set. This standard was both strengthened and expanded in 1996. In

Table 2.1 Chinese National Ambient Air Quality Standard (GB 3095-2012) vs. WHO AQG

		Chinese AQC(μg/m^3)		
Pollutants	Averaging time	Grade I	Grade II	WHO AQG (μg/m^3)
$PM_{2.5}$	Annual	15	35	10
	24-h	35	75	25
PM_{10}	Annual	40	70	20
	24-h	50	150	50
SO_2	Annual	20	60	–
	24-h	50	150	20
	1-h	150	500	200
NO_2	Annual	40	40	40
	24-h	80	80	–
	1-h	200	200	200
O_3	Maximum daily 8-h	100	160	100
	1-h	160	200	
CO	24-h	4000	4000	
	1-h	10,000	10,000	

2000, the standard was updated with less stringent limits for certain pollutants. In February 2012, China released a new ambient air quality standard, GB 3095-2012, which set limits for the first time on $PM_{2.5}$ and Maxium daily 8-h ozone (Table 2.1). Meanwhile, the standard threshold of PM_{10} and oxynitride has also been tightened up.

Current air quality standards include two grades of limit values. Grade I standards apply to special regions such as national parks. Grade II standards apply to all other areas, including urban and industrial areas.

The 24-h and annual $PM_{2.5}$ limit values are set at 75 μg/m^3 and 35 μg/m^3 (Table 2.1). WHO AQG is stricter than the Chinese National Ambient Air Quality Standard. The recommended WHO AQG short-term (24-h) and long-term (annual average) values were 25 μg/m^3 and 10 μg/m^3 for $PM_{2.5}$. The United States published the National Ambient Air Quality Standard for $PM_{2.5}$ in 1997 (24-h average, 65μg/m^3; annual average, 15 μg/m^3), but the Ministry of Environmental Protection of the People's Republic of China did not published the National Ambient Air Quality Standard for $PM_{2.5}$ until 2012.

The 24-h and annual PM_{10} limit values are set at 150 μg/m^3 and 70 μg/m^3. WHO AQG for PM_{10} is lower than the Chinese National Ambient Air Quality Standard (Table 2.1). The recommended WHO AQG short-term (24-h) and long-term (annual average) values were 50 μg/m^3 and 20 μg/m^3 for PM_{10}.

The Chinese National Ambient Air Quality Standard (GB 3095-2012) set by the Ministry of Environmental Protection of the People's Republic of China (MEP) for SO_2, as well as WHO guideline, is shown in Table 2.1. The limit value for the annual and 24-h mean SO_2 concentration are set at 60 μg/m^3 and 150 μg/m^3. Unlike the usual 24-h and annual mean levels, WHO recommends that SO_2 follows a more stringent 10-min and 24-h intervals based on recommendations resulting from epidemiological studies. The yearly guideline is not needed since the 24-h guideline would be sufficient in assuring low annual average level.

The limit value for the annual, 24-h, and 1-h mean NO_2 concentrations is set at 40 μg/m^3, 80 μg/m^3, and 200 μg/m^3, respectively, which was identical to WHO AGQ. A maximum daily 8-h O_3 mean concentration was set at 160 μg/m^3 and 1-h mean was 200 μg/m^3 in China. The WHO AQG for O_3 is a daily maximum 8-h mean concentration of 100 μg/m^3, as shown in Table 2.1. This recommended limit was reduced from the previous level of 120 μg/m^3, based on recent conclusive associations between daily mortality and lower O_3 concentrations [27].

2.3 Characteristics of Major Air Pollutants

2.3.1 *Fine Particle ($PM_{2.5}$)*

2.3.1.1 Characteristics of $PM_{2.5}$ Pollution

China is one of the countries worst hit by $PM_{2.5}$ pollution. According to the global map of $PM_{2.5}$ published by NASA [16], the $PM_{2.5}$ pollution in north and east of China is the most serious, which was higher than that in India (Fig. 2.1). Recently, following the change of energy consumption structure, the pollution characteristics of particulate matters in China has changed from "coal smoke pollution" to "complex pollution," which is characterized by high $PM_{2.5}$ and O_3 concentration.

According to the report of the China air quality database, the CNAAQS limit value for $PM_{2.5}$ was exceeded in many cities of China in 2015. $PM_{2.5}$ has the highest percentage of exceedance among monitored pollutants. In 2015, the annual $PM_{2.5}$ concentrations in 338 cities were 11–125 μg/m^3 with an average of 50 μg/m^3, which was 0.43 times higher than the Chinese National Ambient Air Quality Standard (35 μg/m^3). The exceedances occurred in 77.5% of the case in all the monitoring

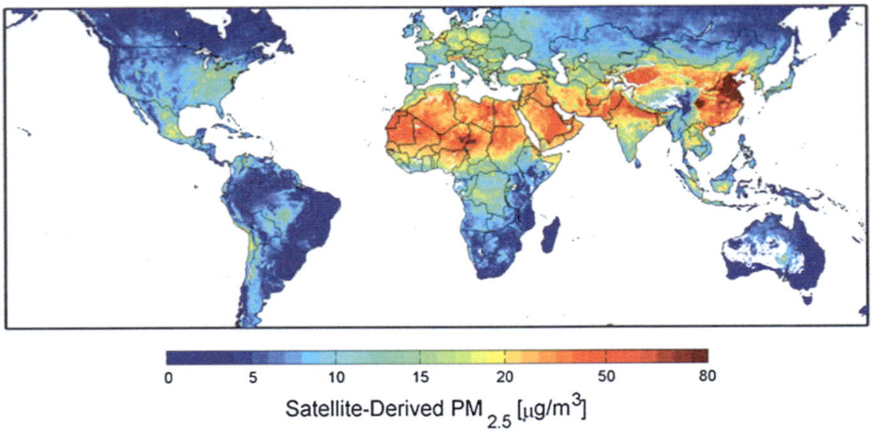

Fig. 2.1 Global satellite-derived map of $PM_{2.5}$ averaged over 2001–2006 [16]

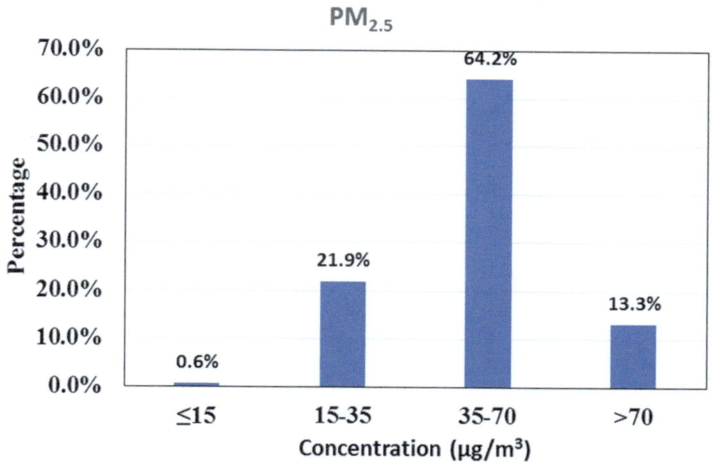

Fig. 2.2 Percentage of cities at different $PM_{2.5}$ concentration levels in 338 cities, 2015 (Data sources: MEP [13])

cities (Fig. 2.2). None of the cities were compliant with annual $PM_{2.5}$ World Health Organization (WHO) air quality guideline (10 μg/m^3). The proportion of the number of days exceeded the Chinese National Ambient Air Quality Standard was about 17.5% [13].

$PM_{2.5}$ has obvious spatial and temporal distributions related to the patterns of source emissions, chemical reaction mechanism, regional transport, and other meteorological conditions (such as dry and wet deposition). Figure 2.3 shows the spatial distribution of $PM_{2.5}$ in 2015 in China. As shown in Fig. 2.3, higher $PM_{2.5}$ concentration is mainly concentrated in Beijing-Tianjin-Hebei region, the north and middle part of Shandong province, the south and middle part of Henan province, and most of Hubei province. Generally, $PM_{2.5}$ annual concentration in the northern region was much higher than in the southern region. A number of studies have revealed that the higher concentrations in the northern region were related to the emissions from fossil fuel combustion and biomass burning. The colder north burns much more coal for winter heating and has more heavy industry, which emits a large amount of particulate matter [5].

With respect to seasonal variation, $PM_{2.5}$ has higher concentration in winter than that in other seasons and the lowest appeared in the summer (Fig. 2.4). The highest seasonal average concentrations were less than twice the lowest average values. Although the low temperature in the winter limited the secondary formation of particles, more frequent occurrences of the stagnant weather conditions caused the accumulation of atmospheric particles and high concentration episodes. Lower concentrations were observed in summer as particulate matters are washed out due to wet deposition.

Because many effective measures have been carried out to improve the air quality, $PM_{2.5}$ annual concentration in China has decreasing trend according to observation data in the recent 3 years (see Fig. 2.5). The annual average of $PM_{2.5}$ in 74 key

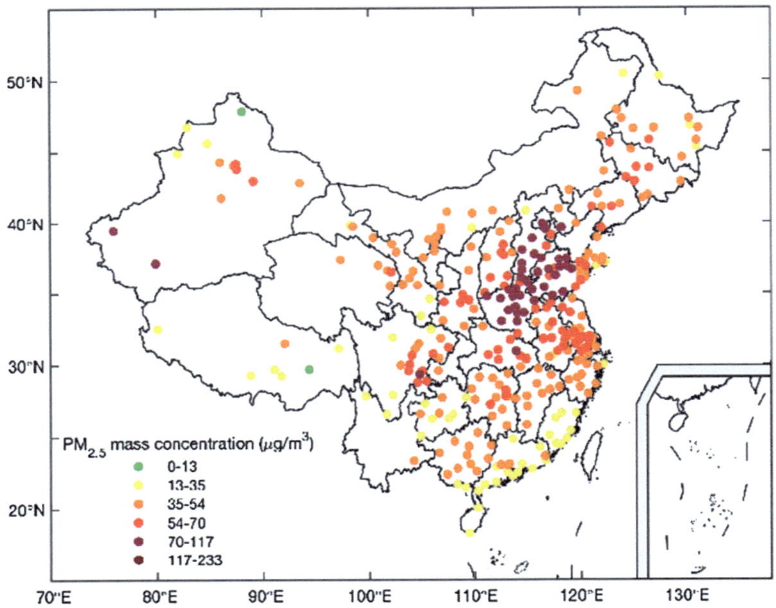

Fig. 2.3 Spatial distribution of $PM_{2.5}$ annual average concentrations in 2015 (Data sources: National air pollution monitoring network in China)

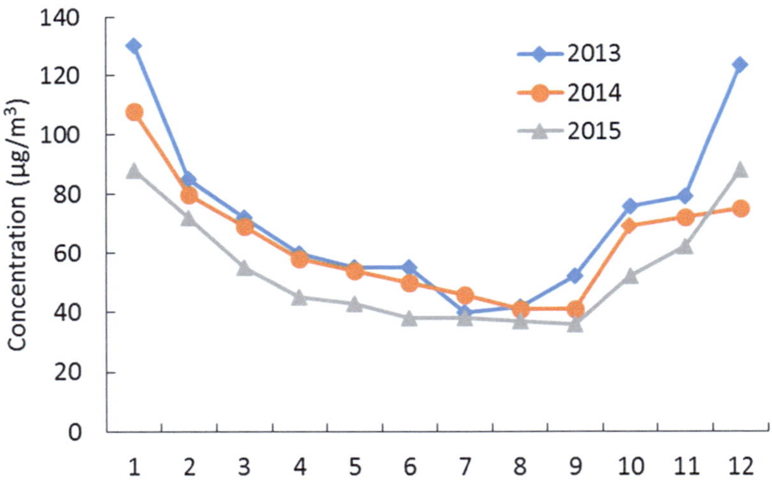

Fig. 2.4 Monthly variations of $PM_{2.5}$ in 2013, 2014, and 2015 (Data sources: National air pollution monitoring network in china)

cities was 72 μg/m^3 in 2013, and it decreased to 55 μg/m^3 in 2015. The government has taken measures and $PM_{2.5}$ has decreased over the recent years; however, $PM_{2.5}$ in most Chinese cities is still far above the Chinese National Ambient Air Quality Standard (GB 3095-2012). In China, the $PM_{2.5}$ has large portion of PM_{10} with 50–85%.

Fig. 2.5 Annual variation trends of $PM_{2.5}$ in 74 key cities in the recent 3 years (Data sources: MEP [11–13])

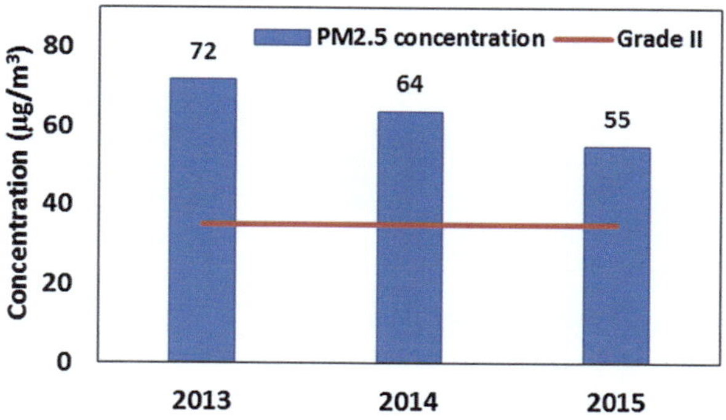

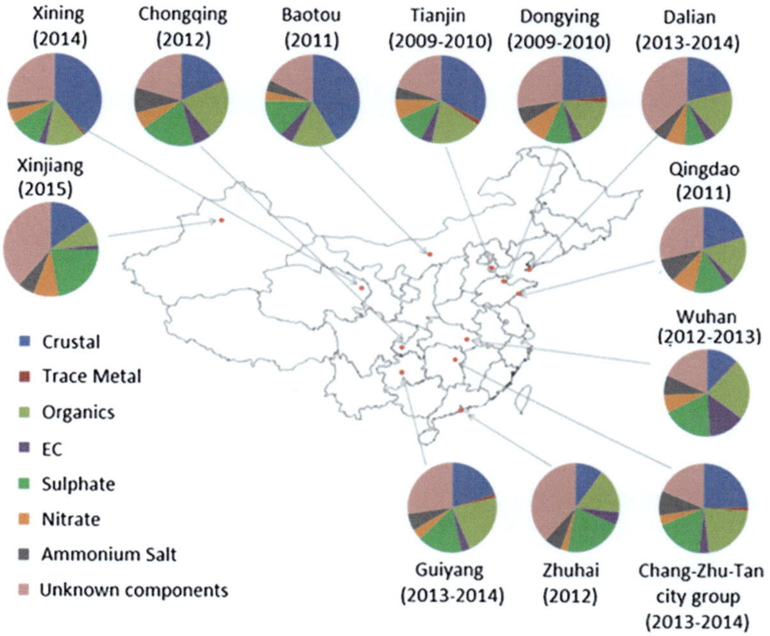

Fig. 2.6 $PM_{2.5}$ speciation in China [1]

2.3.1.2 Chemical Composition and Source Apportionment of $PM_{2.5}$

Particulate matter originated from both primary emission sources and reaction of precursor gases, such as sulfur dioxide (SO_2), nitrogen oxides (NO_X), ammonia (NH_3), and volatile organic compounds (VOCs). The main precursor gases NH_3, SO_2, and NOx react in the atmosphere to form ammonium, sulfate, and nitrate compounds. These compounds form new particles in the air or condense onto preexisting ones and form so-called secondary inorganic aerosols. Figure 2.6 shows the chemical composition of $PM_{2.5}$ in most cities of China. The chemical composition of $PM_{2.5}$ is varied at different cities, which is related with pollution sources and

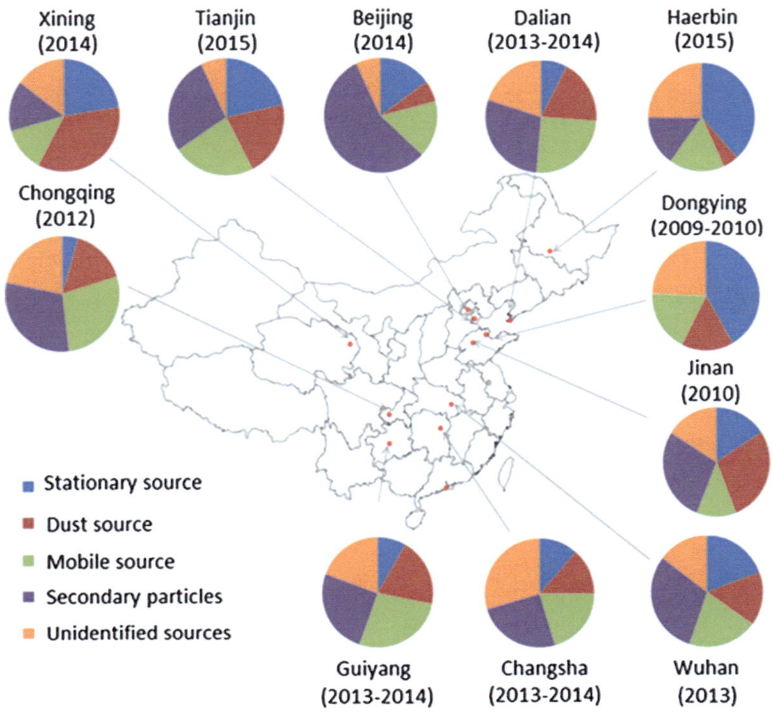

Fig. 2.7 The major sources of $PM_{2.5}$ in many Chinese cities [1]

meteorological conditions. In general, the crustal elements and organic matter are major species of $PM_{2.5}$. Secondary particles, such as sulfate, nitrate, and ammonium salt, have higher fractions in the eastern cities. Yang et al. [30] and He et al. [5] also find secondary ions, organic carbons, and crustal material that are the main components in urban and rural sites of China. This result indicated that there are more local formation/production and regional transport of the secondary aerosols in the eastern region, thus more intensive characteristic of "complex atmospheric pollution" compared to the western region.

$PM_{2.5}$ can be emitted directly from selected sources (primary PM), such as combustion and industry, or generated by gas-to-particle conversion in the atmosphere (secondary PM). Figure 2.7 shows the major sources of $PM_{2.5}$ in many Chinese cities. From it we can see that source contribution rates are varied in different cities. In generally, secondary particles formation and motor vehicle exhaust were the main sources of $PM_{2.5}$ in megacities (such as Beijing, Wuhan, and Chongqing). $PM_{2.5}$ in western cities (such as Xining) was influenced mainly by dust. The contribution of stationary sources, including coal combustion and industrial emissions, shows a downward trend from north to south. During the haze pollution events, a large fraction of $PM_{2.5}$ was secondary species, that is, secondary organic aerosol (SOA) and secondary inorganic aerosol (SIA, sulfate, nitrate, and ammonium). The contribution of primary particulate to $PM_{2.5}$ was small [8].

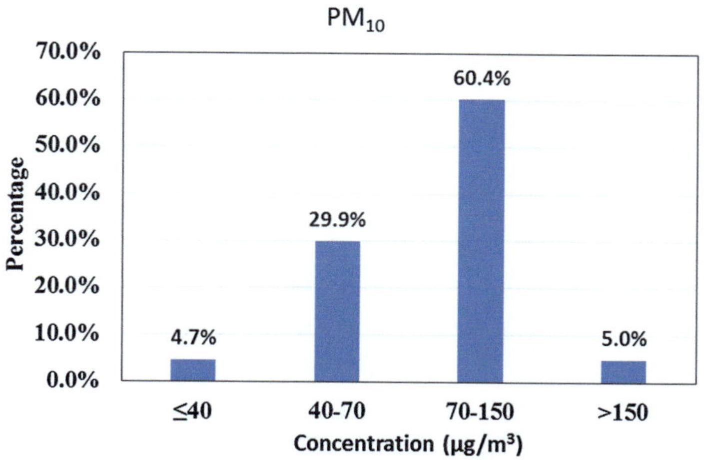

Fig. 2.8 Percentage of cities at different PM_{10} concentration levels in 338 cities, 2015 (Data sources: MEP [13])

2.3.2 *Inhalable Particulate Matter (PM_{10})*

2.3.2.1 Characteristics of PM_{10} Pollution

In China, PM_{10} remains an important pollutant. In 2015, PM_{10} concentrations at 65.4% of the monitoring cities were found to be exceeded than the CNAAQS (Fig. 2.8). The annual PM_{10} concentrations in 338 cities were 24–357 μg/m^3 with an average of 87 μg/m^3, which exceeded the Chinese National Ambient Air Quality Standard (70 μg/m^3). Days of daily concentrations exceeding the air standard was about 12.1% of all monitoring days.

Figure 2.9 shows the spatial distribution of PM_{10} in 2015. As shown in Fig. 2.9, PM_{10} annual concentration in the northern region was much higher than that in the southern region. The higher PM_{10} concentrations in the northern region were related to the influence of dust-sand.

The trends of PM_{10} in the recent 3 years in 74 key cities were calculated based on the officially reported data (Fig. 2.10). Although PM_{10} annual concentration also was decreasing trend, it was still far above the Chinese National Ambient Air Quality Standard (GB 3095-2012). The annual average of PM_{10} in 74 key cities was 118 μg/m^3 in 2013, and it decreased to 93 μg/m^3 in 2015 which was about 33% higher than the Chinese grade II standards.

PM_{10} has also obvious seasonal variation, showing the concentrations in winter were higher than that in other seasons (Fig. 2.11). The highest concentrations appeared in December and January. Lower concentrations were observed in July and August, which was related with the frequency of rain.

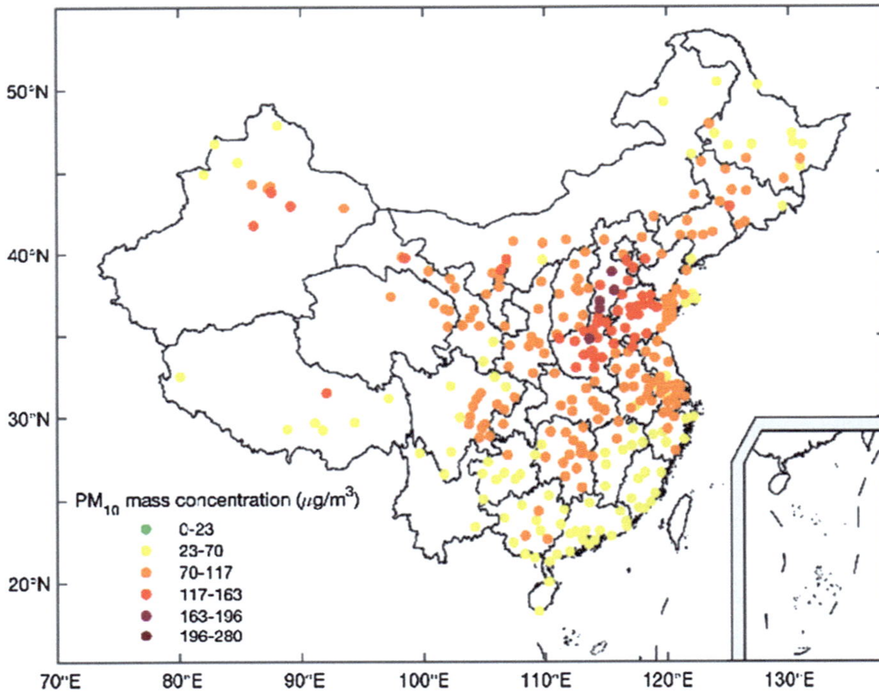

Fig. 2.9 Spatial distribution of PM_{10} annual average concentrations in 2015 (Data sources: National air pollution monitoring network in China)

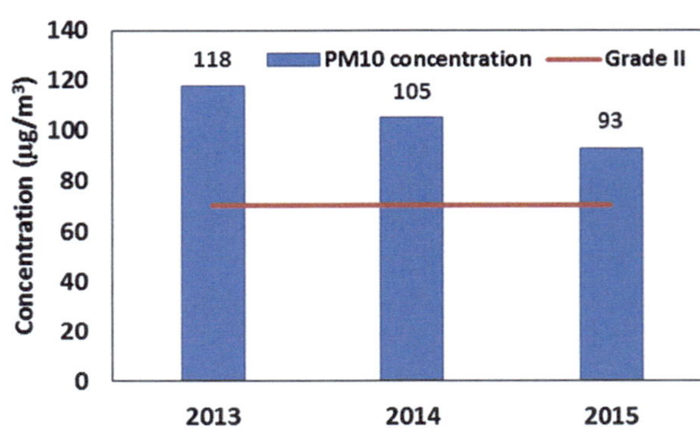

Fig. 2.10 Annual variation tendency of PM_{10} in 74 key cities of China (Data sources: MEP [11–13])

2.3.2.2 Chemical Composition and Source Apportionment of PM_{10}

Soil dust was the first abundant component for PM_{10} in most cities of China. And the secondary aerosol was the second important component. Carbonaceous matter has also important contribution to PM_{10} mass concentration [2, 21].

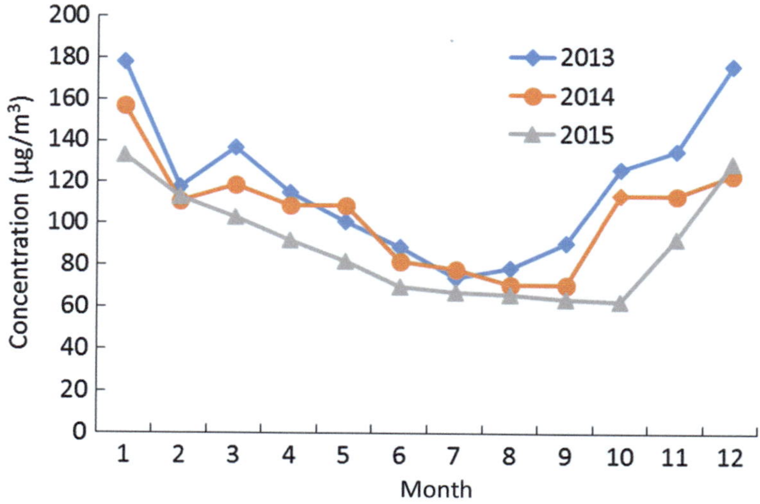

Fig. 2.11 Monthly variations of PM_{10} in 2013, 2014, and 2015 (Data sources: National air pollution monitoring network in china)

Figure 2.12 shows the major sources of PM_{10} in some Chinese cities. It can be found that dust was the main source for PM_{10}. Stationary source and mobile source also have important contribution to PM_{10} in the northern cities (such as Dalian, Shenyang, and Harbin).

2.3.3 *Ozone (O_3)*

2.3.3.1 Characteristics of O_3 Pollution

O_3 is a strong oxidant, formed from the reactions of precursors (VOCs, NO_X, and so on) and sunlight. The major health effect of O_3 is its effect on the respiratory systems. O_3 is the main component of photochemical smog. In 1974, the first photochemical smog events in China appeared in the Xigu Industrial District of Lanzhou City. Photochemical smog events have also appeared in some suburban regions. Photochemical smog, high O_3, and NO_X concentrations have gradually emerged into China's three city clusters (Beijing-Tianjin-Hebei region, the Yangtze River Delta, and Pearl River Delta).

Although O_3 annual concentrations in the recent 3 years were lower than the CNAAQS, the average annual O_3 concentrations were increasing during 2013–2015 (as Fig. 2.13 shows), indicating that O_3 pollution should be paid more attention. Year-to-year differences in the O_3 levels are also induced by meteorological variations.

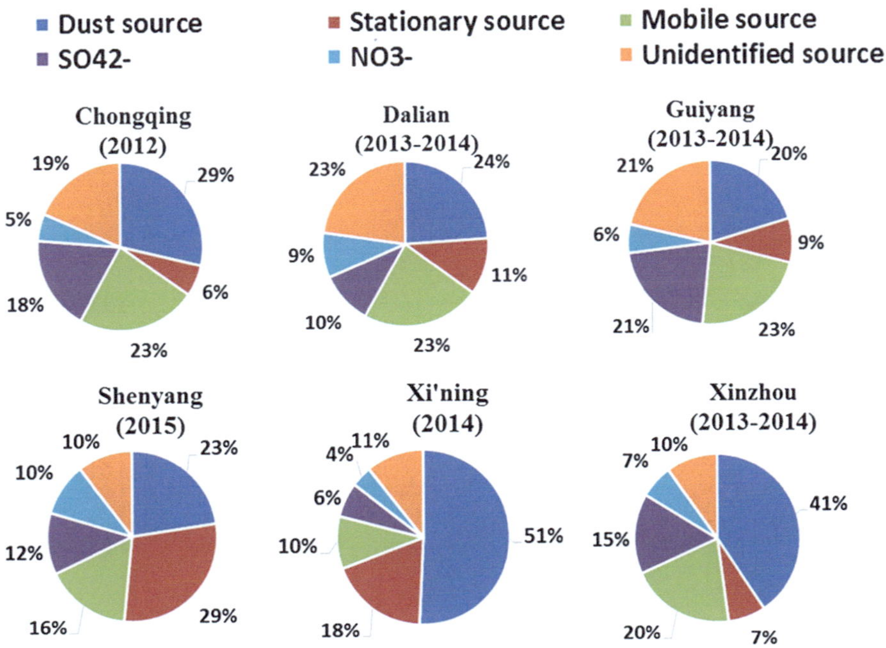

Fig. 2.12 The major sources of PM_{10} in many Chinese cities

Fig. 2.13 Annual variation tendency of O_3 in 74 key cities of China (Data sources: MEP [11, 12, 13]

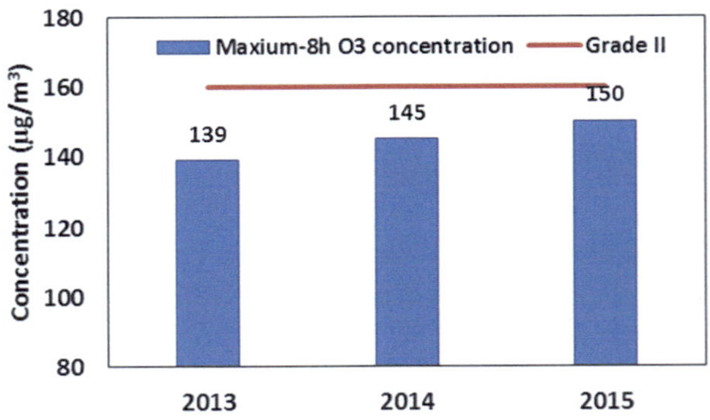

In 2015, the 90% of O_3 maximum daily 8-h mean concentrations in 338 cities were 62–203 μg/m³ with an average of 134 μg/m³. Sixteen percent of the total cities did not meet the CNAAQS (Fig. 2.14). Days of daily concentrations exceeding the air standard was about 4.6% of all monitoring days.

Differences in the distribution of O_3 precursor emission sources and climatic conditions in Europe result in considerable regional differences in O_3 concentrations. Higher ozone concentrations are observed, in general, in summer months as it is formed by photochemical reactions of NO_x and VOCs. Ozone concentrations tend to peak in early to midafternoon in areas where there is strong photochemical activity. The values indicate that ozone levels are within CNAAQS.

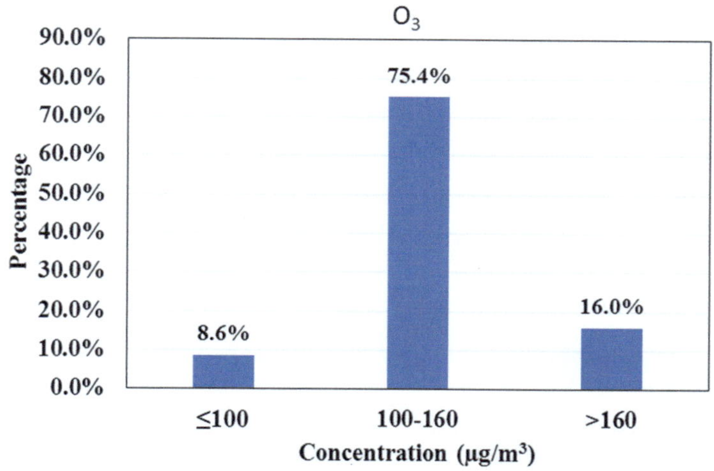

Fig. 2.14 Percentage of cities at different O_3 concentration levels in 338 cities, 2015 (Data source: MEP [13])

2.3.3.2 Sources and Formation of O_3

VOCs and NO_X emissions from motor vehicle were the major precursor gases of O_3 formation. Shao et al. [20] found that alkenes contribute a large fraction of VOC activity with 75%. The formation of O_3 is VOC-limited in urban areas of China and NO_X-limited in nonurban areas [7, 24, 31]. The influence of biogenic VOCs on O_3 formation was minor [19]. Heterogeneous NO_2 could increase the concentration. The influence of the reaction of NO_3 and N_2O_5 on O_3 was unimportant [29]. CH_4 and CO also play a role in O_3 formation in certain environments. Tie et al. [24] reported that oxidation of CO contributed to 54% of the total O_3 production in eastern region of China.

Both local formation and regional transport contributed to O_3 concentrations. Wang et al. [25] reported that the contribution from regional transport was about 17.8% of O_3 concentration in PRD. Tang [22] found that 40% of O_3 concentration in Beijing was from southern and southeastern regions.

Meteorological conditions have also a major influence on O_3 formation. High O_3 concentration was related to the occurrence of high-pressure synoptic systems [28].

2.3.4 Sulfur Dioxide (SO_2)

SO_2 has greatly contributed to acid rain and has adverse effects on ecosystems and the respiratory system [26]. It is also the main precursor to formation of particulate matter.

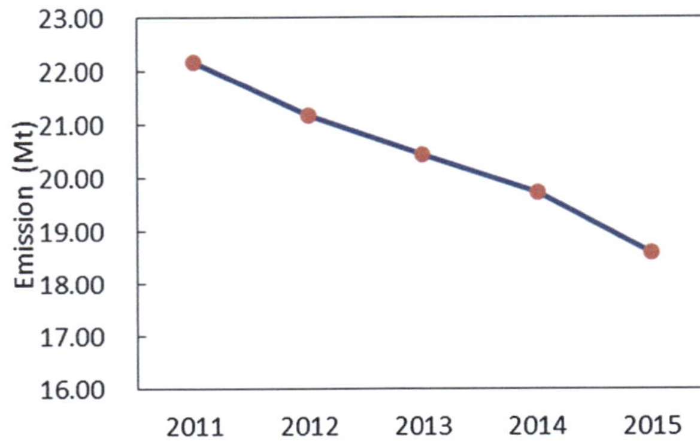

Fig. 2.15 Emission of SO_2 in China from 2011 to 2015 (Data sources: MEP [9–13])

Global SO_2 emissions have been dramatically reduced from 121 Tg to 103 Tg during the period of 1990–2010. Figure 2.11 gives the emission trend of SO_2 from 2011 to 2015 in China. Following the emission control legislations, SO_2 emissions have been decreasing dramatically. National emission of SO_2 in 2011 was about 22.17 Mt/year and it decreased to 18.59 Mt/year in 2015. In the period 2011–2015, SO_2 emission decreased by 16% (Fig. 2.15).

SO_2 is emitted primarily from fuels containing sulfur burning. The main anthropogenic emissions of SO_2 in China are derived from industrial sources (including power plant, domestic heating, and industrial production processes), and contribution from urban life source was little. As reported in "Annual Report of Environmental Statistics" [11], SO_2 emission in 2014 was about 19.7 Mt, and the contribution of industrial source and urban life source was about 88% and 12%, respectively.

With respect to the spatial distribution of SO_2 emission, Cao et al. (2010) found that the SO_2 emission in Shandong province, Hebei province, and Shanxi province was the highest, which was related to large consumption of coal in these regions [3]. The SO_2 emission in the western region (Qinghai, Xizang, and Gansu provinces) was relatively little.

In 2015, the annual SO_2 concentrations in 338 cities were 3–87 $\mu g/m^3$ with an average of 25 $\mu g/m^3$. The average concentration was lower than the Chinese National Ambient Air Quality grade II standard (60 $\mu g/m^3$). SO_2 concentrations at 3.3% of the monitoring cities were found to exceed the CNAAQS (Fig. 2.16). Days of daily concentrations exceeding the air quality standard was about 0.7% of all monitoring days.

Figure 2.17 shows the spatial distribution of SO_2 annual concentration in 2015. Maximum SO_2 annual concentrations were found in northern regions, especially in North China and Inner Mongolia region. It may be related to coal heating in China. The SO_2 concentration in southern regions was relatively lower.

Figure 2.18 shows the annual variation of SO_2 concentrations in 74 key cities of China. It is clear that the average annual values for 74 key cities show a decline trend in the recent 3 years, and all annual SO_2 concentrations stayed below the grade II standard value, indicating that the measures taken to control SO_2 pollution were effective.

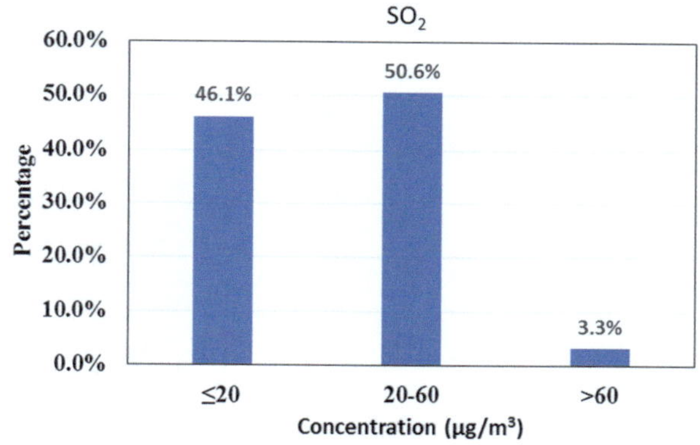

Fig. 2.16 Percentage of cities at different SO_2 concentration levels in 338 cities, 2015 (Data sources source: MEP 11)

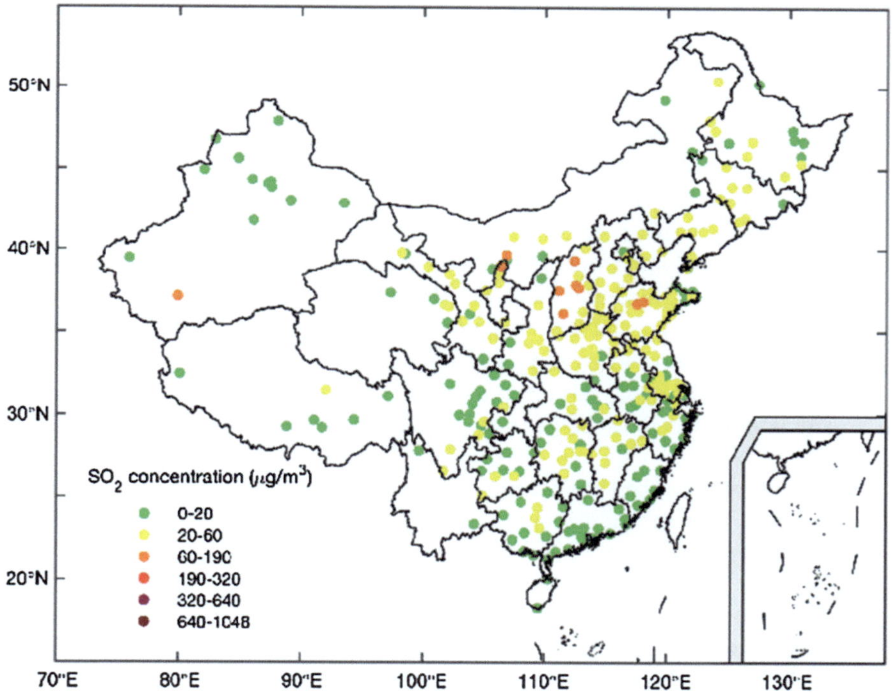

Fig. 2.17 Spatial distribution of SO_2 annual average concentrations in 2015 (Data sources: National air pollution monitoring network in China)

SO_2 shows the highest concentration in the winter and the lowest in the summer, which is due to the effects of emission sources and meteorological conditions. The energy structure was based on coal in China. Energy for heating is mainly coal in winter and these coals contain high-sulfur fraction over 0.5%, which cause the higher emission in winter. In addition, slow winds and shallow mixing layers occur more frequently in winter, trapping the pollutants near the surface and leading to high concentrations.

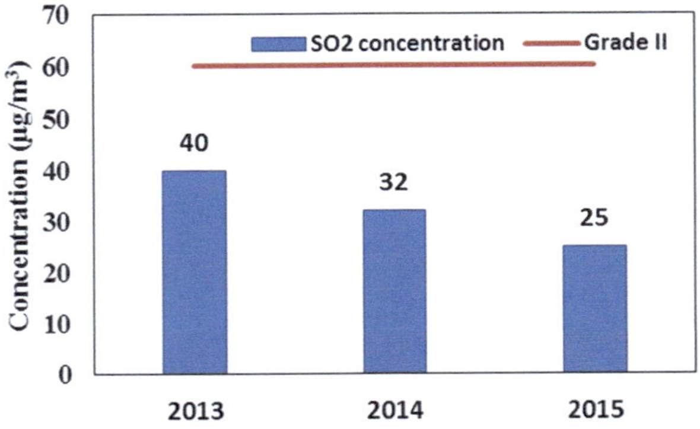

Fig. 2.18 Annual variation tendency of SO_2 in 74 key cities of China (Data sources: MEP [11–13])

2.3.5 *Nitrogen Oxides (NO_X)*

Vehicle is the main contributor of NO_X. Although the vehicle population increased by about 14.9% per year during 2007–2015 in China, the NO_X emissions have decreased obviously by about 23% during 2011–2015. This indicated that the control measure for NO_X was effective (Fig. 2.19).

In many developed counties, the main emission sources of NO_X are mobile vehicles. In the United States, 57.5% of NO_X emission was from mobile sources, with fuel combustion and industrial processes only account for 24.2% and 8.4%, respectively. However, in China, the main NO_X emissions sources are industrial sources (including power plants and industrial production) and motor vehicles. As reported in the "Annual Report of Environmental Statistics" [11], NO_2 emission in 2014 in China was about 20.8 Mt, and the emissions from industrial source accounted for 67% and motor vehicle accounted for 30%. Urban life source only accounted for 3%.

Emissions of NO_X vary significantly by province owing to factors such as population, energy sources, and economic base. In China, NO_2 emission in Shandong province, Jiangsu province, Hebei province, and Guangdong province was the highest, which was related to large consumption of coal in these regions [9].

Figure 2.20 gives the annual variation tendency of NO_2 in the recent 3 years. It can be observed that NO_2 was decreasing during 2013–2015. The NO_2 concentrations in 2013 and 2014 were 44 μg/m³ and 42 μg/m³, respectively. It decreased to 39 μg/m³ in 2015 and was lower than the CNAAQS grade II standard.

In 2015, the annual NO_2 concentrations in 338 cities were 8–63 μg/m³ with an average of 30 μg/m³. The average of NO_2 annual concentration did not exceeded the Chinese National Ambient Air Quality Standard and WHO AQG (40 μg/m³). NO_2 concentrations at 81.7% of the monitoring cities were found to be lower than the CNAAQS (Fig. 2.21), and only 19.3% of the monitoring cities exceeded the CNAAQS. The number of days with NO_2 concentrations exceeding the CNAAQS occupied only 1.6% of all monitoring days.

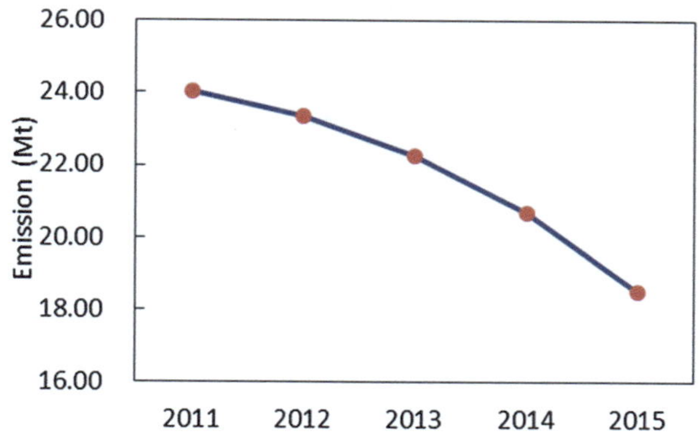

Fig. 2.19 Emission of NO_X in China from 2011 to 2015 (Data sources: MEP [9–13])

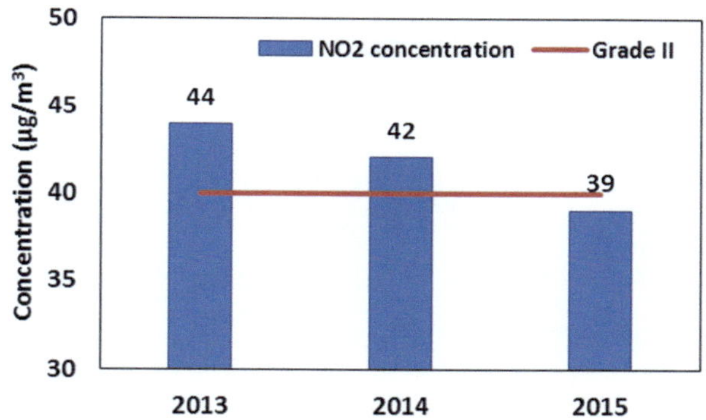

Fig. 2.20 Annual variation tendency of NO_2 in 74 key cities of China (Data sources: MEP [11–13])

Spatial distribution of NO_2 showed that the concentration in northern regions was higher than that in southern regions (Fig. 2.22). The NO_2 concentrations in North China, Pearl River Delta, and Urumqi City were the highest.

2.4 Conclusion

With the development of economy and industries, air pollution is getting more and more serious in China. Although reduced emissions have improved air quality in China, heavy pollution events occurred frequently.

$PM_{2.5}$ was the major pollutant in China. Although $PM_{2.5}$ annual concentration in China has decreasing trend, the CNAAQS limit value for $PM_{2.5}$ exceeded in large parts of China. The exceedances occurred in 77.5% of the case in all the monitoring cities in 2015. Beijing-Tianjin-Hebei region, the north and middle part of Shandong province, the south and middle part of Henan province, and most of Hubei province have higher $PM_{2.5}$ concentrations. The crustal elements and organic matter are major species of $PM_{2.5}$. Secondary particles formation and motor vehicle exhaust were the main sources of $PM_{2.5}$ in megacities.

Fig. 2.21 Percentage of cities at different NO_2 concentration levels in 338 cities, 2015 (Data source: MEP [13])

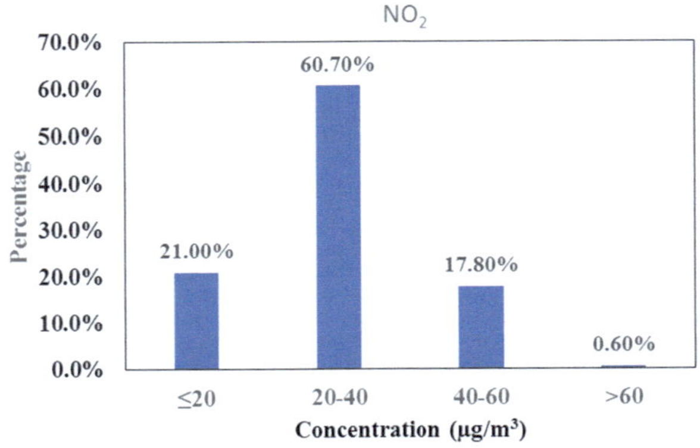

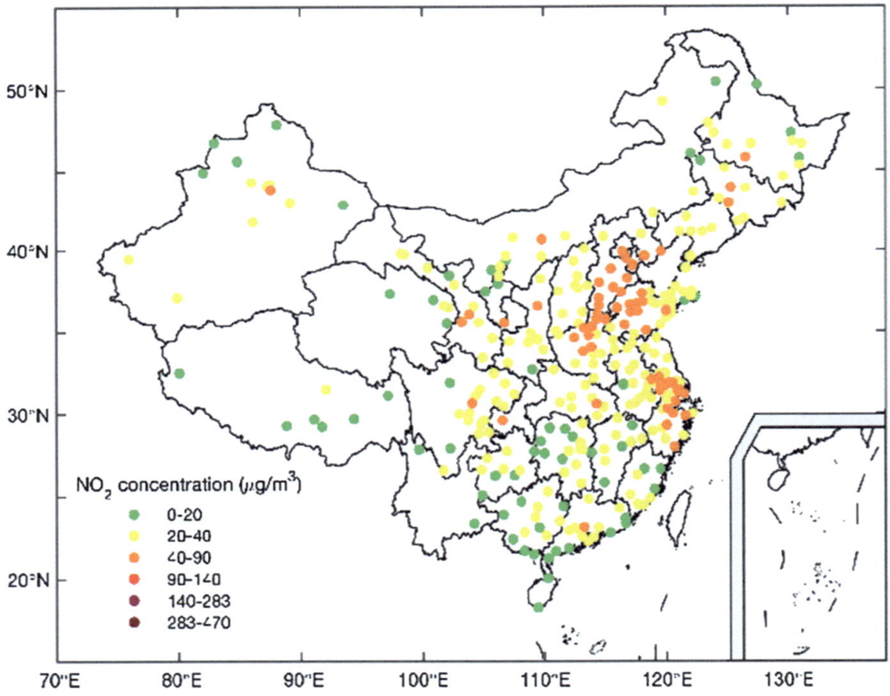

Fig. 2.22 Spatial distribution of NO_2 annual average concentrations in 2015 (Data sources: National air pollution monitoring network in China)

PM_{10} also is an important pollutant in China. In 2015, PM_{10} concentrations at 65.4% of the monitoring cities were found to be exceeded the CNAAQS. PM_{10} annual concentration in the northern region was much higher than that in the southern region, which related to the influence of dust on northern cities. Soil dust was the first abundant component for PM_{10} in most cities of China.

Although O_3 annual concentrations in the recent 3 years were lower than the CNAAQS, the average annual O_3 concentrations were increasing during 2013–2015. In 2015, the 90% of O_3 maximum daily 8-h mean concentrations in 338 cities were 62–203 μg/m^3 with an average of 134 μg/m^3. VOCs and NO_X emissions from motor vehicle was the major precursor gases of O_3 formation. The formation of O_3 is VOC-limited in urban areas of China and NO_X-limited in nonurban areas.

The average annual values of SO_2 for 74 key cities shows a decline trend in the recent 3 years, and all annual SO_2 concentrations stayed below the grade II standard value. NO_X also has decreasing trend and it decreased to 39 μg/m^3 in 2015, which was lower than the CNAAQS grade II standard. These results indicated that the measures taken to control SO_2 and NO_X pollution were effective.

Acknowledgments The authors would like to thank the support from the National Basic Research Program of China (Grant No. 2011CB503801) and the Central Level, Scientific Research Institutes for Basic R&D Special fund business (Grant No. 2016YSKY-023).

References

1. Bai Z, Han J, Azzi M. Insights into measurements of ambient air $PM_{2.5}$, in China. Trends Environ Anal Chem. 2017;13:1–9.
2. Cao J, Shen Z, Chow JC, Qi G, Watson JG. Seasonal variations and sources of mass and chemical composition for PM_{10} aerosol in Hangzhou, China. Particuology. 2009;7(3):161–8.
3. Cao G, An X, Zhou C, Ren Y, Tu J. Emission inventory of air pollutants in China. China Environ Sci. 2010;30(7):900–6.
4. Ciocoo A, Thompson D. A follow-up on Donora ten years after: methodology and findings. Am J Public Health. 1961;51:155–64.
5. He K, Huo H, Zhang Q. Urban air pollution in China: current status, characteristics and progress. Annu Rev Energy Environ. 2002;27(1):397–431.
6. Hsu A, Alexandre N, Cohen S, et al. 2016 environmental performance index. New Haven: Yale University; 2016. Available: www.epi.yale.edu
7. Huang JP, Fung JCH, Lau AKH, Qin Y. Numerical simulation and process analysis of typhoon-related ozone episodes in Hong Kong. J Geophysical Res. 2005;110(D5):755–65.
8. Huang RJ, Zhang Y, Bozzetti C, Ho KF, Cao JJ, Han Y, Daellenbach KR, Slowik JG, Platt SM, Canonaco F, Zotter P, Wolf R, Pieber SM, Bruns EA, Crippa M, Ciarelli G, Piazzalunga A, Schwikowski M, Abbaszade G, Schnelle-Kreis J, Zimmermann R, An Z, Szidet S, Baltenperger U, Haddad IE, Prevot ASH. High secondary aerosol contribution to particulate pollution during haze events in China. Nature. 2014;514(7521):218–22.
9. MEP (Ministry of Environmental Protection of the People's Republic of China). Chinese environmental condition report in 2011. 2012.
10. MEP (Ministry of Environmental Protection of the People's Republic of China). Chinese environmental condition report in 2012. 2013.
11. MEP (Ministry of Environmental Protection of the People's Republic of China). Chinese environmental condition report in 2013. 2014.
12. MEP (Ministry of Environmental Protection of the People's Republic of China). Chinese environmental condition report in 2014. 2015.
13. MEP (Ministry of Environmental Protection of the People's Republic of China). Chinese environmental condition report in 2015. 2016.
14. MEP (Ministry of Environmental Protection of the People's Republic of China). Annual report of environmental statistics of 2014. 2016.

15. MEP (Ministry of Environmental Protection of the People's Republic of China). China vehicle environmental management annual report, 2016.
16. NASA. New map offers a global view of health-sapping air pollution. 2010. https://www.nasa.gov/topics/earth/features/health-sapping.html
17. NBSC (National Bureau of Statistics of China). Statistical communique of 2015 China's economic and social development, people's daily, 2016, Mar 1 sect 010.
18. NBSC (National Bureau of Statistics of China). China statistic yearbook-2016. China statistics Press; 2017.
19. Shao M, Zhao M, Zhang Y, Peng L, Li J. Biogenic vocs emissions and its impact on ozone formation in major cities of China. J EnvironSci Health. 2000;35(10):1941–50.
20. Shao M, Fu LL, Liu Y, Lu SH, Zhang YH, Tang XY. Major reactive species of ambient volatile organic compounds (VOCs) and their sources in Beijing. Sci China Series D—Earth Sci. 2005;48(z2):147–54.
21. Sun Y, Zhuang G, Tang AA, Wang Y, An ZS. Chemical characteristics of PM2.5 and PM10 in haze-fog episodes in Beijing. Environ Sci Technol. 2006;40(10):3148.
22. Tang X. The characteristics of urban air pollution in China. In: Fritz JJ, editor. Urbanization, energy, and air pollution in China. Washington, DC: The National Academies Press; 2004. p. 47–54.
23. Tiao G, Box G, Hamming W. Analysis of los angeles photochemical smog data: a statistical overview. J Air Pollut Control Assoc. 1975;25(3):260–8.
24. Tie X, Brasseur GP, Zhao CS, Granier C, Massie S, Qin Y, Wang P, Wang G, Yang P, Richter A. Chemical characterization of air pollution in Eastern China and the Eastern United States. Atmos Environ. 2006;40(14):2607–25.
25. Wang X, Carmichael G, Chen D, Tang Y. Impacts of different emission sources on air quality during March 2001 in the Pearl River Delta (PRD) region. Atmos Environ. 2005;39(29):5227–41.
26. WHO (World Health Organization). Review of evidence on health aspects of air pollution – REVIHAAP Project, Technical 617 Report. World Health Organization, Regional Office for Europe, Copenhagen, Denmark. 2013.
27. WHO (World Health Organization). Ambient (outdoor) air quality and health, fact sheet No 313, updated September 2016, World Health Organization http://www.who.int/mediacentre/factsheets/fs313/en/
28. Xu J, Zhu Y. Some characteristics of ozone concentrations and their relations with meteorological factors in Shanghai. Atmos Environ. 1994;28(20):3387–92.
29. Xu J, Zhang YH, Wang W. Numerical study on the impacts of heterogeneous reactions on ozone formation in the Beijing urban area. Adv Atmos Sci. 2006;23(4):605–14.
30. Yang F, Tan J, Zhao Q, Du Z, He K, Ma Y, Duan F, Chen G, Zhao Q. Characteristics of PM2.5 speciation in representative megacities and across China. Atmos Chem Phys. 2011;11(11):1025–51.
31. Zhang J, Wang T, Chameides WL, Cardelino C, Kwok J, Blake DR, Ding A, So KL. Ozone production and hydrocarbon reactivity in Hong Kong, Southern China. Atmos Chem. Phys. 2007;6(5):557–73.

Chapter 3
Human Exposure Assessment for Air Pollution

Bin Han, Li-Wen Hu, and Zhipeng Bai

Abstract Assessment of human exposure to air pollution is a fundamental part of the more general process of health risk assessment. The measurement methods for exposure assessment now include personal exposure monitoring, indoor-outdoor sampling, mobile monitoring, and exposure assessment modeling (such as proximity models, interpolation model, air dispersion models, and land-use regression (LUR) models). Among these methods, personal exposure measurement is considered to be the most accurate method of pollutant exposure assessment until now, since it can better quantify observed differences and better reflect exposure among smaller groups of people at ground level. And since the great differences of geographical environment, source distribution, pollution characteristics, economic conditions, and living habits, there is a wide range of differences between indoor, outdoor, and individual air pollution exposure in different regions of China. In general, the indoor particles in most Chinese families comprise infiltrated outdoor particles, particles generated indoors, and a few secondary organic aerosol particles, and in most cases, outdoor particle pollution concentrations are a major contributor to indoor concentrations in China. Furthermore, since the time, energy, and expense are limited, it is difficult to measure the concentration of pollutants for each individual. In recent years, obtaining the concentration of air pollutants by using a variety of exposure assessment models is becoming a main method which could solve the problem of the increasing number of individuals in epidemiology studies.

Keywords Air pollution • Personal exposure • Exposure measurement • Exposure model

---

B. Han • Z. Bai (✉)
State Key Laboratory of Environmental Criteria and Risk Assessment, Chinese Research Academy of Environmental Sciences, Beijing 100012, China

Atmospheric Chemistry & Aerosol Division, Chinese Research Academy of Environmental Sciences, Beijing, China
e-mail: baizp@craes.org.cn

L.-W. Hu
Guangzhou Key Laboratory of Environmental Pollution and Health Risk Assessment, Department of Preventive Medicine, School of Public Health, Sun Yat-sen University, 74 Zhongshan 2nd Road, Yuexiu District, Guangzhou 510080, Guangdong, China

G.-H. Dong (ed.), *Ambient Air Pollution and Health Impact in China*, Advances in Experimental Medicine and Biology 1017, DOI 10.1007/978-981-10-5657-4_3

3.1 Introduction

Assessment of human exposure to air pollution is a fundamental part of the more general process of health risk assessment, which is the process by which risk increase for human health on exposure to air-polluted substances can be estimated. The effects of a specific pollutant on health can be completely different depending whether exposure is acute or chronic or continuous or intermittent, so the assessment of total air pollution exposure is essential and crucial for a proper determination of possible links between air pollution and health effects.

The human contact with a pollutant may occur through a variety of source, and to assess the total exposure, each source should be taken into account. Generally, three sources can be defined: outdoor, indoor, and workplace. In China, workplace environment is regulated separately and managed through industrial hygiene practices. In this circumstance, the microenvironment of air pollution always included two main sources to the public: indoor and outdoor.

Human exposure to air pollution may be evaluated by direct or indirect methods. Direct methods are carried out by personal exposure measurement, either with pumped or diffusive sampling, or by measurements of biological markers; we can also call it "personal exposure monitoring." Indirect methods involve calculation of human exposure, by combination of concentrations at various locations and time spent by people in each specific environment, such as the indoor or outdoor.

In this chapter, we will describe the current state of air pollution assessment in China.

3.2 Principal Definition and Framework for Assessment of Human Exposure to Air Pollution

3.2.1 Exposure Science

Exposure science plays an important role in the development of epidemiology, toxicology, and risk assessment. It provides critical information for protecting human and ecosystem health. It also can play an effective role in other fields, such as environmental regulation and legislation and urban and ecosystem planning. Exposure science links human and ecologic behavior to environmental processes in such a way that the information generated can be used to mitigate or prevent future adverse effects.

In Fig. 3.1, the conceptual framework from pollution sources to final health outcomes identifies and links the core elements of exposure science: sources of stressors, environmental intensity (such as air pollutant concentrations), time activity and behavior, contact of stressors and receptors, and outcomes of contact.

Figure 3.2 shows the role of upstream human and natural factors in determining which stressors are mobilized and transported to key receptors. This figure indicates

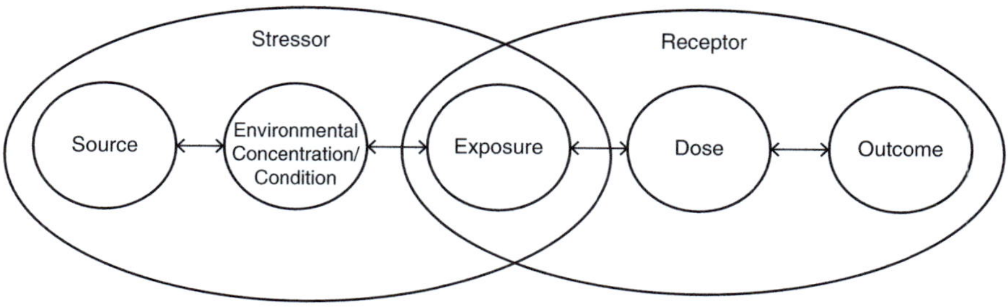

Fig. 3.1 The classic environmental health conception [82]

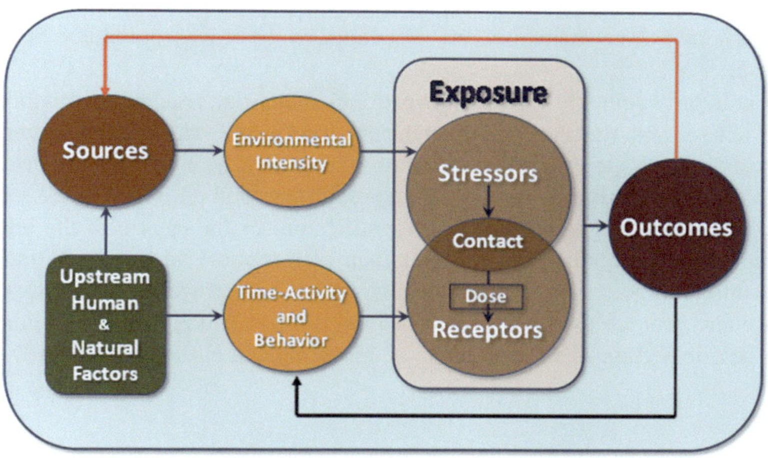

Fig. 3.2 Core elements of exposure science [33]

the role of behavior of receptors and time in changing the contact that results from environmental intensities that affect exposure. It reveals both external and internal exposure; however, it still retains the concept that exposure is measured at certain borderline between receptor and emission source. In addition, the dose is the amount of certain material, which passes through or has some effects across the borderline to get into the contact with the target system, organ, or even cell and, then, produces a corresponding outcome. For example, a dose in a tissue (e.g., blood) can be used as an exposure to another tissue of blood perfusion. Figure 3.2 identifies the feedbacks inherent in exposure science. For example, consider how the behavior of a patient or organism changes and affects exposure. The outcome may also influence the source, just as a person with an environment-mediated infectious disease becomes a source of water supply pathogens [33].

3.2.2 Air Pollution Exposure Assessment in Environmental Epidemiology

Human exposure to air pollutants is related with various health and nuisance effects. Acute biological effects include outcomes, such as an increase of existing disease (e.g., increased frequency and severity of asthma), acute respiratory system infections (e.g., increased respiratory disease ratio in children), transient pulmonary function, and allergic reactions. Chronic health outcomes include long-term decrements in lung growth, chronic obstructive pulmonary disease (e.g., bronchitis), cancer, neurobehavioral changes, and heart disease. However, common effects of most population are nuisance effects. These are often acute, including harmful scents; eyes, nose, and throat irritation; and coughing, which is a symptomatic breathing response.

Air pollutants found in various occupational, industrial, residential, outdoor, and public access and transport microenvironments include extensive and complex chemicals in both gaseous and particulate phase. Ideally, air pollutants involving adverse health or nuisance effects will be identified, and an exposure assessment protocol will be designed. Often, the identification of the causes of the effect is confounded by air quality factors other than conventional air pollutants (such as temperature, humidity, noise, or lighting). In practice, however, exposure to a class of pollutants, source categories, or a proxy pollutant usually must be addressed when specific pollutants are not able to be recognized or easily measured [81].

3.2.3 Framework of Air Pollution Exposure Assessment

The pathways by which a person is exposed to specific environmental pollutants include inhalation, ingestion, and/or dermal absorption, which usually depend on the pollutants exposed to. Air pollution exposure can be regarded as the process of exposure to harmful substances in the air that can be inhaled into the organism. Based on the early framework proposed by Nieuwenhuijsen et al. [83], the basic process of air pollution exposure is shown in Fig. 3.3 [128].

In Fig. 3.3, air pollutants usually originate from outdoor and indoor pollution sources. If we try to accurately evaluate personal exposure, we should consider both environmental (outdoor background) and indoor exposure. Methods for evaluating individual exposure include personal monitoring, biological monitoring (e.g., biomarkers), and environmental monitoring or modeling. The first two are commonly used to assess indoor exposure and environmental exposure, while the latter one is more widely used in separate environmental exposures.

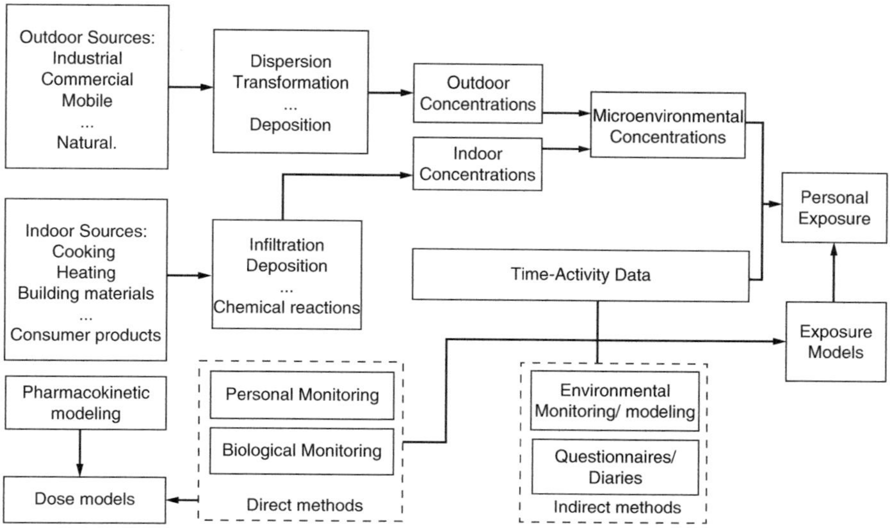

Fig. 3.3 A framework of air pollution exposure and its possible assessment approaches [128]

3.3 Exposure Measurement

3.3.1 Personal Exposure Monitoring

In theory, personal exposure to any certain air pollutant is defined as the concentration of the given air pollutant in inhaled air entering the nose or mouth. Practically, it is measured by sampling air pollutant with a personal exposure monitor (PEM) worn by a person and sampling from a point near the breathing zone (but not affected by exhaled breath).

There are many suitable measurement methods to monitoring air pollution; three examples are as follows:

1. Direct-Reading Devices
The concentration of air pollution can be determined with the use of portable direct-reading instruments. The advantages of this kind instrument are that they usually have data logging capabilities and can store data for many days. Direct calculation of concentrations with different averaging times and statistical analysis of the data can be implemented. The disadvantage is that the date may be less accurate for some specific air pollution.

2. Pumped or Passive Samplers
Sampling and analytical methods generally involve taking a sample of air and analyzing it subsequently in the laboratory. The air can be sampled by a sealable container, made with inert materials, like glass bottles, plastic bags, or inertized stainless steel containers.

Otherwise, the airborne pollutants can be concentrated by drawing air with pumps through a collecting medium where they are captured. Typically, a known volume of air is pumped through a collector such as a filter (glass and quartz fiber filters, Teflon and polycarbonate membranes, etc.), a chemical solution, or an adsorbent bed for a known period of time. The performances of these samplers are influenced mainly by airflow rate, contacting surface area, the presence of interfering compounds, breakthrough phenomena, and usually the temperature and humidity. Sampling and analytical methods are available for hundreds of pollutants.

As exposure evaluation of airborne particles is one of the main contents of the air pollution health effect study, we take the monitoring of particle matters as an example here. Specifically, PEMs for particulate matter use measurement techniques similar to those employed for measuring ambient particulate matter.

But different from the ambient PM monitoring, PEMs must be worn by studying participants, so they must be quiet, compact, and battery operated. Because of these requirements, the type of pumps and the total sample volume that can be collected by the PEMS were limited. So generally, small sample volumes limit personal exposure measurements to PM mass and a few elements detected by X-ray fluorescence (XRF). And in most studies, $PM_{2.5}$ and PM_{10} have not been collected concurrently. The most used PEMs now rely on filter-based mass measurement of a particle size fraction (PM_{10} or $PM_{2.5}$) with battery-operated pumps to integrate a 24- or 12-h period at flow rates of 2–4 L/min.

Sampling and analytical methods can also rely on passive sampling devices, thus avoiding the use of pumps. They are also called diffusive samplers, because the diffusion of molecules in air is the driving force of the sampling process. In passive samplers, the target compounds migrate toward the absorbing/reacting medium by diffusing through a membrane.

Qualitative and quantitative analyses of the trapped air pollutants are usually performed by spectrophotometry, gas or liquid chromatography, mass spectrometry, and so on after extraction of the compounds by solvent desorption or by heating (thermal desorption). Take the PM as an example, the collected filters are first subjected to gravimetric analysis and then to chemical analysis (such as sulfate, nitrate, chloride, ammonium, metals, elemental carbon/organic carbon (EC/OC), polycyclic aromatic hydrocarbons (PAHs), etc.)

Care should be taken to follow proper sampling protocol, such as that it must be aware of the limitations involved with the use of sampling instruments as well as the inadequacies of the chemical analysis techniques. For example, the measurements of PM may not be accurate if the sampling is conducted in wetter winter rather than the drier summer and there may be excessive loss of semi-volatiles if the sampling is conducted under very hot conditions ($T > 35$ °C). These might result in underestimate of the actual mass concentrations, and the error might further disturb the overall chemical mass balance.

3. Real-Time Monitoring

Modern technology has facilitated real-time monitoring of particulates and gases. Instrumental systems available for the real-time or semi-real-time measurement of air pollutants increased considerably. For instance, optical particle counters (OPC), scanning mobility particle sizer (SMPS), tapered element oscillating microbalance (TEOM), ultrafine particle counters (UPC), etc., can measure the number of particles and/or real time their aerosol size distribution. In addition, portable gas chromatographs (GC) are now available to monitor volatile organic compounds (VOCs) and other key gaseous pollutants and can provide real-time information about their concentrations. In Table 3.1, there are some typologies of professional and accurate instruments available for real-time measuring of several properties of airborne particles.

Since people spend most of the time (70–90% of the total time) in the indoor environment (such as the family, the workplace, and so on), many exposure studies have shown that there would be significant differences among the concentrations of fixed monitoring sites and outdoor, indoor, and personal levels. Therefore, detail studying the personal exposure both from the indoor and outdoor environments and their relationships of air pollutants (components) can help to understand and correct the exposure error caused by human activities and air penetrating behavior and improve the authenticity and accuracy of the results of epidemiological studies.

Table 3.1 Real-time instrument for different purpose of particulate matter monitoring

Monitoring purpose	Instrument	References
Number of particles	Optical particle counters (OPC)	[62]
	Condensation particle counters (CPC)	[22]
Distribution of the aerodynamic size of particles	Scanning mobility particle sizers (SMPS)	[101]
	Aerodynamic particle sizers (APS)	[101]
	Optical particle sizers (OPS)	[20]
	Differential mobility particle sizers (DMPS)	[57]
Particle mass concentration	Tapered element oscillating microbalance (TEOM)	[87]
	Beta attenuation mass monitor (BAM)	[37]
	Continuous aerosol mass monitor (CAMM)	[45]
	Nephelometer	[8]
Active surface area of the nanoparticles	Nanoparticle surface area monitor (NSAM)	[100]
Black carbon (BC)	Multiangle absorption photometer (MAAP)	[102]
	Particulate soot absorption photometer (PSAP)	[102]
Number concentration of UFC ultrafine particles (UFP)	Ultrafine particle counters (UPC)	[13]

3.3.2 Indoor-Outdoor Sampling

At present, a large number of related studies made use of the air pollutant concentrations from fixed monitoring sites to assess the relationship between air pollutants and health. However, in general, it has been found that population exposure might exceed the average urban concentration and is higher indoors than outdoors. In the case of $PM_{2.5}$ and polychlorinated biphenyls (PCBs), it has been demonstrated that the average concentration indoors turns out to be high than the outdoor average ambient level in some studies [15, 73, 123]. Therefore, some recent studies have criticized outdoor/ambient concentration for being poor indicators of the actual personal exposure levels of the general population to ambient air pollutants. This is true since most of the people spend nearly 90% of their time indoors.

Meanwhile, personal exposure measurement, where pollution is measured using personal wearable devices, can better quantify observed differences and better reflect exposure among smaller groups of people at ground level. It is also considered to be the most accurate method of pollutant exposure assessment.

At present, this methodology is increasingly used but is still less common than scenario-based methods because of costs involved in their collection and because necessary data, measurement devices, and models are not available to provide interpolation of the health implications for all the situations and populations of interest with sufficient temporal and spatial specificity.

So in practical, overall considering neither the monitoring costs affordable nor technically feasible in epidemiology studies, it is better to collect and carefully analyze a small number of well-chosen samples taken at key locations to provide a reliable index of environmental conditions, such as the indoor and outdoor microenvironment. The results obtained by these samples could be used along with exposure models and statistics to derive estimates on the exposure of the overall population, that is, the "microenvironmental" methods, which combined measurements in important microenvironments (e.g., inside the residence (indoors) and outdoors in the community) with data on time-activity patterns.

The measurement methods to monitoring indoor and outdoor air pollution were similar to the personal exposure monitoring, e.g., direct-reading devices, sampling and analytical methods with either pumped or passive sampling, and real-time monitoring. There are also central site-based outdoor air monitoring stations.

In China, there were still needs of systematic study on indoor/outdoor and individual exposure characteristics. Since the great differences of geographical environment, source distribution, pollution characteristics, economic conditions, and living habits, there is a wide range of differences between indoor, outdoor, and individual air pollution exposure in different regions of China. Take the children as an example, there are 26.8% of Chinese children exposed to the indoor environment of using solid fuel cooking or heating; 13.6% of Chinese children take their main activities within the oil, petrochemical, and other industrial enterprises surrounding in the area of 1 km; and 14.6% of them take their main activities within the traffic arteries surrounding in the area of 50 m. These different environmental characteristics of pollutants make the relationship between the indoor, outdoor, and individual air pollution exposure quite significantly different in children of China compared with the other countries [77].

Therefore, to understand the relationship between indoor and outdoor particle concentration in China, there are three widely used parameters just as that in the other country, i.e., indoor/outdoor (I/O) ratio, penetration factor, and infiltration factor [21, 98].

1. I/O ratio

I/O ratio directly represents the relationship between indoor and outdoor concentrations of pollutants, which is very easy to understand and widely used. I/O ratio is defined as:

$$\mathrm{I/Oratio} = \frac{C_{\mathrm{in}}}{C_{\mathrm{out}}}$$

where C_{in} and C_{out} are the indoor and outdoor pollutant's concentration, respectively.

2. Infiltration factor (F_{in})

F_{in} represents the equilibrium fraction of ambient particles that penetrates indoors and remains suspended. It avoids the mixture with indoor particle sources. F_{in} isdefined as $F_{\mathrm{in}} = \frac{\mathrm{aP}}{a+K'}$, which based on the steady-state and zero indoor particle emission rate as case of Eq. 3.1:

$$V\frac{\mathrm{dC_{in}}}{\mathrm{dt}} = \mathrm{aPVC_{out}} - \mathrm{aVC_{in}} - \mathrm{KVC_{in}} + \dot{S}, \tag{3.1}$$

where V is the volume of the room, t is time, a is the air exchange rate due to infiltration, P is the penetration factor, K is the particle deposition rate, and $\dot{S}$ is the indoor particle emission rate. It should be noted that the resuspension of particles is neglected in this equation. All the parameters except V and a are a function of both time and particle size [21, 69].

When the mechanical ventilation system is used, the filter efficiency can affect Eq. 3.1. Thus the generalized definition of F_{in} can be expressed as:

$$C_{\mathrm{in}} = F_{\mathrm{in}} C_{\mathrm{out}} + C_{\dot{S}}$$

where $C_{S}^{\cdot}$ is the indoor particle concentration which is contributed by indoor sources.

3. Penetration factor (P)

Penetration factor is defined as the fraction of particles in the infiltration air that passes through the building shell. It is the most relevant parameter for the penetration mechanism through cracks. The penetration factor could be calculated by experiment data measured in both real buildings and laboratories. Here, we mainly introduce some equations in real buildings such as:

$$C_{\text{in}}\left(t\right)=\left[C_i-\frac{\dot{S}+\text{aPVC}_{\text{out}}}{\left(a+K\right)V}\right]e^{-\left(a+K\right)t}+\frac{\dot{S}+\text{aPVC}_{\text{out}}}{\left(a+K\right)V}=\left[C_i-C_f\right]e^{-\left(a+K\right)t}+C_f$$

where $C_i = C_{\text{in}}(t = 0)$ and C_f = Cin(t=). When there are no particle sources $\left(\dot{\text{S}}=0\right)$ such as the measurement was conducted at night, and the particle decay curve and air exchange rate could be determined, such as the studies by Tung et al. [107], Thatcher et al. [106], and Chao et al. (2001)[19], then P can be calculated by:

$$P=\frac{\left(a+K\right)C_f}{\text{aC}_{\text{out}}}$$

However, when indoor particle emissions cannot be avoided, the method developed by Long et al. [70] which is based on the linear regression approach can be used. After measuring the indoor and outdoor particle concentrations under different conditions, $C_{\dot{S}}$ can be obtained. Then, according to $\text{I / Oratio}=\frac{\text{aP}}{a+K}+\frac{\dot{S}}{\left(a+K\right)\text{VC}_{\text{out}}}$, a linear expression can be rewritten as:

$$\frac{C_{\text{out}}}{C_{\text{in}}-C_{\dot{S}}}=\frac{K}{P}\left(\frac{1}{a}\right)+\frac{1}{P}$$

Based on this equation, after measuring air exchange rates under different conditions, the penetration factor can be easily obtained.

In China, the degree of human exposure to indoor air pollutants has not yet been well characterized, as well as the relationship between indoor and outdoor. This is principally because indoor air contaminants arise from multiple sources and their concentration levels and constituents differ greatly from those found in outdoor air with temporal and spatial variations. Take the study in Tian, China, for example, Xu et al. [122] collected daily personal exposure, residential indoor/outdoor, and community central site PM filter samples in an elderly community during the non-heating and heating periods in 2009, and mass concentrations of the particulate compounds were estimated based on the chemical analysis results of particulate species. The results showed that the infiltration factors (Fin) of particulate compounds were different during different seasons; in summer the Fin of PM_{10} is 0.74 ± 0.31, while in winter the Fin of PM_{10} is 0.44 ± 0.22. And the PM mass observed during the heating period could be well represented through chemical mass reconstruction. So to characterize the relationship of human exposure to indoor and outdoor air pollutants, a list of typical personal exposure, indoor air pollution, and outdoor/ambient air pollution monitoring studies in China is gathered in Table 3.2.

Studying the personal, indoor, and outdoor relationships of air pollutants (components) can help to understand and correct the exposure error caused by penetrating behavior and improve the authenticity and accuracy of the results of epidemiological studies in China. As studies showed in Table 3.2, in general, the indoor particles comprise infiltrated outdoor particles, particles generated indoors, and a few secondary organic aerosol (SOA) particles in most Chinese families, and in most cases, outdoor particle pollution concentrations are a major contributor to indoor concentrations, though the I/O ratio <1. Several studies have also found that outdoor-generated $PM_{2.5}$ showed stronger relationships with indoor $PM_{2.5}$ exposure. However, as indoor air pollution of particles was determined by many factors, such as the season, the building physics, heating, ventilation, and air conditioning system features, filter configurations, etc., more considerations and researches should be taken.

3.3.3 *Mobile Monitoring*

During the past decades, several air quality monitoring techniques were developed following the continuously improvements in air pollution researches, thus allowing the collection and the analysis of high-quality data. Nowadays, airborne PM is constantly monitored by national environmental agencies, which have organized large network of monitoring. However, these monitoring sites and devices are typically stationary, large, sparsely deployed and expensive. Moreover, fixed stations are not sufficient to monitor adequately the air quality in large urban areas and cannot characterize the spatial variations of air pollutants. Indeed, the concentration and chemical composition of PM in urban areas present considerable differences at a local scale depending on the emission sources, topography, and weather conditions [78]. The limited number of available monitoring stations installed in the urban areas affects the resolution of air pollution maps/forecast models that are based on datasets generated by these devices [44, 55], which are also the only ones recognized by agencies and public institutions.

For the reasons above, mobile air quality monitoring is attracting an increasingly growing interest [76, 88, 108]. With the advent of portable instruments providing high-time resolution and real-time data, the approach to air quality monitoring has deeply changed. A wide typology of professional and accurate instruments is now available for real-time measuring of several properties of airborne particles, such as OPC, SMPS, and CAMM, as shown in Table 3.1. These professional instruments have been used on different platforms (i.e., car, van, pedestrian, bicycle, tram, airship) for mobile monitoring of the various properties of PM. In particular, mobile air quality monitoring is mainly carried out through motor vehicles that can be equipped with voluminous and moderately heavy instruments (e.g., [61, 65, 86, 115, 118]). Table 3.3 summarized the main articles on mobile monitoring of PM.

However, there still exists some limitation on mobile monitoring. First, to derive time-representative pollutant maps associated to different traffic and meteorological

Table 3.2 Character of personal exposure, indoor air pollution, and outdoor/ambient air pollution monitoring studies in China

Studies	Time	Location	Population	Air pollutants	Monitoring	Main outcome
Xu et al. (2014) [121]	2009–2011	Tianjin, China	Two elderly communities, 181 elderly people	PM_{10}; $PM_{2.5}$; OC; EC; PAH	Personal/indoor/ outdoor	F_{in} of PM_{10}(mean ± SD) were In summer: 0.74 ± 0.31 In winter: 0.44 ± 0.22
Han et al. (2015) [39]	Non-heating and heating period, 2009	Tianjin, China	12 individual and 36 homes	PAHs	Personal/indoor/ outdoor	The average I/O ratios were lower than 1 both in heating and non-heating period
Cao et al. (2005) [15]	March and April 2004	Hong Kong, China	Six residences (two roadside, two urban, and two rural)	$PM_{2.5}$; OC; EC	Indoor/outdoor	The average I/O ratios of 24 h $PM_{2.5}$, OC, and EC were 1.4, 1.8, and 1.2, respectively
Deng et al. (2016) [31]	2014	Hong Kong, China	Five kindergartens	$PM_{2.5}$; BC; etc.	Indoor/outdoor	The O/I ratios were BC, 0.37–2.4; formaldehyde, 2.4; acetaldehyde, 1.9
Chan (2002) [17]	March–December	Hong Kong, China	Student' office	Respirable suspended particulate (RSP); NO_x	Indoor/outdoor	The average I/O ratios of RSP and NO_x were 0.7887 and 0.5623, respectively
Chao et al. (2001)	April 1998–January 1999	Hong Kong, China	Five homes	$PM_{2.5}$	Indoor/outdoor	The I/O ratios were within the range of 0–1
Wang et al. (2006) [116]	August 2004–September 2004	Guangzhou, China	Four hospitals	PM_{10}; $PM_{2.5}$; 18 target elements	Indoor/outdoor	The average I/O ratios were $PM_{2.5}$: 1.01(0.56–1.38) PM_{10}: 0.89(0.45–1.21)
Huang et al. (2005, 2007) [47, 48]	July and August 2004	Guangzhou, China	Nine residences	$PM_{2.5}$; OC; EC; 18 element	Indoor/outdoor	The I/O ratios of 18 element ranged from 0.4974 to 2.0497
Ke et al. (2011) [58]	March 25–April 5, 2010	Guangzhou, China	A primary school	$PM_{2.5}$	Indoor/outdoor	The I/O ratio = 0.825(0.66–2.22)

Wang et al. (2014) [114]	2010	Guangzhou, China	Elementary school and industrial downwind community	$PM_{2.5}$ and its components	Indoor/outdoor	The I/O ratios of $PM_{2.5}$ were
						At school: 1.0 ± 0.5
						In community: =1.0 ± 0.2
Chen et al. (2008) [23]	October 2004 and April 2005	Guangzhou and Fushan, China	14 homes, 16 workplaces, and 10 other indoor microenvironments	Polybrominated diphenyl ethers (PBDEs)	Indoor/outdoor	The concentrations of $\sum_{10}$PBDEs were
						In-house: 125.1–2,877 pg m^{-3}
						The office: 181.3–8,315 pg m^{-3}
						The other workplace: 322.1–2,437 pg m^{-3}
						Outdoor air: 203.2–2,426 pg m^{-3}
Wang et al. 2016 [113]	2014–2015 winter	Yangtze River Delta, China	Four residential dwellings	$PM_{2.5}$	Indoor/outdoor	The I/O ratios and F_{in} were 0.876 and 0.867, respectively
Shi et al. (2015) [99]	2013–2014	Beijing, China	180 representative residences	CO_2 multizone network airflow model (CONTAM)	Indoor/outdoor distributions by CO_2	The annual average F_{in} range from 0.02 to 0.82 h^{-1} with a median value of 0.16 h^{-1}
Huang et al. (2015) [50]	Non-heating season 2013	Beijing, China	41 urban residences having 1- to 8-year-old children	$PM_{2.5}$	Indoor /outdoor/ ambient	The median indoor/ambient ratio of $PM_{2.5}$ mass concentration was 0.62
Qi et al. (2017) [93]	Non-heating and heating seasons 2015	Beijing, China	13 households	$PM_{2.5}$	Indoor/outdoor (ambient)	The mean I/O ratios were
						Non-heating: 0.88 ± 0.56
						Heating: 0.53 ± 0.35
Han et al. (2015) [40]	Non-heating and heating periods 2013–2014	Beijing, China	A three-bedroom apartment at the 15th floor of a 20-story apartment building	$PM_{2.5}$	Indoor/outdoor (ambient)	$PM_{2.5}$ concentrations in ambient air were
						Heating: 82 ± 67 mg/m^3
						Non-heating: 86 ± 53 mg/m^3
						In indoor air were
						Heating: 59 ± 33
						Non-heating: 55 ± 30 mg/m^3

(continued)

Table 3.2 (continued)

Studies	Time	Location	Population	Air pollutants	Monitoring	Main outcome
Xu et al. (2011) [123]	July 2007–August 2008	Beijing, China	*Ca* 20 homes of 60 aged people	$PM_{2.5}$	Indoor/outdoor	The daily mean concentrations of outdoor and indoor $PM_{2.5}$ were 85.5 ± 48.7 μg/m^3 and 110.5 ± 77.5 μg/m^3, respectively
Pu et al. (2015) [92]	July–September 2013	Beijing, China	40 families in urban area distributed in 6 districts	Carbonyls	Indoor/outdoor	The total concentrations of carbonyls were 136.0 ± 77.5 μg/m^3
Liu et al. (2013) [68]	November–December 2009	Beijing, China	210 residential homes	Acetone, acrolein, and xylenes	Indoor/outdoor	The I/O ratios of acetone, acrolein, and xylenes were <1.0
Yao et al. (2015) [126]	January–February 2013	Jinan, China	A student lab and a yard	$PM_{2.5}$; perchlorate; chlorate	Indoor/outdoor	The I/O ratios of PM $_{2.5}$, chlorate, and perchlorate were 0.40–0.91, 0.05–0.70, and 0.12–16.47, respectively
Han et al. (2016) [41]	Winter and summer 2008	Shanghai, China	20 homes and 20 offices as well as 10 outdoor sites	PBDE	Indoor (home and office)/outdoor	The median I/O ratios of Σ_{15}PBDEs and BDE − 209 were
						In summer: 6.4 (1.0–11.5) and 0.41 (0.06–1.58)
						In winter: 2.7 (1.0–24.6) and 0.17 (0.02–0.25)
Lei et al. (2016) [64]	December 2014–March 2015	Shanghai, China	51 graduate students aged 22–28 years at a university in Xuhui District in Shanghai	$PM_{2.5}$; BC	Personal/indoor/outdoor (ambient)	The concentrations of $PM_{2.5}$ were
						Personal: 110.0 μg m^{-3}(average)
						Ambient: 94.5 μg m^{-3} (median)

Li et al. (2016) [67]	February–March 2013 and September 2013	Lanzhou, China	53 households during the heating seasons and 54 households during non-heating seasons	$PM_{2.5}$	Indoor (kitchen/bedroom)/outdoor	The average concentrations of $PM_{2.5}$ in the kitchen, bedroom, and outdoor were
						Heating: 125 ± 51, 119 ± 64, and 328 ± 104 μg/m^3
						Non-heating: 80 ± 67, 80 ± 50, and 80 ± 49 μg/m^3
Chen et al. (2017) [28]	May, 2011 (non-heating period)	Shanxi, China	84 households from four villages	PAHs	Personal/microenvironments (kitchen, bedroom, outdoor)	Total concentrations of nPAHs and oPAHs were
						Personal: 0.47 ± 0.28 and 18 ± 22 ng/m^3
						Kitchen: 2.2 ± 2.5 and 230 ± 520 ng/m^3
						Living room: 0.73 ± 0.48 and 61 ± 120 ng/m^3
						Outdoor: 0.33 ± 0.21 and 9.9 ± 1.8 ng/m^3
Hu et al. (2014) [53]	August 2008–February 2009	Xuanwei and Fuyuan, China	163 nonsmoking female heads of households enrolled from 30 villages	$PM_{2.5}$	Personal/indoor	Geometric means (geometric standard deviation) of $PM_{2.5}$ concentrations in Xuanwei and Fuyuan were
						Personal: 166 (2.0) and 146 (1.9) μg/m^3
						Indoor air: 162 (2.1) and 136 (2.0) μg/m^3
Li (1994) [66]	–	China	Three domestic environments in residential and industrial communities	Total suspended particulate (TSP), PM_{10}	Indoor/outdoor	The average I/O was 0.60

Table 3.3 Summary table of the main articles on mobile monitoring reporting information on the instrumentation, platform, and measured parameters

Study	Platform	Instrumentations	Measured parameters	Objectives
Wang et al. (2009) [115]	Van	Scanning mobility particle sizer – TSI DMA 3081	Size distribution D = 15–673 nm	Verify the effectiveness of pollution control policies adopted for the 2008 Beijing Olympic Games
		Optical particle counter – GRIMM Dust Monitor 1.108	Size distribution D = 0.3–20 μm	
		Nanoparticle surface area monitor/TSI 3550	Active surface area	
		Multiangle absorption	Black carbon position	
		Photometer – Thermo Model 5012		
Zwack et al. (2011) [129]	Pedestrians	TSI, P-Track 8525	Ultrafine particles (UFP)	Discern contributions of traffic within street canyons by regression techniques
		TSI, Dust Track 820	$PM_{2.5}$	
Elen et al. (2013)[32]	Bicycle	TSI, P-Track 8525	UFP, PM_1, $PM_{2.5}$, PM_{10}, TSP, BC	Mobile air quality
		GRIMM Dust Monitor 1.108		Monitoring in small areas
		AE51, AethLabs MicroAeth		
Castellini et al. (2014) [16]	Cabin of urban metro system	Customized optical particle counter (OPC)	Size distribution (0.28–10 μm)	Monitoring air quality in urban environment
Frick and Hoppel (2000) [36]	Airship NRL DMA size spectrometer	PMS optical particle counter	Size distribution (0.005–0.6 μm)	Effect of ship's exhaust on marine boundary layer
		TSI condensation particle counter	Size distribution (0.25–23.5 μm)	
			Size distribution (>0.003 μm)	

(continued)

Table 3.3 (continued)

Study	Platform	Instrumentations	Measured parameters	Objectives
Riley et al. (2106) [94]	Van	TSI portable CPC (ethanol-based) model 3007	Particle number (PN) conc., 10 nm–1 μm	Aircraft turbine emission
Hudda et al. (2014) [51]		AethLabs MicroAeth AE51 model year 2010	BC	
		EcoChem PAH analyzer, model PAS 2000	Particle-bound polycyclic aromatic hydrocarbons (PB-PAH) and elemental carbon	
		Garmin GPSMAP 76CSx	Location, speed	
		2B Technologies Model 401-410	NO_X	
		Aerodyne CAPS NO_2 monitor	NO_2	
Chan et al. (2017) [18]	Three SCANIA 43-seat tour buses	Air quality data was recorded using the Smart eHome Wireless Indoor Air Quality (IAQ) Monitoring System	The CO_2 concentrations and temperature conditions	CO_2 concentrations and temperature changes within high-passenger occupied tour buses
Wang et al. (2016) [117]	Vehicle	Cavity Enhanced Differential Optical Absorption Spectroscopy (CE-DOAS) an on-vehicle, real-time system, consists of an Aethalometer, a vehicle exhaust analyzer and a portable global positioning system (GPS) device	BC	The primary vehicular emissions and the ambient air quality
Hu et al. (2011) [52]	Vehicle	Vehicular sensor networks (VSNs)	CO_2	Equip fewer nodes on cars to achieve fine-grained monitoring
		A Zig Bee-based prototype		

conditions, a large amount of data are required [86, 108]. The need for a high number of repeated measurements to assess the temporal and spatial variability of air pollutants has been recently raised by Peters et al. [89]. However, the minimum number of measurements to get a reliable representation of the variability of pollutants is not straightforward [109]. Essentially, the number of repeated measurements is a function of the desired spatial resolution [108]. An appropriate background correction of the final must be also considered [108].

Second, the chemical characterization of PM is another crucial issue and is mandatory to identify the origin of specific components, i.e., their emission sources. Currently, the chemical features of airborne particles are determined with different analytical techniques able to perform major and trace elemental analysis and anion and cation speciation analysis, e.g., XRF, ICP-MS, PIXE, INAA, AAS, IC, and AC. Although instruments allowing real-time measurements of the chemical composition of PM have been developed and are currently available (e.g., ion trap mass spectrometer, aerosol chemical speciation monitor, thermal desorption chemical ionization mass spectrometry), at present they still cannot be easily extended to mobile platforms mainly due to their dimensions (e.g., [43, 90, 95, 103, 112, 125]).

Third limitations for the integration of these instruments are the low concentrations of some chemical components in PM that would require long acquisition times. As an example, Voisin et al. [112] reported the application of thermal desorption chemical ionization mass spectrometry to the analysis of ammonium sulfate nanoparticles with a sensitivity to ambient mass concentration of 50 pg/m^3 using an aerosol sampling flow of 8 L/min and a collection time of 10 min.

A major demand in PM mobile monitoring is the development of compact (i.e., suitable for different mobile platforms) and high-performance instruments for real-time chemical analysis. Still a critical issue is the sensitivity to aerosol mass concentration.

Currently, the real-time chemical characterization of aerosol allows a time resolution of tens of minutes and a maximum sensitivity to aerosol concentration of tens of pg/m^3 [29, 112].

3.4 Exposure Assessment Modeling

Exposure assessment is one of the important components in the assessment on health effects caused by air pollution. The inclusion of more individual samples may reduce the error of health assessment; however, the increase in the number of individuals brings in the monitoring problem. Since the time, energy, and expense are limited, it is difficult to measure the concentration of pollutants for each individual. Therefore, it is the main method to obtain the concentration of air pollutants by using the exposure assessment model. In recent years, a variety of models have been developed by scientists around the world and gradually improved and developed. In this section, we introduced four common used exposure assessment models for reference and supply the applications of these models. For some other newly developed models, we use Sect. 4.5 "Other Models" to give brief introduction.

3.4.1 Proximity Models

The proximity model represents one of the most basic approaches to distinguish air exposure within a city. This approach hypothesizes that the distances from the sources are related with health outcomes of population and explains the relationship between air pollution and health effects. The model is simple, is cost-effective, and can be run on GIS software.

The model was used in Europe, the United States, the United Kingdom, and Canada from 1997 to 2003 to assess the volume of traffic and the distance from the road in urban scale. In practice, different types of roads are often categorized and measured to differentiate traffic volumes. Janssen et al. [54] measured the concentrations of $PM_{2.5}$, NO_2, and benzene along 24 major roads inside and outside the school within the range of 400 m. Wyler et al. [120] calculated the hourly average traffic volume of the roads around the participant's home based on the traffic list. Venn et al. [110] used traffic activity guidelines to measure the sum of road traffic per 1 km^2. English et al. [34] used the traffic pollution emission model to estimate the exposure in different radii around the home.

In China, Deng et al. [30] showed children living close to the main road were more influenced by the exposure to air pollutants. Huang and Guo [49] reviewed the assessment methods of vehicle exhaust pollution exposure, and this model can be applied to assess the vehicle exhaust exposure.

Although the proximity approach provides a direct application for long-term exposure, there are still some limitations. The disadvantage is that it can only be judged qualitatively, but not quantitatively. The result is more susceptible to some cofounders, such as the impact of wind direction, terrain factors, and misclassification [55].

3.4.2 Interpolation Model

The interpolation model is based on deterministic and stochastic geostatistical modeling techniques. Contaminants will be distributed at the various monitoring sites in the study area. In general, estimates are also obtained in a grid, and the entire grid is distributed over the studying area, so that continuous contaminant concentrations can be obtained. In general, the kriging model is the most suitable interpolation method in the field of atmospheric pollution, and its best linear unbiased estimator can estimate the variables of any point [14].

The main advantage of this method is that the predicted value of the unknown point and the corresponding standard error (kriging difference) can be obtained at the same time. These standard errors make the prediction uncertainty of the unknown point be quantified, and the credibility of interpolation is low. The interpolation method relies on the continuous spatial distribution of data. Apart from the noise and error of data, spatial dependence includes two kinds of data, one is global data such as global trend and the other is local distance [4, 14]. Other methods, such as Spline and IDW based on deterministic or geographic algorithms, allow reasonable estimation of pollutant concentrations at unknown points. These methods are easier to apply and are more suitable for monitoring areas with fewer outlets and larger distances.

In North America and Europe, there are already many differential applications in cities. Pikhart et al. [91] used geostatistics to estimate SO_2 levels in small areas. Mulholland et al. [80] used the universal kriging method to analyze the temporal and spatial distribution of ozone in 20 counties of Atlanta. Abbey et al. [1] used the inverse distance weighting method to calculate the relative risk of exposure. The advantage of the interpolation method over the adjacent model is that it utilizes real contaminant measurement data to reliably quantify exposure level differences. Kriging is more problematic than other interpolation methods in that the local variability is greater and the choice is better, but regional marginal representation is still a problem. The ordinary kriging method assumes that there is no global change in the singularity of the data, also known as the static hypothesis [80]. Based on this assumption, the pollutant change at each point in the study area is based on the distance between two points and the direction of the line. Although this technique has been able to detect similar errors, an estimation error is generated if this assumption is violated [91]. The universal kriging is an expansion of ordinary kriging, incorporating trend functions to account for spatial variations in pollutants.

The disadvantage of the geostatistical difference is the availability of monitoring data. This model requires intensive monitoring network, according to research accuracy and pollutant change scale, the local pollutant emission source, the size of the estimated error, the topography of the area, and the main meteorological conditions of comprehensive judgment, the city area usually 10–100 monitoring points. In general, government-monitoring sites are sparse, and the selected sites are susceptible to industrial or severe traffic pollution. Selecting a government-monitoring network would generally smooth the pollution level map more than the actual level, and a relatively small data point would have a relatively large estimation bias. This problem is particularly severe for pollutants such as NO_x that are subject to large variations in small areas. If you seek more site data, it will come with high cost or short sampling time. So the researchers can only select one to two kinds of representative pollutants that are easy to monitor. Applying the geostatistical model requires a mastery of GIS and statistical software, such as ESRI's ArcGIS model.

In China, there are similar studies, such as Fan et al. [35] and Meng and Ma [74]. However, their researches are limited to the introduction of methods and monitoring data processing, but no health effects were mentioned, which meant they didn't combine pollutant data with health data together to get the relationship between air pollutants and health. In a recent study, to assess the benefits of the program "Air Pollution Prevention and Control Action Plan (Action Plan)" in Beijing-Tianjin-Hebei (BTH) Region, Chen et al. [26] estimated $PM_{2.5}$ concentration on a grid with a spatial resolution of 1×1 km^2 using Voronoi spatial interpolation method. And by combining with the exposure-response function between $PM_{2.5}$ concentration and health endpoints based on Environmental Benefits Mapping and Analysis Program

(BenMAP), the economic loss is assessed, which provides scientific basis for implementation of air pollution control strategies. They also use this method to assess the population exposure to $PM_{2.5}$ for mortality in China [27].

3.4.3 Air Dispersion Models

Diffusion models are generally based on the Gaussian equation, using emission sources, meteorological conditions, and topographic data to assess spatial exposure to air pollution [5, 7]. Data on emission, meteorological parameters, and topography were used to estimate spatial exposure of air pollution. In recent years, dispersion models have been applied together with GIS. The combination analyzes both information from traditional monitoring systems and data concerning the population distribution in the studying area. With the supplement of data concerning the topography of the studying area, models of road networks, and traffic volume observation, a more realistic solution of the problem is formed [55].

Such models have been well used in many studies, such as Bartonova et al. [5] predicting TSP and NO_x in 1999 and Bellander et al. [7] and Nyberg et al. [85] establishing a NO_x prediction model. The establishment of a diffusion model requires the data of pollutants, meteorological conditions, and emission sources. Contaminant data, i.e., background data, are commonly used for routine monitoring of government-monitoring stations in the vicinity of the study area. Meteorological data generally include wind speed, wind direction, ambient temperature, solar irradiance, and atmospheric stability. Data of emission sources are generally grouped into two categories: one is human-made stationary sources such as domestic heating or plant emissions. Information on the emission factors (e.g., annual emissions, smoke height, diameter, temperature, vertical discharge velocity) of each source of pollution, as well as the type and location of the equipment, are collected [46]. With the improvement of the accuracy of the model, only simple update data are needed to predict the concentration of pollutant at the required time and place. Hruba et al. [46] established the US EPA's long-term industrial source complex model with the data of 151 pollution sources and two residential areas in the Banska Bystrica area of the Slovakia center and estimated the distribution of airborne particulate contamination by this model. The results of this model are used effectively in studies of air pollution in Europe and children's respiratory health. Gualtieri and Tartaglia [38] estimated the concentration of air pollutants using the traffic volume of the road network, i.e., assessing the distribution of pollutants in typical road contaminated areas with the Gaussian model. From 1990 to 2000, a large number of studies using diffusion model combined with GIS assessed and predicted urban NO_x, $PM_{2.5}$, and SO_2. Some domestic scholars also used the diffusion model to predict the concentration of urban air pollutants.

In China, Ma et al. [71] used the Gaussian point source diffusion model and the kriging interpolation model to analyze the spatial distribution of SO_2 concentration in Lanzhou City. Both could reflect the spatial distribution of SO_2 accurately but still had a certain distance from the measured value. In addition, Yang et al. [124] applied the AERMOD diffusion model with the pollutant data of Shenyang. Meanwhile, the diffusion model also has some shortcomings: (1) the relatively complex data input, (2) the inaccurately estimated diffusion method, (3) more testing data for cross-validation, and (4) estimated errors due to temporal variation. Sampling interval difference would appear during the input of pollutant not related to traffic or significantly related to diesel vehicles. In addition, point-and-surface emissions are generally reported with annual averages, whereas heavy-oil vehicles do not increase in hourly traffic. The hourly meteorological data will also bring the error. In addition, the overloaded data model program had higher requirement on GIS professional proficiency and computer performance. Therefore, it is not suitable for forecast in large areas, and the current application is relatively small. Huang and Guo [49] incorporated motor vehicle emissions data into the model to estimate the changes in motor vehicle tail gas over time and space, including motor vehicle emissions, the number of motor vehicles, the main type of each type standard emissions, motor vehicle speed, road network status, and other information.

In general, more studies applied diffusion model in China, compared with other models. However, these studies only focused on the methodologies and regional data validation but were insufficient on the population health data. Therefore, the interdisciplinary cooperation between environmental sciences and public health should be encouraged; thus we can both simulate the exposure level and estimate health effects of pollutants.

3.4.4 Land-Use Regression Models

The land-use regression model is based on the land-use and traffic information around the observed point to predict pollutants. The regression model is an effective way to evaluate the exposure assessment of traffic-related pollutants by using the pollutant concentration at one point as the dependent variable and the land-use data at that point as the independent variable. The land-use regression methodology tries to make prediction on the concentrations of certain pollutant at a given site based on surrounding geographic variables, such as land-use type, traffic characteristics, etc. More specifically, this method uses measured concentrations of certain pollutants at given location or monitoring site as the dependent variable (response variable) and geographic information within areas around location (we call them buffers) as the independent variables or predictors for measured concentrations [11, 12, 63]. The method entails the use of least-squares regression modeling to predict pollution surfaces based on pollution monitoring data and existing exogenous independent variables [55].

Briggs et al. [11] firstly used the land-use regression model to simulate small-scale changes in the Small Area Variations in Air Quality and Health (SAVIAH) study, to study the influence on children's respiratory disease by long-term exposure at mean air pollution levels. As the study population included thousands of children, sampling of air pollution at each location was less feasible. Thus, 80 sampling points were planned in each city to ensure sufficient sample density to build an effective model. The regression equation R^2 (0.79–0.87) was obtained for the average annual NO_2 concentration. Thereafter, the regression model was widely used, especially in Europe and North America, and most of the application were at large multi-city region, with a few at small-scale urban areas. Other studies have been used throughout the country, such as the Netherlands and the United Kingdom [104]. The ATMOSPHERE project simulated the concentration of pollutants at the scale of 1 × 1 km across Europe [9].

Land-use regression is based on the use of different monitoring data to build the model and usually in the large-scale study often used routine monitoring network data [2, 42]. Currently, only a handful of research used data from routine monitoring sites [6, 10, 11, 79, 96, 104]. For most urban studies, there are few regular monitoring points and cannot meet the need of establishing a small-scale model of outdoor air change. Therefore, passive sampling methods were adopted to meet the required quantity. Madsen et al. [72] found that 40 monitoring sites and 80 monitoring sites had no significant effect on the model results, in Oslo urban study. Jerrett et al. [56] reported that the model established with 65 sites was similar to that with 94 sites. In general, linear regression is used to calculate the regression model, and the model residuals are independent. When there is no spatial autocorrelation, the general least squares is also used to establish the model [56, 72, 97]. If there is a correlation between the residuals of the land-use regression model [9], the generalized kriging model (the combination of linear regression and the regular kriging model) was combined. The current land-use regression model has been used to model the spatial distribution of NO_2, NO_x, $PM_{2.5}$, EC in $PM_{2.5}$, Soot, and VOCs. The interpretation of the model is usually around 60–70%. Compared with other methods, the land-use model has lower cost, but this method is limited to the areas with similar land use, weather, and traffic conditions. Therefore, only a few studies have been successfully applied when the model is applied to other areas with different land use and different terrain. Briggs et al. [12] successfully applied the model developed in Huddersfield at other times in the city, as well as three other UK cities. However, when applying the regression coefficients in Amsterdam in the study of Hamilton district in Canada, Briggs et al. [12] did not achieve the desired results, as no significant spatial variability appeared near the roads and underestimation/overestimation existed in other areas. Even in regions with similar geographical conditions and traffic conditions, intensive monitoring data are still required. Therefore, land-use model is a type of semiquantitative model with empirical relation, which is different with the diffusion model, and the limitation in migrated application which makes it can only be applied in the vicinity of areas with similar variables.

In China, the application of LUR in the health assessment of environmental exposure is relatively late. In 2003, Zhang and Wang [127] used the evolution of land use and cover to conduct the spatial analysis and probe the impact of land use on regional atmospheric environment in Shanghai. Chen et al. [24] firstly developed a multiple linear land-use regression model through the NO_2 and PM_{10} of conventional monitoring in Tianjin and found that the R^2 were 0.74 and 0.72 in heating season and 0.61 and 0.49 in non-heating season for NO_2 and PM_{10}, respectively. After that, they also established the models of SO_2, NO_2, and PM_{10} and obtained the better predictive effect which indicated the R^2 were 0.78, 0.89, and 0.84, respectively [25]. Wu et al. [119] combined the spatial-temporal prediction model and land-use information to predict the $PM_{2.5}$ levels in Beijing and found that the temporal variation of $PM_{2.5}$ was greater than the spatial variation and the adjusted R^2 was 0.43–0.65. Recently, Meng et al. [75] successfully established the land-use regression model for the NO_2 concentrations in Shanghai from 2008 to 2011. The model showed that the R^2 was 0.75 and superior to the results of kriging model and inverse distance weight model.

3.4.5 Other Models

Because LUR model is only used to single city or area, it is difficult to apply to the other cities or regions. Therefore, more scholars used the data from high-quality remote sensing satellites to improve the application of LUR in larger areas. Novotny et al. [84] combined the land-use information and NO_2 concentrations of ground and satellite to predict the NO_2 concentration of the entire United States. The prediction accuracy was 30 m, and the correlation of predicted value with monitoring value can reach 0.78. And the model can be used to any sites which have sufficient road network and monitoring data of pollutants. Vienneau et al. [111] also used the monitoring data from more than 1,500 points and the satellite data of NO_2 and $PM_{2.5}$, and combined with the land-use information to predict the NO_2 and PM_{10} levels of the whole 100 m grid in Western Europe. The prediction model R^2 for NO_2 and PM_{10} was increased by 0.05 and 0.11. Knibbs et al. [60] also combined the satellite data and generalized estimating equation model to predict the concentrations on a national scale of Australia. They found that the interpretations of annual and monthly NO_2 were 81% and 76%, respectively.

In order to conduct long-term exposure studies of prospective cohorts, the spatial-temporal model was used to predict the temporal and spatial variations of pollutants. The principle of the model is that the regional pollutant concentrations are divided into several time and space domain according to local geographic information data, meteorological data, and monitoring data of pollutants. The concentrations of every domain vary with the space and time. In the context of fully consideration of various uncertain factors, the Bayesian function algorithm was used to obtain the concentration in the location of each object. The indoor concentrations could be calculated through the building characteristic and permeability parameter simulation. Then the exposure concentration of each subject was weighted based on the indoor and outdoor activity time [105]. In recent years, spatial-temporal model has been used in some studies [3, 59].

References

1. Abbey DE, Nishino N, McDonnell WF, Burchette RJ, Knutsen SF, Beeson WL, Yang JX. Long-term inhalable particles and other air pollutants related to mortality in nonsmokers. Am J Respir Crit Care Med. 1999;159(2):373–82.
2. Adam-Poupart A, Brand A, Fournier M, Jerrett M, Smargiassi A. Spatiotemporal modeling of ozone levels in Quebec (Canada): a comparison of kriging, Land-Use Regression (LUR), and combined Bayesian maximum entropy-LUR approaches. Environ Health Perspect. 2014;122(9):970–6.
3. Adar SD, Sheppard L, Vedal S, Polak JF, Sampson PD, Roux AVD, Budoff M, Jacobs DR, Barr RG, Watson K, Kaufman JD. Fine particulate air pollution and the progression of carotid intima-medial thickness: a prospective cohort study from the multi-ethnic study of atherosclerosis and air pollution. PLoS Med. 2013;10(4):e1001430.
4. Bailey TC, Gatrell AC. Interactive spatial data analysis. Essex: Longman Scientific & Technical; 1995.
5. Bartonova A, Clench-Aas J, Gram F, Gronskei KE, Guerreiro C, Larssen S, Tonnesen DA, Walker SE. Air pollution exposure monitoring and estimation part V. Traffic exposure in adults. J Environ Monit. 1999;1(4):337–40.
6. Beelen R, Hoek G, Fischer P, van den Brandt PA, Brunekreef B. Estimated long-term outdoor air pollution concentrations in a cohort study. Atmos Environ. 2007;41(7):1343–58.
7. Bellander T, Berglind N, Gustavsson P, Jonson T, Nyberg F, Pershagen G, Jarup L. Using geographic information systems to assess individual historical exposure to air pollution from traffic and house heating in Stockholm. Environ Health Perspect. 2001;109(6):633–9.
8. Brauer M, Bartlett K, RegaladoPineda J, PerezPadilla R. Assessment of particulate concentrations from domestic biomass combustion in rural Mexico. Environ Sci Technol. 1996;30:104–9.
9. Briggs D, Aaheim A, Dore C, Hoek G, Petrakis M, Shaddick G. Air pollution modelling for support to policy on health and environmental risks in Europe. Final Report Section 6. Imperial College, London. EVK2-2002-00176. 2005. Available from: http://www.apmosphere.org.
10. Briggs DJ. The use of GIS to evaluate traffic-related pollution. Occup Environ Med. 2007;64(1):1–2.
11. Briggs DJ, Collins S, Elliott P, Fischer P, Kingham S, Lebret E, Pryl K, VanReeuwijk H, Smallbone K, VanderVeen A. Mapping urban air pollution using GIS: a regression-based approach. Int J Geogr Inf Sci. 1997;11(7):699–718.
12. Briggs DJ, de Hoogh C, Guiliver J, Wills J, Elliott P, Kingham S, Smallbone K. A regression-based method for mapping traffic-related air pollution: application and testing in four contrasting urban environments. Sci Total Environ. 2000;253(1–3):151–67.
13. Brock CA, Schroder F, Karcher B, Petzold A, Busen R, Fiebig M. Ultrafine particle size distributions measured in aircraft exhaust plumes. J Geophys Res-Atmos. 2000;105:26555–67.
14. Burrough PA, McDonnell R, McDonnell RA, Lloyd CD. Principles of geographical information systems. Oxford: Oxford University Press; 2015.
15. Cao JJ, Lee SC, Chow JCL, Cheng Y, Ho KF, Fung K, Liu SX, Watson JG. Indoor/outdoor relationships for $PM_{2.5}$ and associated carbonaceous pollutants at residential homes in Hong Kong -case study. Indoor Air. 2005;15:197–204.
16. Castellini S, Moroni B, Cappelletti D. PMetro: measurement of urban aerosols on a mobile platform. Measurement. 2014;49:99–106.
17. Chan AT. Indoor-outdoor relationships of particulate matter and nitrogen oxides under different outdoor meteorological conditions. Atmos Environ. 2002;36:1543–51.
18. Chan KL, Wang S, Liu C, Zhou B, Wenig MO, Saiz-Lopez A. On the summer time air quality and related photochemical processes in the megacity Shanghai, China. Sci Total Environ. 2017;580:974–83.

19. Chao CYH, Tung TC. An empirical model for outdoor contaminant transmission into residential buildings and experimental verification. Atmos Environ. 2001;35:1585–96.
20. Chazette P, Randriamiarisoa H, Sanak J, Couvert P, Flamant C. Optical properties of urban aerosol from airborne and ground-based in situ measurements performed during the Etude et Simulation de la Qualite de l'air en Ile de France (ESQUIF) program. J Geophys Res-Atmos. 2005;110:D02206.
21. Chen C, Zhao B. Review of relationship between indoor and outdoor particles: I/O ratio, infiltration factor and penetration factor. Atmos Environ. 2011;45:275–88.
22. Cheng YS. Condensation particle counters. In: Aerosol measurement. Wiley; 2011. p. 381–392.
23. Chen LG, Mai BX, Xu ZC, Peng XC, Han JL, Ran Y, Sheng GY, Fu JM. In- and outdoor sources of polybrominated diphenyl ethers and their human inhalation exposure in Guangzhou, China. Atmos Environ. 2008;42(1):78–86.
24. Chen L, Bai ZP, Kong SF, Han B, You Y, Ding XA, Du SY, Liu AX. A land use regression for predicting NO_2 and PM_{10} concentrations in different seasons in Tianjin region, China. J Environ Sci (China). 2010;22(9):1364–73.
25. Chen L, Wang YM, Li PW, Ji YQ, Kong SF, Li ZY, Bai ZP. A land use regression model incorporating data on industrial point source pollution. J Environ Sci (China). 2012;24(7):1251–8.
26. Chen L, Shi M, Gao S, Li S, Mao J, Zhang H, Sun Y, Bai Z, Wang Z. Assessment of population exposure to PM2.5 for mortality in China and its public health benefit based on BenMAP. Environ Pollut. 2017;221:311–7.
27. Chen L, Shi M, Li S, Gao S, Zhang H, Sun Y, Mao J, Bai Z, Wang Z, Zhou J. Quantifying public health benefits of environmental strategy of $PM_{2.5}$ air quality management in Beijing-Tianjin-Hebei region, China. J Environ Sci (China). 2017;57:33–40.
28. Chen Y, Du W, Shen G, Zhuo S, Zhu X, Shen H, Huang Y, Su S, Lin N, Pei L, Zheng X, Wu J, Duan Y, Wang X, Liu W, Wong M, Tao S. Household air pollution and personal exposure to nitrated and oxygenated polycyclicaromatics (PAHs) in rural households: influence of household cooking energies. Indoor Air. 2017;27(1):169–78.
29. Crippa M, DeCarlo PF, Slowik JG, Mohr C, Heringa MF, Chirico R, Poulain L, Freutel F, Sciare J, Cozic J, DiMarco CF, Elsasser M, Nicolas JB, Marchand N, Abidi E, Wiedensohler A, Drewnick F, Schneider J, Borrmann S, Nemitz E, Zimmermann R, Jaffrezo J-L, Pevot ASH, Baltensperger U. Wintertime aerosol chemical composition and source apportionment of the organic fraction in the metropolitan area of Paris. Atmos Chem Phys. 2013;13:961–81.
30. Deng F, Wang X, Su H, Wu S, Lv H, Guo X. Personal exposure to $PM_{2.5}$ of children living near traffic road and the influencing factors in Beijing. J Environ Health. 2009;09:762–5.
31. Deng WJ, Zheng HL, Tsui AK, Chen XW. Measurement and health risk assessment of $PM_{2.5}$, flame retardants, carbonyls and black carbon in indoor and outdoor air in kindergartens in Hong Kong. Environ Int. 2016;96:65–74.
32. Elen B, Peters J, Poppel MV, Bleux N, Theunis J, Reggente M, Standaert A. The Aeroflex: a bicycle for mobile air quality measurements. Sensors (Basel). 2012;13(1):221–40.
33. Eisenberg JN, Lei X, Hubbard AH, Brookhart MA, Colford JM. The role of disease transmission and conferred immunity in outbreaks: analysis of the 1993 Cryptosporidium outbreak in Milwaukee, Wisconsin. Am J Epidemiol. 2005;161(1):62–72.
34. English P, Neutra R, Scalf R, Sullivan M, Waller L, Zhu L. Examining associations between childhood asthma and traffic flow using a geographic information system. Environ Health Perspect. 1999;107(9):761–7.
35. Fan M, Chen C, Chi T. Research and application on the continuity of monitoring data using kriging interpolation. Comput Eng Appl. 2005;09:210–2.
36. Frick G, Hoppel W. Airship measurements of ship's exhaust plumes and their effect on marine boundary layer clouds. J Atmos Sci. 2000;57:2625–48.
37. Gobeli D, Schloesser H, Pottberg T. Met one instruments BAM-1020 beta attenuation mass monitor US-EPA PM2.5 federal equivalent method field test results. In: The Air & Waste Management Association (A&WMA) Conference, Kansas City, MO. Citeseer; 2008.

38. Gualtieri G, Tartaglia M. Predicting urban traffic air pollution: a GIS framework. Transp Res Part D-Transp Environ. 1998;3(5):329–36.
39. Han B, Bai Z, Liu Y, You Y, Xu J, Zhou J, Zhang J, Niu C, Zhang N, He F, Ding X. Characterizations, relationship, and potential sources of outdoor and indoor particulate matter bound polycyclic aromatic hydrocarbons (PAHs) in a community of Tianjin, Northern China. Indoor Air. 2015;25(3):320–8.
40. Han Y, Qi M, Chen Y, Shen H, Liu J, Huang Y, Chen H, Liu W, Wang X, Liu J, Xing B, Tao S. Influences of ambient air $PM_{2.5}$ concentration and meteorological condition on the indoor $PM_{2.5}$ concentrations in a residential apartment in Beijing using a new approach. Environ Pollut. 2015;205:307–14.
41. Han WL, Fan T, Xu BH, Feng JL, Zhang G, Wu MH, Yu YX, Fu JM. Passive sampling of polybrominated diphenyl ethers in indoor and outdoor air in Shanghai, China: seasonal variations, sources, and inhalation exposure. Environ Sci Pollut Res. 2016;23:5771–81.
42. Hart JE, Yanosky JD, Puett RC, Ryan L, Dockery DW, Smith TJ, Garshick E, Laden F. Spatial modeling of PM_{10} and NO_2 in the continental United States, 1985–2000. Environ Health Perspect. 2009;117(11):1690–6.
43. Held A, Rathbone GJ, Smith JN. A thermal desorption chemical ionization ion trap mass spectrometer for the chemical characterization of ultrafine aerosol particles. Aerosol Sci Technol. 2009;43:264–72.
44. Hoek G, Beelen R, de Hoogh K, Vienneau D, Gulliver J, Fischer P, Briggs D. A review of land-use regression models to assess spatial variation of outdoor air pollution. Atmos Environ. 2008;42:7561–78.
45. Hopke PK, Liu W, Han YJ, Yi SM, Holsen TM, Cybart S, Milligan M. Measured summertime concentrations of particulate components, Hg0, and speciated polycyclic aromatic hydrocarbons at rural sites in New York State. Environ Pollut. 2003;123(3):413–25.
46. Hruba F, Fabianova E, Koppova K, Vandenberg JJ. Childhood respiratory symptoms, hospital admissions, and long-term exposure to airborne particulate matter. J Expo Anal Environ Epidemiol. 2001;11(1):33–40.
47. Huang H, Lee S, Cao J, Zou C, Chen X, Fan S. Characterization of indoor outdoor organic and elemental carbon in $PM_{2.5}$ during summer in Guangzhou city. Acta Sci Circumst (in Chinese). 2005;25(9):1242–9.
48. Huang H, Li S, Cao J, Zou C, Chen X, Fan S. Characterization and source of element components in indoor and outdoor $PM_{2.5}$ during summer and winter in Guangzhou City. Acta Sci Circumst (in Chinese). 2007;23(4):383–8.
49. Huang J, Guo X. Research progress in exposure assessment methods of vehicle exhaust pollution. Environ Chem. 2011;01:138–42.
50. Huang LH, Pu ZN, Li M, Sundell J. Characterizing the indoor-outdoor relationship of fine particulate matter in non-heating season for urban residences in Beijing. PLoS One. 2015;10:0138559.
51. Hudda N, Gould T, Hartin K, Larson TV, Fruin SA. Emissions from an international airport increase particle number concentrations 4-fold at 10 km downwind. Environ Sci Technol. 2014;48:6628–35.
52. Hu SC, Wang YC, Huang CY, Tseng YC. Measuring air quality in city areas by vehicular wireless sensor networks. Mobile applications: status and trends. J Syst Softw. 2011;84:2005–12.
53. Hu W, Downward GS, Reiss B, Xu J, Bassig BA, Dean Hosgood H, Zhang LL, Seow WJ, Wu GP, Chapman RS, Tian LW, Wei FS, Vermeulen R, Lan Q. Personal and indoor PM2.5 exposure from burning solid fuels in vented and unvented stoves in a rural region of China with a high incidence of lung cancer. Environ Sci Technol. 2014;48(15):8456–64.
54. Janssen NAH, van Vliet PHN, Aarts F, Harssema H, Brunekreef B. Assessment of exposure to traffic related air pollution of children attending schools near motorways. Atmos Environ. 2001;35(22):3875–84.
55. Jerrett M, Arain A, Kanaroglou P, Beckerman B, Potoglou D, Sahsuvaroglu T, Morrison J, Giovis C. A review and evaluation of intraurban air pollution exposure models. J Expo Anal Environ Epidemiol. 2005;15(2):185–204.

56. Jerrett M, Arain MA, Kanaroglou P, Beckerman B, Crouse D, Gilbert NL, Brook JR, Finkelstein N, Finkelstein MM. Modeling the intraurban variability of ambient traffic pollution in Toronto, Canada. J Toxicol Environ Health A. 2007;70(3–4):200–12.
57. Keady PB, Quant FR, Sem GJ. Differential mobility particle sizer: a new instrument for high-resolution aerosol size distribution measurement below 1 μm. TSI Quarterly. 1983;9:3–11.
58. Ke Z, Wang J, Zhen J, Lai S, Zhang Y. $PM_{2.5}$ exposure assessment of school children at a primary school in Guangzhou, China. China. Environ Sci (in Chinese). 2011;31(10):1618–24.
59. Keller JP, Olives C, Kim SY, Sheppard L, Sampson PD, Szpiro AA, Oron AP, Lindstrom J, Vedal S, Kaufman JD. A unified spatiotemporal modeling approach for predicting concentrations of multiple air pollutants in the multi-ethnic study of atherosclerosis and air pollution. Environ Health Perspect. 2015;123(4):301–9.
60. Knibbs LD, Hewson MG, Bechle MJ, Marshall JD, Barnett AG. A national satellite-based land-use regression model for air pollution exposure assessment in Australia. Environ Res. 2014:135204–11.
61. Kolb CE, Herndon SC, McManus JB, Shorter JH, Zahniser MS, Nelson DD, Jayne JT, Canagaratna MR, Worsnop DR. Mobile laboratory with rapid response instruments for real-time measurements of urban and regional trace gas and particulate distributions and emission source characteristics. Environ Sci Technol. 2004;38(21):5694–703.
62. Kulkarni P, Baron PA, Willeke K. Aerosol measurement: principles, techniques, and applications. New York: Wiley; 2011.
63. Lebret E, Briggs D, van Reeuwijk H, Fischer P, Smallbone K, Harssema H, Kriz B, Gorynski P, Elliott P. Small area variations in ambient NO_2 concentrations in four European areas. Atmos Environ. 2000;34(2):177–85.
64. Lei X, Xiu G, Li B, Zhang K, Zhao M. Individual exposure of graduate students to $PM_{2.5}$ and black carbon in Shanghai, China. Environ Sci Pollut Res Int. 2016;23(12):12120–7.
65. Levy I, Mihele C, Lu G, Narayan J, Hilker N, Brook JR. Elucidating multipollutant exposure across a complex metropolitan area by systematic deployment of a mobile laboratory. Atmos Chem Phys. 2014;14:7173–93.
66. Li CS. Relationships of indoor/outdoor inhalable and respirable particles in domestic environments. Sci Total Environ. 1994;151(3):205–11.
67. Li TX, Cao SZ, Fan DL, Zhang YQ, Wang BB, Zhao XG, Leaderer BP, Shen GF, Zhang YW, Duan XL. Household concentrations and personal exposure of $PM_{2.5}$ among urban residents using different cooking fuels. Sci Total Environ. 2016;548–549:6–12.
68. Liu QY, Liu YJ, Zhang MG. Personal exposure and source characteristics of carbonyl compounds and BTEXs within homes in Beijing, China. Build Environ. 2013;61:210–6.
69. Li Y, Chen Z. A balance-point method for assessing the effect of natural ventilation on indoor particle concentrations. Atmos Environ. 2003;37:4277–85.
70. Long CM, Suh HH, Catalano PJ, Koutrakis P. Using time- and size resolved particulate data to quantify indoor penetration and deposition behavior. Environ Sci Technol. 2001;35:2089–99.
71. Ma A, Hu H, An X, Gao F, Wang Q2. Using GIS to simulate distribution of ground concentration of urban air pollution. Urban Environment and Urban Ecology. 2007;02:32–4. 38
72. Madsen C, Carlsen KCL, Hoek G, Oftedal B, Nafstad P, Meliefste K, Jacobsen R, Nystad W, Carlsen KH, Brunekreef B. Modeling the intra-urban variability of outdoor traffic pollution in Oslo, Norway – a GA(2)LEN project. Atmos Environ. 2007;41(35):7500–11.
73. Melymuk L, Bohlin-Nizzetto P, Kukučka P, Vojta Š, Kalina J, Čupr P, Klánová J. Seasonality and indoor/outdoor relationships of flame retardants and PCBs in residential air. Environ Pollut. 2016;218:392–401.
74. Meng J, Ma X. Kriging analysis and its application on urban atmospheric environmental pollution. Math Pract Theory. 2002;02:309–12.
75. Meng X, Chen L, Cai J, Zou B, Wu CF, Fu QY, Zhang Y, Liu Y, Kan HD. A land use regression model for estimating the NO_2 concentration in Shanghai, China. Environ Res. 2015:137308–15.

76. Merbitz H, Fritz S, Schneider C. Mobile measurements and regression modeling of the spatial particulate matter variability in an urban area. Sci Total Environ. 2012;438:389–403.
77. Ministry of Environmental Protection of the People's Republic of China. Report of environmental exposure related activity patterns research of Chinese population (Children). Beijing; 2016.
78. Monn C, Carabias V, Junker M, Waeber R, Karrer M, Wanner HU. Small-scale spatial variability of particulate matter <10 mu m (PM_{10}) and nitrogen dioxide. Atmos Environ. 1997;31:2243–7.
79. Moore DK, Jerrett M, Mack WJ, Kunzli N. A land use regression model for predicting ambient fine particulate matter across Los Angeles, CA. J Environ Monit. 2007;9(3):246–52.
80. Mulholland JA, Butler AJ, Wilkinson JG, Russell AG, Tolbert PE. Temporal and spatial distributions of ozone in Atlanta: regulatory and epidemiologic implications. J Air Waste Manage Assoc. 1998;48(5):418–26.
81. National Research Council. Human exposure assessment for airborne pollutants: advances and opportunities. Washington, DC: National Academies Press; 1991.
82. National Research Council. Exposure science in the 21st century: a vision and a strategy. Washington, DC: National Academies Press; 2012.
83. Nieuwenhuijsen M, Paustenbach D, Duarte-Davidson R. New developments in exposure assessment: the impact on the practice of health risk assessment and epidemiological studies. Environ Int. 2006;32(8):996–1009.
84. Novotny EV, Bechle MJ, Millet DB, Marshall JD. National satellite-based land-use regression: NO_2 in the United States. Environ Sci Technol. 2011;45(10):4407–14.
85. Nyberg F, Gustavsson P, Jarup L, Bellander T, Berglind N, Jakobsson R, Pershagen G. Urban air pollution and lung cancer in Stockholm. Epidemiology. 2000;11(5):487–95.
86. Padro-Martinez LT, Patton AP, Trull JB, Zamore W, Brugge D, Durant JL. Mobile monitoring of particle number concentration and other traffic-related air pollutants in a near-highway neighborhood over the course of a year. Atmos Environ. 2012;61:253–64.
87. Patashnick H, Rupprecht EG. Continuous PM-10 measurements using the tapered element oscillating microbalance. J Air Waste Manage Assoc. 1991;41:1079–83.
88. Patton AP, Perkins J, Zamore W, Levy JI, Brugge D, Durant JL. Spatial and temporal differences in traffic-related air pollution in three urban neighborhoods near an interstate highway. Atmos Environ. 2014;99:309–21.
89. Peters J, Van den Bossche J, Reggente M, Van Poppel M, De Baets B, Theunis J. Cyclist exposure to UFP and BC on urban routes in Antwerp, Belgium. Atmos Environ. 2014;92:31–43.
90. Petit JE, Favez SJ, Crenn V, Sarda-Estève R, Bonnaire N, Močnik G, Dupont JC, Haeffelin M, Leoz-Garziandia E. Two years of near real-time chemical composition of submicron aerosols in the region of Paris using an Aerosol Chemical Speciation Monitor (ACSM) and a multi-wavelength Aethalometer. Atmos Chem Phys. 2015;15:2985–3005.
91. Pikhart H, Bobak M, Gorynski P, Wojtyniak B, Danova J, Celko MA, Kriz B, Briggs D, Elliott P. Outdoor sulphur dioxide and respiratory symptoms in Czech and Polish school children: a small-area study (SAVIAH). Int Arch Occup Environ Health. 2001;74(8):574–8.
92. Pu ZN, Huang LH, Yue Y, Mo JH, Zhang YP. Characteristics of carbonyls in Beijing urban residences: concentrations, source strengths and influential factors. Procedia Eng. 2015;121:2052–9.
93. Qi M, Zhu X, Du W, Chen Y, Chen Y, Huang T, Pan X, Zhong Q, Sun X, Zeng EY, Xing B, Tao S. Exposure and health impact evaluation based on simultaneous measurement of indoor and ambient $PM_{2.5}$ in Haidian, Beijing. Environ Pollut. 2017;220(Pt A):704–12.
94. Riley EA, Gould T, Hartin K, Fruin SA, Simpson CD, Yost MG, Larson T. Ultrafine particle size as a tracer for aircraft turbine emissions. Atmos Environ. 2016;139:20–9.
95. Ripoll A, Minguillón MC, Pey J, Jimenez JL, Day DA, Sosedova Y, Canonaco F, Prévôt ASH, Querol X, Alastuey A. Long-term real-time chemical characterization of submicron aerosols at Montsec (southern Pyrenees, 1570 m a.s.l.). Atmos Chem Phys. 2015;15:2935–51.

96. Ross Z, English PB, Scalf R, Gunier R, Smorodinsky S, Wall S, Jerrett M. Nitrogen dioxide prediction in Southern California using land use regression modeling: potential for environmental health analyses. J Expo Sci Environ Epidemiol. 2006;16(2):106–14.
97. Ross Z, Jerrett M, Ito K, Tempalski B, Thurston GD. A land use regression for predicting fine particulate matter concentrations in the New York City region. Atmos Environ. 2007;41(11):2255–69.
98. Shi S, Chen C, Zhao B. Modifications of exposure to ambient particulate matter: tackling bias in using ambient concentration as surrogate with particle infiltration factor and ambient exposure factor. Environ Pollut. 2017;220(Pt A):337–47.
99. Shi SS, Chen C, Zhao B. Air infiltration rate distributions of residences in Beijing. Build Environ. 2015;92:528–37.
100. Shin WG, Pui DYH, Fissan H, Neumann S, Trampe A. Calibration and numerical simulation of nanoparticle surface area monitor (TSI model 3550 NSAM). In: Maynard AD, Pui DYH, editors. Nanotechnology and occupational health. Dordrecht: Springer; 2007. p. 61–9.
101. Sioutas C, Abt E, Wolfson JM, Koutrakis P. Evaluation of the measurement performance of the scanning mobility particle sizer and aerodynamic particle sizer. Aerosol Sci Technol. 1999;30:84–92.
102. Slowik JG, Cross ES, Han JH, Davidovits P, Onasch TB, Jayne JT, Williams LR, Canagaratna MR, Worsnop DR, Chakrabarty RK, Moosmüller H, Arnott WP, Schwarz JP, Gao RS, Fahey DW, Kok GL, Petzold A. An inter-comparison of instruments measuring black carbon content of soot particles. Aerosol Sci Technol. 2007;41:295–314.
103. Smith D, Španěl P, Herbig J, Beauchamp J. Mass spectrometry for real-time quantitative breath analysis. J Breath Res. 2014;8(2):027101.
104. Stedman JR, Vincent KJ, Campbell GW, Goodwin JWL, Downing CEH. New high resolution maps of estimated background ambient NOx and NO_2 concentrations in the UK. Atmos Environ. 1997;31(21):3591–602.
105. Szpiro AA, Sampson PD, Sheppard L, Lumley T, Adar SD, Kaufman JD. Predicting intraurban variation in air pollution concentrations with complex spatio-temporal dependencies. Environmetrics. 2010;21(6):606–31.
106. Thatcher TL, Lunden MM, Revzan KL, Sextro RG, Brown NJ. A concentration rebound method for measuring particle penetration and deposition in the indoor environment. Aerosol Sci Technol. 2003;37:847–64.
107. Tung TCW, Chao CYH, John B. A methodology to investigate the particulate penetration coefficient through building shell. Atmos Environ. 1999;33:881–93.
108. Van den Bossche J, Peters J, Verwaeren J, Botteldooren D, Theunis J, De Baets B. Mobile monitoring for mapping spatial variation in urban air quality: development and validation of a methodology based on an extensive dataset. Atmos Environ. 2015;105:148–61.
109. Van Poppel M, Peters J, Bleux N. Methodology for setup and data processing of mobile air quality measurements to assess the spatial variability of concentrations in urban environments. Environ Pollut. 2013;183:224–33.
110. Venn A, Lewis S, Cooper M, Hubbard R, Hill I, Boddy R, Bell M, Britton J. Local road traffic activity and the prevalence, severity, and persistence of wheeze in school children: combined cross sectional and longitudinal study. Occup Environ Med. 2000;57(3):152–8.
111. Vienneau D, de Hoogh K, Bechle MJ, Beelen R, van Donkelaar A, Martin RV, Millet DB, Hoek G, Marshall JD. Western European land use regression incorporating satellite- and ground-based measurements of NO_2 and PM_{10}. Environ Sci Technol. 2013;47(23):13555–64.
112. Voisin D, Smith JN, Sakurai H, McMurry PH, Eisele FL. Thermal desorption chemical ionization mass spectrometer for ultrafine particle chemical composition. Aerosol Sci Technol. 2003;37:471–5.
113. Wang F, Meng D, Li X, Tan J. Indoor-outdoor relationships of $PM_{2.5}$ in four residential dwellings in winter in the Yangtze River Delta, China. Environ Pollut. 2016;215:280–9.
114. Wang J, Lai S, Ke Z, Zhang Y, Yin S, Zheng J. Exposure assessment, chemical characterization and source identification of $PM_{2.5}$ for school children and industrial downwind residents in Guangzhou, China. Environ Geochem Health. 2014;36:385–97.

115. Wang M, Zhu T, Zheng J, Zhang RY, Zhang SQ, Xie XX, Han YQ, Li Y. Use of a mobile laboratory to evaluate changes in on-road air pollutants during the Beijing 2008 summer Olympics. Atmos Chem Phys. 2009;9:8247–63.
116. Wang X, Bi X, Sheng G, Fu J. Hospital indoor $PM_{10}/PM_{2.5}$ and associated trace elements in Guangzhou, China. Sci Total Environ. 2006;366:124–35.
117. Wang Y, Xing Z, Zhao S, Zheng M, Mu C, Du K. Are emissions of black carbon from gasoline vehicles overestimated? Real-time, in situ measurement of black carbon emission factors. Sci Total Environ. 2016b;547:422–42.
118. Westerdahl D, Fruin S, Sax T, Fine PM, Sioutas C. Mobile platform measurements of ultrafine particles and associated pollutant concentrations on freeways and residential streets in Los Angeles. Atmos Environ. 2005;39:3597–610.
119. Wu JS, Li JC, Peng J, Li WF, Xu G, Dong CC. Applying land use regression model to estimate spatial variation of $PM_{2.5}$ in Beijing, China. Environ Sci Pollut Res Int. 2015;22(9):7045–61.
120. Wyler C, Braun-Fahrlander C, Kunzli N, Schindler C, Ackermann-Liebrich U, Perruchoud AP, Leuenberger P, Wuthrich B. Exposure to motor vehicle traffic and allergic sensitization. Epidemiology. 2000;11(4):450–6.
121. Xu J, Bai Z, You Y, Zhou J, Zhang J, Niu C, Liu Y, Zhang N, He F, Ding X. Residential indoor and personal PM_{10} exposures of ambient origin based on chemical components. J Expo Sci Environ Epidemiol. 2014;24(4):428–36.
122. Xu J, Zhang N, Han B, You Y, Zhou J, Zhang J, Niu C, Liu Y, He F, Ding X, Bai Z. Assessment on personal exposure to particulate compounds using an empirical exposure model in an elderly community in Tianjin, China. Sci Total Environ. 2016;572:1080–91.
123. Xu MM, Jia YP, Li GX, Pan XC. Evaluation of personal integrated exposure to fine particle in a community in Beijing. J Environ Health. 2011;28(11):941–3.
124. Yang H, Ma Y, Zhang Y. Application of atmospheric diffusion model system in Shenyang Liaoning urban and rural. Environ Sci Technol. 2005;01:7–9.
125. Yang M, Reilly PTA, Boraas KB, Whitten WB, Ramsey JM. Real-time chemical analysis of aerosol particles using an ion trap mass spectrometer. Rapid Commmun Mass Sp. 1996;10:347–51.
126. Yao L, Yang LX, Chen JM, Toda K, Wang XF, Zhang JM, Yamasaki D, Nakamura Y, Sui X, Zheng LF, Wen L, Xu CH, Wang WX. Levels, indoor–outdoor relationships and exposure risks of airborne particle-associated perchlorate and chlorate in two urban areas in Eastern Asia. Chemosphere. 2015;135:31–7.
127. Zhang H, Wang X. Urban land use/cover dynamics of shanghai metropolitan area and its potential impact on local air environment. J Fudan University (Nat Sci). 2003;06:925–9.
128. Zou B, Wilson JG, Zhan FB, Zeng Y. Air pollution exposure assessment methods utilized in epidemiological studies. J Environ Monit. 2009;11(3):475–90.
129. Zwack LM, Paciorek CJ, Spengler JD, Levy JI. Characterizing local traffic contributions to particulate air pollution in street canyons using mobile monitoring techniques. Atmos Environ. 2011;45:2507–14.

Chapter 4
Ambient Air Pollution and Biomarkers of Health Effect

Di Yang, Xuan Yang, Furong Deng, and Xinbiao Guo

Abstract Recently, the air pollution situation of our country is very serious along with the development of urbanization and industrialization. Studies indicate that the exposure of air pollution can cause a rise of incidence and mortality of many diseases, such as chronic obstructive pulmonary disease (COPD), asthma, myocardial infarction, and so on. However, there is now growing evidence showing that significant air pollution exposures are associated with early biomarkers in various systems of the body. In order to better prevent and control the damage effect of air pollution, this article summarizes comprehensively epidemiological studies about the bad effects on the biomarkers of respiratory system, cardiovascular system, and genetic and epigenetic system exposure to ambient air pollution.

Keywords Air pollution • Inflammation • Oxidative stress • Genetic • Epigenetic

4.1 Introduction

It is well known that air pollutants can cause serious damage to the human body. Air pollutants are a complex mixture of particulate matter (PM) [125] and gases including ozone (O_3), nitrogen dioxide (NO_2), sulfur dioxide(SO_2) [21], polycyclic aromatic hydrocarbons (PAHs), and volatile organic compounds (VOCs). One group of PM was identified with aerodynamic diameter ≤2.5μm ($PM_{2.5}$), with a characteristic of smaller molecular diameter and large surface areas absorbing many heavy mental, and with the capability of entering the deeper respiratory tract and crossing the lung capillary network, leading to severe damage to various systems of the body. Recently, the air pollution situation of our country is very serious because of urbanization and modernization. Especially since January 2013, the incidence and scope of smog and haze in China gradually have increased greatly; the serious air

D. Yang • X. Yang • F. Deng (✉) • X. Guo
Department of Occupational and Environmental Health Sciences, School of Public Health, Peking University, No. 38 Xueyuan Road, Beijing, China
e-mail: lotus321321@126.com

G.-H. Dong (ed.), *Ambient Air Pollution and Health Impact in China*, Advances in Experimental Medicine and Biology 1017, DOI 10.1007/978-981-10-5657-4_4

pollution of the Beijing-Tianjin-Hebei region spreads its effect to national land area of about 1/10 and affected population of more than 400 million. According to the Beijing Municipal Environmental Protection Bureau data, the concentration of $PM_{2.5}$ in many monitoring stations exceeded 700 μg/m^3 in Beijing during haze period.

Air pollutants are associated with numerous health effects and can cause a variety of diseases, including asthma, COPD [30], myocardial infarction [17], and so on, and can increase morbidity and mortality. The Global Burden of Diseases, Injuries, and Risk Factors Study 2010 indicated that $PM_{2.5}$ induced 1.2 million early deaths and 25 million disability-adjusted life-years in China in 2010 [75].

We should have a clear understanding of the pathogenic mechanisms of pollutants to better prevent and control the damage effects on the human body by air pollutants. The commonly acknowledged ways include inflammation, oxidative stress, and genetic and epigenetic pathways. In this review, we attempt to summarize the bad effects on the biomarkers of respiratory system, cardiovascular system, and genetic and epigenetic system exposure to ambient air pollution from the epidemiological studies in china.

4.2 Inflammation

Inflammation is one of the most critical mechanisms of the body disease caused by the exposure of air pollutants. As one of the most serious air pollutants, $PM_{2.5}$, with a characteristic of smaller molecular diameter and capability of entering the deeper respiratory tract and absorbing on the cell surface, could induce the occurrence of inflammation through the stimulation of the transition metal and organic matter absorbed in its surface and increase inflammatory factor (interleukin, hydrogen sulfide (H_2S)), leading to a further pathophysiological change, and then finally make the occurrence of diseases.

4.2.1 Inflammation-Related Biomarkers of Respiratory System Exposure to Ambient Air Pollution

4.2.1.1 Exhaled Nitric Oxide (eNO)

Nitric oxide (NO), as an important physiological transmitter, is involved in the occurrence and development of many diseases. The eNO of a normal person is an extremely small amount (<25 ppb) of metabolites of L-arginine with the effect of nitric oxide synthase (NOS) in the airway epithelial cell. Inducible nitric oxide synthases (iNOS) induced by the pulmonary inflammation can promote the massive production of NO to make an anti-inflammatory effect. So, the change of the eNO level can reflect the airway inflammatory response of the human body followed by the simulation of pollutants. As an effective noninvasive sampling method, eNO is the most commonly used indicator of respiratory inflammation in the epidemiologic study. Most studies show that the level of eNO of different subjects (the healthy or patients, children, or adults) increases to varying degrees with the increasing exposure level of pollutants.

A panel study of 35 healthy college students as subjects found that an interquartile range (IQR) increase in $PM_{2.5}$ was associated with an increase in eNO of 11.3% (95% confidence interval (CI): 0.6%, 23.0%) [19]; the use of an air purifier can effectively decrease elevated eNO levels caused by air pollution; studies found that the mean $PM_{2.5}$ concentration in rooms with a true air purifier was greatly reduced to 41.3 mg/m^3, 57% lower than the concentration of the sham group, and eNO level was decreased significantly by 17.0% (95% CI: 3.6%, 32.5%) in geometric mean. In addition, some studies have drawn the opposite conclusion. A panel study of 75 Fudan University healthy college students as subjects found that an interquartile range increase (0.3 mg/m^3) of 2-h carbon monoxide (CO) exposure was associated with a decrease in eNO of 10.6% (95% CI: 6.3%, 14.9%) [140]. In two-pollutant model, the association remained when controlling for the concomitant exposure. Studies indicated that the effects of different pollutants on the level of eNO in the subjects may be different, and a further study is needed in the future.

There were studies which found that air pollution can also cause increased levels of eNO in patients with chronic obstructive pulmonary disease (COPD). A panel study of 23 COPD patients as subjects in Beijing found that interquartile range increases in $PM_{2.5}$, a particulate matter with aerodynamic diameter $\leq$10 μm (PM_{10}), and SO_2 were associated with maximum increases in eNO of 13.6% (95% CI: 4.8%, 23.2%), 9.2% (95% CI: 2.1%, 16.8%), and 34.2% (95% CI: 17.3%, 53.4%), respectively [119]. A similar conclusion was reached in a study of 18 COPD patients that is an interquartile range increase in $PM_{2.5}$ was associated with an increase in eNO of 13.13% [84]. Some further studies found that the contribution of the different components in $PM_{2.5}$ is different. A panel study of 30 COPD patients as subjects found that interquartile range increases in $PM_{2.5}$ at lag 1 day were associated with increases in eNO of 13.3%, and a further component analysis of $PM_{2.5}$ found that interquartile range increases in four components of $PM_{2.5}$ (organic carbon (OC), element carbon (EC), nitrate ions (NO_3^-), and ammonium (NH_4^+)) at lag 1 day were associated with increases in eNO of 16.93%, 8.97%, 18.26%, and 11.42%, respectively [18]. The above research results indicated that eNO can serve as an important biomarker of inflammation in patients with COPD after the exposure of air pollutants.

In addition, as the sensitive population, children without developed systems are even more susceptible to have an elevated level of eNO after the exposure of traffic-related air pollution. A panel study of 36 preschool children (28 healthy and 8 asthma) as subjects found that interquartile range increases in $PM_{2.5}$ and BC were associated with increases in eNO of 16.6% (95% CI:14.1%, 19.2%) and 18.7 (95% CI:15.0%, 22.5%), respectively [69]. In two-pollutant model, the association between black carbon (BC) and eNO became stronger after the adjustment of $PM_{2.5}$.

There were no significant effects on the levels of eNO by BC and $PM_{2.5}$ on the asthma children and the healthy children. All of the above studies indicated that air pollution had a serious effect on the children.

4.2.1.2 Exhaled Carbon Monoxide (eCO)

As an important cell messenger molecule, CO plays a key role in the human physiological processes. With the occurrence of the inflammatory reaction and the action of the reduced coenzyme II – nicotinamide adenine dinucleotide phosphate (NADPH) and cytochrome P450 reductase and molecular oxygen – heme oxygenase (HO) catalyzes the degradation of heme to biliverdin, CO, and iron to play an anti-inflammatory, antioxidant effect together. In recent years, CO has been served as a marker of respiratory inflammation in many studies, and a cross-sectional study found that the odds ratios (OR) and 95% CI of airflow obstruction (forced expiratory volume in 1 s (FEV1)/forced vital capacity (FVC) ratio <0.7) for never smokers with eCO at the level of 7–14 ppm and >14 ppm, compared with those having eCO <7, were 1.38 (95% CI: 1.31, 1.45) and 1.65 (95% CI:1.52, 1.80), respectively [130].

Coal burning is the main source of indoor air pollution, and the incomplete combustion of coal can produce CO of which exposure (especially in rural areas) can induce the elevated level of eCO of resident. A cross-sectional study of 510,000 residents as subjects found that the eCO levels of residents using coal as fuel were much higher than those using other fuels (natural gas and wood) in rural areas and urbans. The average levels of eCO of subjects using natural gas, wood, and coal as fuel were 20.4 ppm, 17.0 ppm, and 32.1 ppm, respectively [130]. There were other cross-sectional studies reaching a similar conclusion that the relative coefficient between the independent variable coal and wood fuel was 0.9 in the fuel type and health outcomes regression model and that the relative coefficient between the independent variable coal stove and wood stove was 0.12 in the stove type and health outcomes regression model [86].

The exposure time of cooking stoves is an important factor influencing the level of eCO in the coal-burning exposure. A study found that a 30-min increase in daily exposure to cooking stoves was associated with a 1.04 ppm increase in exhaled CO (95% CI: 1.02, 1.06) [86]. Another study found that the eCO level of subjects sampled in winter was higher than that sampled in any other season in rural areas, which may be related to the more times of cooking indoor in winter in rural areas [130]. The above studies show that environmental CO exposure is the main cause of elevated level of eCO.

4.2.1.3 Exhaled Hydrogen Sulfide (FeH_2S)

Endogenous H_2S, as a novel gas transmitter, may be involved in the pathophysiological processes of COPD by regulating inflammatory responses. A previous study demonstrated that the sputum H_2S levels in patients with acute COPD exacerbation were higher than those in patients with stable COPD, and sputum-to-serum H_2S ratio was positively correlated with sputum neutrophils and levels of inflammatory biomarkers (interleukin-6 and interleukin-8) in sputum and serum. A panel study of 23 COPD patients as subjects took FeH_2S as an inflammatory biomarker of respiratory and obtained the hourly average concentrations of ambient air pollutants $PM_{2.5}$, PM_{10}, SO_2, and NO_2 from Beijing Municipal Environmental Monitoring Center, collecting repeated exhaled breath samples every 5–7 days simultaneously to explore the exposure-response relationship between pollutants and FeH_2S [119]. The study found that the 24-h average concentration of $PM_{2.5}$, PM_{10}, and SO_2 was correlated significantly with FeH_2S and IQR increases in $PM_{2.5}$ (76.5 μg/m^3, 6-day), PM_{10} (75.0 μg/m^3, 6-day), and SO_2 (45.7 μg/m^3, 7-day) were associated with increases in FeH_2S of 11.4% (95% CI: 4.6%, 18.6%), 7.8% (95%CI:2.3%, 13.7%), and 18.1% (95% CI: 5.5%, 32.2%), respectively. Estimates are adjusted for temperature, relative humidity, age, BMI, sex, long-term time trend, and day of the week. The results indicated that FeH_2S can be used as an inflammatory marker in COPD patients.

4.2.1.4 Fractional Exhaled Nitrite and Nitrate (FeNO)

NO can be formed with the action of nitric oxide synthase (NOS) on substrate L-arginine in airway epithelial cells. Though NO is recognized as one of the markers of reactive airway inflammation, some researchers prefer to use fractional exhaled nitrite and nitrate, the ultimate metabolic products of NO, to represent the FeNO because of the high concentration of NO in the nasal cavity.

There are different degrees of increase in the level of the fractional exhaled nitrite in subjects along with the increasing level of air pollutants. A panel study of 125 healthy adults as subjects in the 2008 Beijing Olympic Games during which the air quality was improved evidently found that exhaled breath condensate (EBC) nitrite was positively associated with almost all pollutants at one or more lags. Across the lags, the largest effect estimates per IQR increase in pollutant concentration were 21.9% for $PM_{2.5}$, 20.5% for SO_2, 15.7% for CO, and 22.2% for NO_2. A further analysis of the constituent of $PM_{2.5}$ found that EBC nitrite was significantly associated with CO, EC, and sulfate radical (SO_4^{2-}) [53].

There is a big difference between the lag time of ultrafine particle (UFP) and $PM_{2.5}$ in the largest lag effect, and the effect of UFP on the EBC nitrite is bigger than $PM_{2.5}$. Another research investigated the association between UFP and EBC nitrite using the same survey mentioned above and found that the largest percentage change in EBC nitrite associated with an interquartile range increase in UFPs at lag 6 was 25.64 (95% CI: 16.12, 35.94) [38]. The difference in lag effect time indicated that there was a big difference in the process of deposition, clearance, and translocation in the lung between UFP and $PM_{2.5}$. Compared with $PM_{2.5}$, UFPs appear to be cleared more slowly and retained longer within the lung after deposition. The effect of $PM_{2.5}$ on the EBC nitrite was reduced in the double-pollutant model after adjustment of other pollutants. However, the effect of UFP remained, indicating that the effect of UFP on the EBC nitrite was more remarkable.

4.2.2 *Inflammation-Related Biomarkers of Cardiovascular System Exposure to Ambient Air Pollution*

4.2.2.1 Count Changes of Different Types of Cells

The white cell, as the human body "guard," can transform and cross the capillary wall to surround and phagocytize foreign materials when bacteria and viruses invade the organism. The count of white blood cells will be higher than normal, following the occurrence of inflammation. Red blood cell, a kind of cell with the largest number in the blood, is the most important media of delivering oxygen to the human body, the count of which will decrease obviously with the occurrence of massive hemorrhage and severe tissue damage.

Many studies proved that $PM_{2.5}$ can trigger the inflammation reaction of the human body. However, some studies found that both the counts of the white blood cell and red blood cell decrease along with the increase of the concentration of $PM_{2.5}$, the mechanism of which needs a further study. In order to investigate the exposure-response relationship, a study chose 107 male polices (exposure group) and 101 officers (control group) as subjects and monitored the personal exposure level of $PM_{2.5}$ continuously for 24 h and then detected the plasma lymphocyte level by blood sampling at the end of the monitoring [136]. The results found that the 24-h average concentration of $PM_{2.5}$ (115.40 ± 46.25 μg/m^3) of the exposure team is significantly higher than the control team (74.96 ± 41.13 μg/m^3). On the contrary, the proportion of lymphocytes to white blood cells of the exposure team is significantly lower than that of the control team in the plasma. Another study of 142 children as subjects, exploring the effect of vehicle exhaust on the cellular immune function of children, drew the similar conclusion that the parameters of lymphocyte subpopulation (CD3+, CD4+, CD8+, CD4+/CD8+) of students in the heavy traffic pollution area is obviously lower than that of students in the low traffic pollution area [135].

A study exploring the relationship between the change of the white blood cell in the human body and the improvement in the Beijing Olympics found that the count of the white blood cell of subjects during the Olympics is significantly lower than that of subjects before the Olympics [59]. In addition, some study found that the 24-h average concentration of $PM_{2.5}$ was negatively related with the counts of subjects in the 0–3 lag effects.

4.2.2.2 Inflammatory Cytokine (Interleukin)

Cytokines are a kind of small molecular proteins with a biological activity and can be synthetized and secreted by immune cells and nonimmune cells (endothelial cells, epithelial cells) in response to a stimulus. Interleukins including interleukin (IL)-1, IL-6, IL-8, and IL-10 were important cytokines secreted by monocytes and lymphocytes and were the most common inflammation markers in the study.

Most studies found that air pollutants can induce elevated levels of IL-6 of subjects. There was a study of 371 residents of some communities in Shanghai using portable monitor (the plastic tube was fixed onto the participants' collar using a clamp) to record $PM_{2.5}$ exposure from 8:00 am to 6:00 pm and detecting the IL-6 level by blood sampling. The results found that the participants living within 50 m to the major road compared with those living more than 200 m away have 8.39 times higher IL-6 [58]. A cohort study, choosing 81 BC-exposed male workers (the exposure group) and 104 non-exposed male workers (the control group), found that there was a significant difference in the mean concentration of IL-6 between the exposure group and the control group, which was 188.32 (95% CI: 46.13, 643.16) pg/mL and 27.51 (95% CI: 2.16, 180.18) pg/mL, respectively [132].

However, there is no consistent conclusion of the effect of air pollutants on the IL-10 of subjects. A panel study of 35 type 2 diabetes patients in Shanghai, exploring the exposure-response relationship between the inflammation marker-interleukin and air particulate matter, detected the IL-10 of subjects six times repeatedly and monitored the mass concentration of $PM_{2.5}$ and PM_{10} and the number concentration with size distributions between 0.25 and 10 μm [107]. The results showed that the percentage increase in IL-10 associated with an IQR increase in 2-h average (particulate number concentration) $PNC_{0.25–0.40}$ was 17% (95%CI: 2%, 32%); on the contrary, a panel study of 371 residents showed that the participants living within 50 m to the major road compared with those living more than 200 m away have 2.54 times lower IL-10. The above results indicated that the effects of air pollutants on IL-10 need a further study.

Some studies exploring the exposure-response relationship between IL-1β and IL-8 in the blood of subjects and BC found that the mean concentration of IL-1β in the exposure group and the control group were 11.88 (95% CI: 1.98, 38.08) pg/mL and 4.16 (95% CI: 0.00, 17.75) pg/mL, respectively [132]. The mean concentration of IL-8 in the exposure group and the control group were 1117.10 (95% CI: 369.36, 373.82) pg/mL and 746.30 (95% CI: 163.55, 1879.01) pg/mL, respectively. In addition, the regression coefficients of independent variables of IL-1β and IL-8 were 0.680 (95% CI: 0.412, 0.949) and 0.659 (95% CI: 0.459, 0.860) in the multiple regression model including age, sex, body mass index (BMI), smoking, alcohol drinking, and BC exposure. The above studies indicated that the BC particles can promote systemic inflammation.

4.2.2.3 Other Inflammatory Cytokine

As an important cellular adhesion molecule, vascular cell adhesion molecule-1 (VCAM-1) that can express with the action of cytokines including IL-1 and tumor necrosis factor (TNF) plays an important role in the process of activation of lymphocytes and inflammation. Though a commonly used biological marker, the conclusions from studies of the healthy and the patients are inconsistent. A panel study of 40 healthy college students in Beijing found that the mean concentration of $PM_{2.5}$ in the Suburban Period and Urban Periods decreased progressively; instead, VCAM-1 showed a higher median level in the Suburban Period. A further chemical constituent analysis found that an interquartile range increase in cadmium (Cd) was associated with a decrease in VCAM-1 of 2.7% (95%CI: 0.5%, 4.8%) [121]. However, a panel study of 35 type 2 diabetes patients found that the percentage increase in VCAM-1 associated with an IQR increase in 2-h average $PNC_{0.25–0.40}$ was 6% (95% CI: 3%, 10%) [107]. Those results indicated that the effect on the VCAM-1 of the healthy and the patients may be different and need a further study.

C-reactive protein (CRP), an acute-phase protein, is produced when the body is under the invasion of germs or there is inflammation in the tissue. The conclusions from studies of the healthy and the patients are inconsistent. A panel study of 40 healthy college students in Beijing found that the mean concentration of $PM_{2.5}$ in the Suburban Period and two Urban Periods decreased progressively [117]. On the contrary, the mean concentration of CRP increased in the suburban period and two urban periods. A study exploring the effects of Beijing air pollution on cardiovascular-related blood parameters of occupational-exposed population (taxi driver) found that there was no statistical significance in terms of the CRP level of female in the study between spring and the less polluting autumn though the CRP level is higher in spring [113]. Instead, the result in man was opposite. However, there were studies showed positive results. A panel study of 35 type 2 diabetes patients found that the percentage increase in VCAM-1 associated with an IQR increase in 2-h average $PNC_{0.25–0.40}$ was 20% (95% CI: 4%, 36%) [107]. Another study of 236 older people found that there was significantly a decrease in the median level of CRP in the subjects between the period vehicle restrictions for the Olympics (3.1 mg/L) and the period before the Olympics (4.25 mg/L) [32].

As an important adhesion molecule mediating adhesion, intercellular adhesion molecule-1 (ICAM-1) plays a key role in the promotion of adhesion of inflammatory sites, immune responses, and inflammatory responses, the concentration of which can increase along with the body inflammation. A panel study of 40 healthy college students in Beijing found that the mean concentration of PM2.5 in the suburban period and two urban periods decreased progressively, and ICAM-1 showed higher median levels in the suburban period and two urban periods, with a concentration of 157.7 ± 136.6, 128.5 ± 108.1, and 141.6 ± 102.6 ng/ml in the three periods, respectively [121]. A further chemical constituent analysis found that an interquartile range increase in Cd and potassium (K) was associated with an increase in ICAM-1 of 28.8% (95% CI: 7.3%, 54.6%) and 23.9% (95% CI: 1.6%, 51.2%), respectively.

As an important inflammatory cytokine, tumor necrosis factor-α (TNF-α) that was mainly produced by mononuclear leucocyte and macrophage can induce the production of IL-6 and enhance the adhesion of neutrophils to vascular endothelium. A panel study of 40 healthy college students in Beijing found that an interquartile range increase in $PM_{2.5}$ was associated with an increase in 7.06% (95% CI: 2.97%, 11.31%) [117]. Another study of traffic polices and the Centers for Disease Control (CDC) staff found that the concentration of TNF-α was significantly higher in the traffic polices than that in the CDC staff, with concentration of 13.69 pg/ml and 8.31 pg/ml, respectively [73]. A cohort study drew a similar conclusion that the concentration of TNF-α of subjects in the exposure group (232.36 pg/mL) was significantly higher than that in the control group (47.75 pg/mL). The regression coefficient of the exposure of BC as independent variable was 1.772 (95% CI:1.436, 2.109) in the regression model of TNF-α as dependent variable, indicating that air pollutants including particulate matter and BC had a good effect on the promotion of the production of TNF-α [132]. A further analysis found that 25 chemical constituents of $PM_{2.5}$ had a positive relationship with TNF-α. An interquartile range increase in zinc (Zn) at 5-day average was associated with an increase in TNF-α of 12.50% (95% CI: 5.88%, 19.54%). The above results show that TNF-α is a reliable indicator of inflammation.

4.3 Oxidative Stress

Oxidative stress was defined as a process that the imbalance of oxidation system and antioxidant system caused by the metabolic disorders or excess of reactive oxygen species (ROS) along the harmful factors induces gene mutation, protein denaturation, and lipid peroxidation and then leads to body damage and the physiological and biochemical dysfunction and metabolic dysfunction of cells. As a pathogenic mechanism of air pollutants, oxidative stress can make a contribution of the development of many diseases.

4.3.1 Oxidative Stress-Related Biomarkers of Respiratory System Exposure to Ambient Air Pollution

4.3.1.1 Malondialdehyde in Exhaled Breath Condensate (EBC)

As a product of lipid peroxidation, malonyldialdehyde (MDA) has long been considered a marker of oxidative stress, the content of which can directly reflect the body oxidative stress. A panel study of 125 healthy college students in the Beijing Olympics measured MDA in exhaled breath condensate of subjects and found that the mean concentration of EBC MDA decreased by 24% significantly from the pre-Olympic to the during-Olympic period and increased by 39% significantly from the during-Olympic period to the post-Olympic period and that the largest increases (in percentage) in EBC MDA associated with each IQR increases in pollutants which occurred at lag 4 for $PM_{2.5}$, at lag 4 for SO_2, at lag 5 for NO_2, and at lag 0 for CO were 11%, 19%, 11%, and 10%, respectively [39]. Another study found there was no significantly relationship in the UFP and EBC MDA, indicating that EBC MDA is vulnerable to be effected by $PM_{2.5}$ [38]. However, there was no statistical significance of the relationship between $PM_{2.5}$ and EBC MDA after the adjustment of SO_2 in the double model, indicating that there was an interaction between $PM_{2.5}$ and SO_2.

4.3.1.2 8-Hydroxydeoxyguanosine (8-OHdG) in Exhaled Breath Condensate

8-isoprostane is the major oxidative stress product in the lung from peroxidation of arachidonic acid catalyzed by free radicals in the body. A panel study of 125 healthy college students in the Beijing Olympics measured 8-OHdG in exhaled breath condensate of subjects and found that the mean concentration of 8-OHdG decreased to 44% (% >detection limit 1.56 pg/ml) from 68% from the pre-Olympic to the during-Olympic period and increased to 74% from the during-Olympic period to the post-Olympic period and that there was an increased odds ratio of a high EBC 8-isoprostanelevel (>75th percentile relative to, 75th percentile values) associated with each IQR increase in all pollutants including $PM_{2.5}$, O_3, NO_2, and CO at multiple lags [38].

4.3.1.3 The pH of Exhaled Breath Condensate

The acidity or acid of exhaled breath condensate (pH) can reflect the pH balance of the surface of airway epithelial cells. The acidification of airway is related with the bronchoconstriction, ciliary dyskinesia, increase of mucus, and damage of airway. So the pH of exhaled breath condensate is considered as a biomarker of oxidative stress in the lung.

A panel study of 125 healthy college students found that there was a significantincrease of 3.5% (95% CI: 2.2%, 4.9%) in EBC pH from the pre-Olympic to the during-Olympic period and a decrease of 4.8% (95% CI: 9.4%, 0.2%) in EBC pH from the during-Olympic period to the post-Olympic period. A further analysis showed that there was a positive relationship between UFP and EBC pH at lag 1 day effect, so as the $PM_{2.5}$ and EBC pH at lag 0–5 day effect [38]. However, three was no statistical significance of the relationship between $PM_{2.5}$, UFP, and EBC pH after the adjustment of SO_2 in the double model, indicating that there was an interaction between $PM_{2.5}$, UFP, and SO_2.

4.3.2 Oxidative Stress-Related Biomarkers of Cardiovascular System Exposure to Ambient Air Pollution

4.3.2.1 The Imbalance of Oxidation System and Antioxidant System

UFP and $PM_{2.5}$ whose surface has transition metals including iron, copper, zinc, and manganese, polycyclic aromatic hydrocarbons, and lipopolysaccharides have the property of active free radicals and simulate the body to produce ROS, leading to the imbalance of oxidation system and antioxidant system. The most common antioxidants causing tissue damage include superoxide dismutase and glutathione peroxidase, which are also considered as the common markers of oxidative damage of body.

The panel study of 40 healthy college students measuring superoxide dismutase and glutathione peroxidase found that an IQR increase of 63.4 mg/m^3 in $PM_{2.5}$ was associated with an increase of 6.3% (95% CI: 0.6%, 12.4%) in extracellular superoxide dismutase (EC-SOD) and an increase of 5.5% (95% CI: 1.3%, 9.8%) in glutathione peroxidase1 (GPX1) at 3-day moving average [120]. A further analysis of constituent showed that two negative ions (nitrate and chloride) and two metals (strontium and iron) were significantly associated with increases in both EC-SOD and GPX1 at 3-day moving average.

Another study of 371 subjects measuring superoxide dismutase and total antioxidant capacity found that the participants living within 50 m to the major road compared with those living more than 200 m away have 1.85 times higher total antioxidant capacity and 1.19 times higher superoxide dismutase [58]. In addition, the study of 170 traffic police (exposed group) and 108 healthy people (control group) as subjects showed that the activity of superoxide dismutase of the exposed group (89.58 ± 14.36) NU/ml was significantly lower than that in the control group (116.83 ± 20.78) NU/ml [128]. The above studies illustrated that the traffic-related air pollution can reduce the body's antioxidant capacity and then promote the body's levels of oxidative stress.

Some studies investigated the relationship between total superoxide dismutase (T-SOD) in the blood and benzene series of air pollutants apart from $PM_{2.5}$. The study of 252 staffs in the gas stations found that the concentration of benzene, toluene, and ethylbenzene of the exposed group (fueling areas) was significantly higher than that in the control group (office area and supermarket area) and that the activity of T-SOD of the exposed group (42.59 ± 6.51) U/ml was significantly lower than that in the control group (49.46 ± 5.92) U/ml [98].

4.3.2.2 Protein Carbonyl

The carbonyl products will increase and the protein function will be loss or even be degraded after the side chain amino acids of protein are oxidized. So the content of carbonyl products can serve as a sensitive index of protein damage of oxidative stress. The blinded crossover study of 17 atopic subjects measuring the content of protein carbonyl explored the combined effect of motor vehicle exhaust and glutathione S-transferase T1 gene (GSTT1) or glutathione S-transferase M1 gene(GSTM1), the main research of which was that the two groups exposed to diesel exhaust (DE) and filtered air by 4 weeks between exposures separately, the order of which was randomized and counterbalanced and then collected the plasma sample of subjects before exposure, 4 h post exposure, and 24 h post exposure [133]. The results showed that only with co-exposure to DE was there a significant increase in carbonyl content 4 h after exposure in the GSTT1 null individuals (0.89 ± 0.74 nmol/mg increase from baseline) compared to GSTT1 present individuals (0.65 ± 0.66 nmol/mg reduction from baseline), stating that there was a combined action of environment and genotype on the content of protein carbonyl of oxidative stress.

4.3.2.3 The Oxidation of Lipoprotein

Low-density lipoprotein (LDL) is a kind of lipoprotein containing abundant cholesterol and cholesterol ester. The ROS produced along with the imbalance of oxidation system and antioxidant system can oxidize LDL and change the particle composition. The oxidation of LDL is the critical step of the occurrence and development of atherosclerosis (AS), so it is a common biomarker of oxidative damage.

The panel study of 40 healthy college students measured the plasma levels of oxidized low-density lipoprotein (ox-LDL) [122]. Daily $PM_{2.5}$ mass samples were collected using SKC sampling systems and analyzed in the laboratory for 32 chemical constituents. The results showed that each IQR increase in $PM_{2.5}$ at lag 1 day, there was a nonsignificant 1.6% (95% CI:0.9%, 4.2%) increase in ox-LDL ($P < 0.10$) and that the integral level of ox-LDL decreased when participants relocated to urban areas from suburban areas. Among the $PM_{2.5}$ chemical constituents, chloride, strontium, and iron showed significant positive associations, whereas secondary organic carbon (SOC) showed a significant inverse association with ox-LDL at 1 day. The results stated that both of $PM_{2.5}$ and its chemical constituents were associated with the change of ox-LDL.

4.3.3 Oxidative Stress-Related Biomarkers of Other System Exposure to Ambient Air Pollution

Because of the characteristics of simple to get and sensitive, urine is also a common sampling to obtain biomarkers. The panel study of 125 healthy college students found that the mean concentration of urinary MDA decreased by 28% from the pre-Olympic to the during-Olympic period and increased by 55% from the during-Olympic period to the post-Olympic period. The largest increases in urinary MDA per IQR increases associated with $PM_{2.5}$ at lag 0, SO_2 at lag 5, NO_2 at lag 1, and CO at lag 0 were 15%, 9%, 13%, and 14%, respectively [38].

4.4 Genetic Biomarkers

The primary biomarkers used to study the genotoxic effects associated with air pollution are ^{32}P-postlabeling, comet assay, urinary 8-OHdG, chromosome aberrations, micronucleus (MN), and sister-chromatid exchange.

4.4.1 DNA Damage

The literature describes three primary assays used to evaluate DNA damage in air pollution-exposed subjects: ^{32}P-postlabeling for DNA adducts, comet assay for DNA damage, and concentration of urinary 8-OHdG.

4.4.1.1 ^{32}P-Postlabeling

DNA adducts are considered as a marker for potential risk of genotoxic effects, which might lead to undesirable outcomes, such as cancer or birth defects. ^{32}P-postlabeling analysis is a highly sensitive method for the detection and measurement of covalent carcinogen adducts, some oxidative DNA lesions, and other DNA modifications [87, 88]. It is suitable for detection of DNA adducts and oxidative lesions in epidemiological studies [34, 104].

One study monitored PAH exposure and DNA adduct in parents of children born with neural tube defects (NTDs) and their matched controls in Shanxi Province, China [78]. To test the hypothesis that parents of children born with a NTD would exhibit a biomarker of exposure at higher levels than the parents of a child with no visible birth defect, venous blood was collected from participants as soon as possible following the birth of the child. PAH concentrations in whole blood were quantified and the frequency of aromatic DNA adducts was measured by ^{32}P-postlabeling in white blood cells (WBCs). Single nucleotide polymorphisms (SNPs) in phase II enzymes were also monitored in an attempt to identify sensitive receptors. The results confirmed significant differences in biomarkers between case and control participants. Both total and carcinogenic PAHs were significantly higher in blood from case mothers compared with control mothers, and the mother's age-adjusted OR for having a child with NTD was 8.7 when their exposure to PAH concentration was above the median concentration. However, bulky DNA adducts exhibited a reversed trend and were significantly higher in both parents of control as compared with NTD cases. The results suggest that PAHs may be a contributing factor to the risk of NTDs and that the lack of a correlation with DNA adducts suggests a possible non-genotoxic mechanism. Besides, the author noted that the PAHs might be used to identify populations exposed to elevated concentrations of combustion by-products.

In the rural county of Xuanwei, Yunnan Province, in southwest China, lung cancer mortality rates are among the highest in China, owing to severe indoor air pollution generated by the smoky coal combustion [15, 77]. An earlier study determined PAH-DNA adduct in brushing cells from fibrobronchoscopy in 30 cases with lung cancer in Xuanwei County and 10 controls in Kunming by ^{32}P-postlabeling assay [127]. They found that PAH-DNA adduct levels in lung cancer patients of Xuanwei County were significantly higher than those in controls ($P < 0.05$). The author suggested that air pollution caused by indoor coal-burning associated directly with the occurrence of lung cancer, and detection of DNA adduct could be used as an indicator for its risk assessment in population.

4.4.1.2 The Comet Assay

The comet assay or single-cell gel electrophoresis was first described in 1984 by Östling and Johanson [83]. A few years later, the method was modified by increasing the pH of electrophoresis to >13 and is the most widely used today. This assay is a sensitive and easy way to detect DNA damage and widely used in genetic toxicity testing, environmental toxicity testing, and molecular epidemiology research [11, 74] and enables the detection of many kinds of DNA damage, including single-strand breaks and double-strand breaks.

In a more recent study, Chu et al. explored the personal 24-h $PM_{2.5}$ exposure levels and DNA damage levels of 328 subjects from Zhuhai, Wuhan, and Tianjin with different $PM_{2.5}$ exposure levels, to explore whether $PM_{2.5}$ exposure and genetic variants contribute to the alteration in DNA damage [22]. Their findings reveal that the DNA damage levels are consistent with the $PM_{2.5}$ exposure levels of each cohort. Besides, genotyping was performed with 241,305 single nucleotide variants (SNVs), and they found that a total of 35 SNVs were consistently associated with DNA damage levels among the three cohorts with pooled P values less than 1.00×10^{-3} after adjustment for age, gender, smoking status, and $PM_{2.5}$ exposure levels, of which 18 SNVs together with gender and $PM_{2.5}$ exposure levels were independent factors contributing to DNA damage. There were three genes significantly associated with DNA damage levels ($P < 0.01$). Gene ontology (GO) analyses indicated that the identified variants were significantly enriched in DNA damage response pathway. The study highlights the importance of genetic variation as well as personal $PM_{2.5}$ exposure in modulating individual DNA damage levels. Another study of traffic policemen in Shenyang City found that the average concentration of $PM_{2.5}$ in field policemen was significantly higher than that in non-field group and DNA damage in field policemen was significantly higher than those in non-field group [72]. They analyzed chemical composition of $PM_{2.5}$ and found that toxic heavy metals, for example, manganese (Mn), zirconium (Zr), copper (Cu), and chromium (Cr), and

PAHs like anthracene, chrysene, and benzo (a) pyrene exist in the $PM_{2.5}$ they collected.

Traffic is also the most important source of PAHs in urban ambient air, and PAHs are ubiquitous constituents of urban airborne particles. One study recruited 91traffic conductors and 53 indoor office workers between May 2009 and June 2011 in Taipei City, Taiwan [52]. To assess the relations between personal exposure to $PM_{2.5}$ and PAHs and the occurrence of DNA strand breaks, they collect breathing-zone particulate PAH samples and urine and blood samples after analyzing work shift of two consecutive days for 1-OHPG, 8-OHdG, and DNA strand breaks, respectively. They found that both urinary 8-OHdG levels and the occurrence of DNA strand breaks in traffic conductors were significantly higher than those in indoor office workers. There was a positive association between particulate PAHs levels and urinary 1-OHPG ($\beta = 0.056$, $P = 0.01$). Besides, urinary 1-OHPG levels showed significant association with urinary 8-OHdG level ($\beta = 0.101$, $P = 0.023$). Their findings suggest that exposure to fine particulates increases DNA damage, and exposure to particulate PAHs might contribute to the induction of oxidative DNA damages.

Recently, diesel engine exhaust (DEE) exposure was associated with the development of lung cancer in a nested case-control study and in a cohort study [2, 95]. PAH is one of the most important carcinogenic components in DEE. To seek early and sensitive biomarkers for prediction of adverse health effects of DEE, one study analyzed the components of DEE particles and examined the genetic and oxidative damages in DEE-exposed workers [33]. In this study, 101 male diesel engine testing workers who were constantly exposed to DEE and 106 matched controls were enrolled, and the components of DEE were analyzed, including $PM_{2.5}$, EC, NO_2, SO_2, and PAHs. They collected post-shift urine samples and analyzed for 1-hydroxypyrere (1-OHP) as an internal exposure marker for DEE. They measured the levels of DNA strand breaks and oxidized purines by comet assay. They found higher levels of $PM_{2.5}$, EC, NO_2, SO_2, and PAHs in the diesel engine testing workshop and significantly higher urinary 1-OHP concentrations in exposed subjects ($P < 0.001$). Compared with controls, the levels of parameters in comet assay were significantly higher in DEE-exposed workers ($P < 0.001$). The findings suggest that DEE exposure mainly induces DNA damage, which might be used as an early biomarker for risk assessment of DEE exposure.

The histone H2A variant H2AX is a key regulator of the DNA damage response. H2AX phosphorylation plays a central role in the response to damaged chromatin. One study collected 307 healthy urban residents in three cities, Zhuhai, Wuhan, and Tianjin, and detected the dust mass concentrations of $PM_{2.5}$and measured the DNA damage levels by comet assay [100]. They found that the DNA damage levels increased along with $PM_{2.5}$ exposure levels with mean percent tail DNA values being 1.36, 1.85, and 2.97, respectively. They also found that three functional SNPs were significantly associated with DNA damage levels (adjusted $P = 0.002$, 0.018, and 0.027, respectively). Significant interactions were observed between certain genetic polymorphisms and $PM_{2.5}$-modulated DNA damage levels. Their finding suggested that genetic variations of H2AX might be associated with the DNA damage levels in urban residents with different exposure to $PM_{2.5}$.

4.4.1.3 8-Hydroxy-Deoxyguanine

8-OHdG could be generated by oxidation of guanine or incorporated during replication or repair as oxidized nucleotides. It is reported that the level of urinary 8-OHdG could serve as a biomarker of early effects of exposure to PM, due to oxidative stress, genotoxicity, and carcinogenicity induced through ambient PM exposure [90]. Previous studies suggest that PM interacts with biological systems through generation of ROS from the surface of particles, organic chemicals, or transition metals [40, 90]. The spectrum of DNA-related oxidative products includes oxidized bases like 8-OHdG [23, 61].

One study evaluated 91 traffic conductors and 53 indoor office workers in Taipei City and found that urinary 8-OHdG levels and the occurrence of DNA strand breaks in traffic conductors significantly exceeded those in indoor office workers [52]. A more recent study investigated the genetic and oxidative damage in DEE-exposed workers and did not find significant differences between DEE-exposed workers and controls in regard to urinary 8-OHdG levels [33]. The reason for this discrepancy might be that the amount of urinary 8-OHdG involves the repair products of oxidized guanine in DNA and the cellular pool from the whole body.

Heavy metals and PAHs are predominating toxic constituents of particulate air pollution that may be related to the increased risk of cardiopulmonary events. One study investigated the oxidative stress induced by PAHs and quinone-structure chemicals in human exposed to heavy traffic volume [111]. They recruited two non-smoking security guards who worked at a university campus gate near a heavily trafficked road and estimated personal exposures to 24 PAHs and anthraquinone (AnQ) in $PM_{2.5}$ and collected the daily pre- and post-work shift spot urines to analyze the 8-OHdG. They found that the average urinary 8-OHdG level was increased more than three times after 8-h work shift than those before the work shift, and all the 24 PAHs and AnQ levels showed positive and significant association with the post-work urinary 8-OHdG levels. Their findings also suggest that $PM_{2.5}$ was attributed to vehicular emission. Another study of children in Guangzhou City found that the children from the polluted area had high urinary levels of urinary monohydroxylated PAH metabolites [2-hydroxynaphthalene (2-OHN), 2-hydroxyfluorene (2-OHF), 2-, 3-, 4-, 9-hydroxyphenanthrene (2-, 3-, 4-, 9-OHPHE), and 1-hydroxypyrene (1-OHP)]. Unexpectedly, the urinary 8-OHdG concentrations in the children from the traffic-polluted area were slightly higher than those in the nonpolluted area (20.87 ± 14.42 μmol/mol creatinine vs. 16.78 ± 13.30 μmol/mol creatinine). The findings suggest that the reason might be that the potential co-exposure of the children to other pollutants affects 8-OHdG concentrations besides PAHs.

One study investigated the effects of the toxic heavy metals [arsenic (As), Cd, Cr, nickel (Ni), and lead (Pb)], and their interactions with PAHs on oxidative stress among coke-oven workers [109]. One thousand three hundred thirty three male workers were recruited in this study. Urinary levels of As, Cd, Cr, Ni, Pb, PAH metabolites, 8-OHdG, and 8-iso-prostaglandin-F2α (8-iso-PGF2α) were determined, and multivariate linear regression models were used to analyze the effects of these metals and their interactions with PAHs on 8-OHdG and 8-iso-PGF2α levels. It was found that urinary As and Ni showed marginal or significant positive linear dose-dependent effects on 8-OHdG, especially among smokers ($\beta = 0.103$ and $P = 0.073$ and $\beta = 0.110$ and $P = 0.002$, respectively). Co-exposure to both high levels of PAHs and heavy metals renders the workers with the highest 8-OHdG and 8-iso-PGF2α (all $P_{interaction} \leq 0.005$). The finding suggested that interaction effects of heavy metals and PAHs could increase the oxidative stress.

The traditional Chinese style of cooking often involves stir-frying and deep-frying, in which oil is usually heated to a high temperature, and some mutagens and human carcinogens such as PAHs, heterocyclic aromatic amines, benzene, and formaldehyde are generated, which may induce DNA damage to the cells and increase susceptibility to lung cancer [20]. One study evaluated 61 Taiwanese male military cooks and 37 office soldiers as the reference group [64]. They collected urine samples pre-shift of the first weekday and post-shift of the fifth workday and measured airborne particulate PAHs in their subjects and concentrations of urinary 1-OHP and urinary 8-OHdG. They found that the level of airborne particulate PAHs in kitchens was significantly higher than those in office areas. The urinary 1- OHP levels among military cooks increased significantly after cooking oil fume exposure, and changes in the levels of 8-OHdG and isoprostane showed a significant relation to the changes in 1-OHP levels.

4.4.2 *Chromosome Damage*

4.4.2.1 The Frequency of Chromosome Aberrations

The frequency of chromosomal aberrations (CA) in peripheral blood lymphocytes has been applied for decades as a biomarker of the early effects of genotoxic carcinogens [1, 6, 80]. It is one of the oldest genotoxicity biomarkers, and it has been used relatively frequently to assess genotoxicity in traffic-exposed subjects [29]. A cohort study on traffic police in Hebi City, Henan Province, investigated CA for genotoxicity assessment of traffic-related air pollution [16]. A group of 45 traffic police and 30 office workers were recruited as the exposure group and control group, respectively. The study found that traffic police had significantly higher frequencies of CA than their respective control group.

4.4.2.2 Micronucleus

An increased frequency of micronucleus (MN) is a marker of genotoxic events and manifestations of chromosomal instability. The MN methodology is simpler than CA analysis, and it has largely replaced CA assay for genotoxicity assessment. The MN assay can be used to detect both chromosome loss and chromosome breaks [36]. Since the MN test is not limited to metaphase cells and has the advantage of allowing rapid screening of large numbers of cells, it is more suitable as a cytogenetic marker than studies involving sister-chromatid exchanges or CA [81].

A potential link between automobile exhaust and genotoxicity was investigated by a study conducted in Lanzhou City [138]. In this study, 78 traffic police working in the field and 57 household register police working was recruited, and their peripheral blood lymphocytes were collected for the frequencies of MN. Compared with the household register police, significant increases of MN were found from the traffic police who is exposed to heavy automobile exhaust ($P < 0.05$, student's t-test). A more recent study among rural and industrial female residents of Shenyang City found that the MN frequency of female residents was higher than their rural control group, suggesting that MN assays might be a sensitive indicator to air pollution-induced genotoxic effects in humans [56].

The indoor air pollution studies also have shown a link between cooking oil fume exposure and the increase of MN level. One study that evaluated 108 subjects engaged in catering trade in Guangzhou City found that the frequency of MN was significantly higher than their control group ($P < 0.01$) and that there was a positive correlation between the levels of MN and length of service ($r = 0.301$, $P < 0.05$) [126].

4.4.2.3 Sister-Chromatid Exchanges

Sister-chromatid exchange (SCE) is the process whereby, during DNA replication, two sister chromatids break and rejoin with one another, physically exchanging regions of the parental strands in the duplicated chromosomes [112]. The frequency of SCE appeared to increase with exposure to various genotoxic carcinogens.

Findings from traffic police study in Lanzhou City, mentioned before, found that the frequency of SCE in peripheral blood lymphocytes from the traffic police was significantly higher than that of the household register police [138]. The findings suggest that the frequency of SCE is suitable for measuring genotoxic effects of traffic-related air pollution. Another study on subjects exposed to inorganic arsenic from the burning of coal in unventilated indoor stoves, a well-known human carcinogen, found that the frequency of SCE was higher in the exposed groups compared with the external control group [129]. The results suggested that long-term exposure may be associated with damage of chromosomes.

4.4.3 Other Biomarkers

4.4.3.1 Telomere Length

Telomeres are complexes of tandem repeats of DNA (5′-TTAGGG-3′) and protein that cap eukaryotic chromosomes and protect chromosome ends from degradation and fusion [5]. It has been reported that short telomeres trigger DNA damage checkpoints, which mediate cellular senescence, and this process may be accelerated by oxidative stress and inflammation [45, 105]. Telomerase, a cellular reverse transcriptase, plays a key role in telomere length (TL) homeostasis, and it uses an internal RNA moiety as a template for the synthesis of telomere repeats [55]. Therefore, TL maintenance is a dynamic process. Evidence is rapidly growing that TL may be affected by air pollution, since PM mass and its metal components have been shown to result in biological changes, including oxidative stress and chronic inflammation, two major determinants of TL.

A study of truck drivers in Beijing demonstrated a significant alteration of blood TL following both short-term and longer exposure to PM [48]. They investigated the relationship of personal and ambient PM exposure with blood TL in 60 truck drivers (highly exposed group) and 60 indoor workers (low exposed group). Personal $PM_{2.5}$ and EC during work hours were measured to the study. Moreover, ambient PM_{10} data on the days of examination, as well as measurements recorded on the days leading up to the examinations, were used, allowing for testing both short- and long-term effects of PM exposure. They observed TL increase associated with personal $PM_{2.5}$ (+5.2%, 95%CI: 1.5%, 9.1%; $P = 0.007$) and EC (+4.9%, 95%CI: 1.2%; 8.8%; $P = 0.01$) and ambient PM_{10} (+7.7%, 95%CI: 3.7%; 11.9%; $P < 0.001$). However, the ambient PM_{10}over the longer-term exposure was significantly associated with shorter TL (−9.9%, 95%CI: −17.6%, −1.5%; $P = 0.02$). The author suggested that the observed TL alterations might participate in the biological pathways of short- and long-term PM effects. Another study of 25 women living in rural Sichuan Province found that women with high exposure to biomass smoke had 43% shorter TL (95% CI: −113%, 28%) than that of women with low exposure, though the results were not statistically significant [93].

Besides, some epidemiological studies focus on the association between air pollution exposure and TL alteration among diabetes patients. A panel study was conducted to investigate TL alteration among 35 type 2 diabetes patients in Shanghai with short-term size-fractionated PM (0.25–10 μm) and gaseous pollutant exposure [123]. They found positive but weak associations between various air pollutants and TL. Taken together, these studies suggest that exposure to PM may induce alteration of TL, which may be a potential biological mechanism for chronic diseases.

4.4.3.2 Mitochondrial DNA

The mitochondrion is the major ATP producer of the mammalian cell and plays a critical role in intracellular respiration and energy metabolism. Mitochondrial DNA (mtDNA) mutations are considered to be associated with a broad spectrum of clinical manifestations, such as cardiac failure, diabetes, and dementia [106]. Compared to human nuclear DNA (nDNA), mtDNA lacks protective chromatin structure, histones, and introns, with less efficient DNA repair systems [67]. Therefore, mtDNA appeared to be particularly vulnerable to ROS-induced damage and has a high mutation rate, resulting in both qualitative and quantitative changes [70]. Each human cell contains between several hundred and over a thousand mitochondria, and each mitochondria carrying 2–10 copies of mtDNA. MtDNA copy number (mtDNAcn) is correlated with the size and number of mitochondria [67]. It has been reported that the amount of mtDNA decreases in peripheral blood correlate with many clinical manifestations, including hypertension and diabetes mellitus [51, 68]. Evidence is rapidly growing that air pollution may affect mtDNAcn.

Previous studies in China on air pollution and mtDNA reported limited and inconsistent results. In Beijing Truck Driver Air Pollution Study, Hou et al. reported decreased blood mtDNAcn in association with increased exposure to element carbon (EC) and ambient PM_{10} exposure, suggesting that mtDNAcn may be influenced by particle exposure [49]. However, another study of 35 type 2 diabetes patients in Shanghai found no significant associations between air pollutants and mtDNAcn [123].

4.5 Epigenetic Biomarkers

Epigenetics investigates heritable changes in gene expression occurring without changes in DNA sequence. Growing evidence suggests links of environmental pollutants with epigenetic variations, including changes in DNA methylation, histone modifications, and microRNAs.

4.5.1 DNA Methylation

Epidemiological studies and laboratory studies have consistently reported that ambient particulate mass and metal components have been linked to broad spectrum of cancer-related changes, including oxidative stress, chronic inflammation, and immunodeficiency. DNA methylation is a heritable trait that is essential for various biological processes and the development in higher organisms. In adult somatic tissues, DNA methylation in the form of 5-methylcytosine (5mC) typically occurs at CpG dinucleotide sites, while non-CpG methylation is prevalent in embryonic stem cells [94].

Previous investigations focused on repeated element sequences from LINE-1and *Alu* families. One study recruited 120 male steel workers from Brescia, Milan, and Beijing to evaluate sensitivity of DNA methylation in differentially evolved LINE, *Alu*, and HERN subfamilies to different types of airborne pollutants [12]. They found that high-exposure group showed subfamily-specific methylation differences compared to low-exposure group. They also found subfamily-specific correlations of methylation with exposure levels, and the effects of the exposures on DNA methylation were dependent on the subfamily evolutionary age, with stronger effects on older LINEs from PM_{10} (P interaction = 0.003) and benzene (P interaction = 0.04), and on younger *Alu* from PM_{10} (P interaction = 0.02). The author suggested that the evolutionary age of repetitive element subfamily determines differential susceptibility of DNA methylation to airborne pollutants.

In recent years, air pollutant-related effect on tandem repeats, a highly represented family of repetitive elements, has been evaluated. In Beijing truck driver air pollution study, methylation of three tandem repeats (*SATα*, *NBL2*, *D4Z4*) of blood samples from truck drivers and office workers were measured [43]. They found an interquartile increase in personal $PM_{2.5}$ and ambient PM_{10} levels was associated with a significant covariate-adjusted decrease in *SATα* and methylation (−1.35% 5mC, $P = 0.01$ and −1.33% 5mC; $P = 0.01$, respectively), and ambient PM_{10} was negatively correlated with *NBL2* methylation in truck drivers (−1.38% 5mC, $P = 0.03$) but not in office workers (1.04% 5mC, $P = 0.13$). The finding suggests that the association of PM exposure with hypomethylation of tandem repeats and tandem-repeat hypomethylation detection in blood samples might identify individuals with biological effects of PM exposure.

It has been reported that 5mC can be oxidized by the enzyme 10–11 translocation (TET) family to generate 5-hydroxymethylcytosine (5hmC) [103]. In contrast to 5mC, associated with gene repression, 5hmC appears to play a role in activating and/or maintaining gene expression [7, 85]. The formation of 5hmC, which can induce demethylation of DNA, may contribute to the dynamics of DNA methylation [79]. A recent study of 60 truck drivers and 60 office workers in Beijing has shown a link between PM_{10}exposure and changes in blood 5hmC level [91]. In this study, PM_{10} was associated with increases in 5hmC of 26.1% in office workers ($P = 0.004$), 20.2% in truck drivers ($P = 0.014$), and 21.9% in all participants combined ($P < 0.001$). However, 5hmC showed no correlations with personal measures of $PM_{2.5}$ and elemental components (family-wise error rate (FWER)>0.05), and 5mC was not correlated with measures of PM_{10}, $PM_{2.5}$, and elemental components. The finding indicates the need to differentiate 5hmC and 5mC in air pollution studies of DNA methylation.

Mitochondria are more autonomous than any other mammalian organelle, with their own genome and mtDNA. It was demonstrated that mitochondrial activity is regulated by modifications of the epigenetic factors, which include DNA methylation. In a study of 40 male participants in Brescia, Milan, and Beijing, Byun et al. found association of mitochondrial MT-TF and MT-RNR1 DNA methylation with metal-rich PM_1 exposure and mtDNA copy number, suggesting that locus-specific mtDNA methylation is correlated to PM exposure and mtDNA damage [13].

4.5.2 MicroRNAs

MicroRNAs (miRNAs) are single-stranded RNAs of 19–25 nucleotides in length, with a regulatory function on target mRNAs by binding to their 3′-UTRs [96]. MiRNAs are essential mediators of a wide range of biological processes, including cellular proliferation, apoptosis, and differentiation, and their dysfunction has been implicated in a broad spectrum of clinical manifestations, such as heart diseases, inflammations, and lung diseases [46, 82, 97].

In Beijing Truck Driver Air Pollution Study, Hou et al. investigated associations of short-term personal $PM_{2.5}$, EC, and ambient PM_{10} exposure with the miRNA profiles in peripheral blood samples from 60 truck drivers and 60 office workers. They observed associations of short-term ambient PM_{10} and EC exposure with the expression of human miRNAs. They also found an interesting association of short-term EC with differential expression of seven viral miRNAs. There are more than 400 viral miRNAs discovered from DNA and RNA viruses, and some of which are detectable in profiles of RNA extracted from human cells [42]. Latent viruses have been found to regulate miRNAs of host cells or to encode their own miRNAs by host cell mechanisms [62]. The above data indicate that latent viral miRNAs are potential mediators of air pollution-associated health effects.

4.5.3 Histone Modification

Histones function both positively and negatively in the regulation of gene expression, which is governed by posttranslational modifications on specific amino acid residues on histones [26]. Modifications on histones include acetylation, phosphorylation, methylation, and ubiquitination [63, 99], and they can be analyzed in different ways. Mass spectrometry has been found to be useful for comparing the global levels of specific modifications and their combinations among different samples [8]. In addition, chromatin immunoprecipitation (ChIP), using an antibody directed against the site-specific modifications, has been applied to analyze the enrichments on specific genome loci [25].

Among histone modifications, histone 3 lysine 27 acetylation (H3K27ac) is a good marker with a putative role in separating active enhancers from their poised counterparts [28]. The relationships between PM exposure and histone modification status were investigated by a study of four Chinese individuals exposed to high or low $PM_{2.5}$. In this study, individual histone modification profiles were compared between different exposure groups and the genome-wide chromatin immunoprecipitation sequencing (ChIP-Seq) data showed a comprehensive differential H3K27ac landscape across the individual genomes, which was associated with high $PM_{2.5}$. Besides, a substantial number of these $PM_{2.5}$-associated differential H3K27ac markers were found in genes involved in immune cell activation.

4.6 Other Air Pollution-Related Subclinical Biomarker

4.6.1 Cardiac Autonomic Function

Previous studies suggest that air pollution exposure may affect autonomic function, which includes altering sympathetic and parasympathetic output, possibly exerting an influence on the stability of atherosclerotic plaques and possibly decreasing HRV [24, 27, 92]. Heart rate variability (HRV) is a noninvasive and sensitive measure of cardiac autonomic function and has been frequently used to assess the cardiac effects of environmental exposures. Reduction of HRV is thought to reflect cardiac autonomic imbalance and worsened cardiovascular prognosis [41].Analysis of HRV is classified into frequency and time domain analyses [102]. Epidemiological studies of both healthy subjects and patients with cardiovascular diseases have demonstrated the link of air pollution with altered autonomic modulation.

Several studies of healthy adult subjects have conducted the relationship between $PM_{2.5}$ exposure and cardiac autonomic modulation. For instance, one study evaluated a panel of 11 young healthy taxi drivers in Beijing and noted association of marked changes in traffic-related $PM_{2.5}$ exposure with altered cardiac autonomic function [116]. The study was conducted before, during, and after the Beijing 2008 Olympic Games to investigate the effect of a wide range of PM exposure, since the unprecedented air pollution control actions focusing on control of traffic pollution throughout the Olympic Games temporarily resulted in marked decrease in the ambient PM air pollution in Beijing. They found that the standard deviation of normal-to-normal (SDNN) intervals decreased by 2.2% (95% CI: –3.8%, –0.6%) with an IQR (69.5 $\mu g/m^3$) increase in the 30-min $PM_{2.5}$ moving average, whereas the results for low-frequency (LF) and high-frequency (HF) powers were similar to those for SDNN. In the same panel, increasing temperature levels was associated with declines in standard deviation of normal-to-normal intervals over different temperature strata and increases in LF and LF/HF ratio in higher temperature range [115]. While some studies have not observed an association, another study of 125 adults in Beijing conducted in pre-, during-, and post-Olympics periods did not find significant changes in HRV indices by period [131]. Additionally, a study of residents in Shanghai documented associations of the increase in personal $PM_{2.5}$ exposure with the decrease in LF and HF in subjects [124]. The reduction in LF and HF reflected the impairment of both sympathetic and parasympathetic nervous systems [14, 66].

However, a contrary conclusion from another panel study of 30 healthy elderly subjects in Beijing reported positive associations between indoor personal $PM_{2.5}$ concentrations and HRV frequency indices [57]. Similarly, a recent study of 371 residents with different residential distance to major roads in Shanghai has showed that the residents living within 50 m to the major road compared with those living more than 200 m away have 1.15 times higher heart rate, 1.06 times lower LF, and 1.05 times lower of HF, which indicate a contribution of long-term exposure to traffic-related air pollution to the development or exacerbation of autonomic modulation impairment [58].Therefore, although air pollution exposure is demonstrated to impair the cardiovascular system, the effects of air pollution on HRV remain unclear.

Some studies focus on the autonomic modulation impairment of air pollutant exposure to patients with cardiovascular diseases and diabetes mellitus. One study of 40 cardiovascular disease patients before and during the 2008 Beijing Olympics found a significant reduction in SDNN and LF in association with exposure to $PM_{2.5}$, black carbon, and nitrogen dioxide but not for sulfur dioxide and zone [54]. Additionally, a study of 53 residents with type 2 diabetes or impaired glucose tolerance in Shanghai noted a significant reduction in SDNN associated with an increase in the number concentration of particles and particles of diameter 100–200 nm

[101]. They also observed a significant association of increased exposure to black carbon, sulfur dioxide, and nitrogen dioxide with HRV reductions. The above data suggest that underlying cardiovascular diseases and diabetes or impaired glucose tolerance may confer reduced autonomic function of heart due to air pollution exposure. Hence, particle interventions to reduce individual exposure and protect susceptible persons are in urgent need. One study that evaluated 98 patients with coronary heart disease walking on a predefined route in central Beijing found that using highly efficient face mask was associated with decreased self-reported symptoms, reduced maximal ST segment depression, and reduced HRV indices [65]. However, heart rate was not influenced by mask use in this study.

4.6.2 Blood Pressure

There has been a growing body of evidence support that air pollution is capable of altering systemic hemodynamics. Short-term exposure has been demonstrated by several studies capable of altering hemodynamic effects, though the results of studies are inconstant. One study that evaluated 125 healthy young adults surrounding the 2008 Beijing Olympics found significant decreases in systolic blood pressure (SBP) from the pre-Olympics to the during-Olympics period and the levels of SBP reversed in the post-Olympics period [131]. However, changes in diastolic blood pressure (DBP) were not observed by period. It is important to note that there have been published studies that have not observed an association between air pollutant exposure and blood pressure. In another study of 201 Beijing residents conducted before, during, and after the Olympics, significant changes on blood pressure over the three periods were not observed, although there was an increase in DBP and reduction in pulse pressure among males during the Olympics [76]. The negative and inverse response results may result from variations in the susceptibility of the subjects evaluated.

Several studies have investigated long-term effects of air pollutant in the development of high blood pressure and hypertension. One study of 24,845 Chinese adults in 11 districts of 3 northeastern Chinese cities from 2009 to 2010 found a significant relation among PM_{10}, SO_2, and O_3 and the prevalence of hypertension [31]. Specifically, the OR for hypertension increased by 1.12 (95% CI: 1.08, 1.16) per 19 $\mu g/m^3$ increase in PM_{10}, 1.11 (95% CI, 1.04, 1.18) per 20 $\mu g/m^3$10 increase in SO_2, and 1.13 (95% CI: 1.06, 1.20) per 22 $\mu g/m^3$ increase in O_3. They also found increases in SBP and DBP associated with an increase in O_3, though these associations were only statistically in men. Consistent with this, the outcome of some other studies also found the adverse association between ambient air pollution and hypertension and positive association with arterial SBP and DBP [134, 139]. Moreover, a panel study of 35 type 2 diabetes mellitus patients from urban community in Shanghai reported that the short-term association of size-fractionated PM and blood pressure strengthened with decreasing diameter of PM, suggesting size patterns of PM in elevating blood pressure [137].

Air pollution-induced BP elevation may be prompted by the components of particle matters. One study investigated changes in BP associated with chemical constituents of particulate air pollution during 460 repeated visits among a panel of 39 university students before and after their natural relocation from a suburban campus to an urban campus with different air pollution levels and contents in Beijing [114]. The results of their study showed positive associations of a subset of $PM_{2.5}$ constituents (organic carbon, elemental carbon, chloride ions, fluoride ions, nickel, zinc, magnesium, lead, and arsenic) with increase in SBP and DBP and negative associations of manganese, chromium, and molybdenum with SBP or DBP. Source appointment for ambient $PM_{2.5}$ was performed in the same panel to evaluate the cardiopulmonary effects associated with source-specific $PM_{2.5}$ concentrations, and results showed significant increases in DBP associated with $PM_{2.5}$ from coal combustion [118]. Moreover, findings from another study of 280 women in rural area of northwestern Yunnan Province where biomass fuels are commonly used show that $PM_{2.5}$exposure from biomass combustion may be a risk factor for increased blood pressure and that black carbon from combustion emissions is more strongly associated with blood pressure than PM mass [3, 4].

4.6.3 Hemostatic Biomarkers

It has been postulated that one of the mechanism to explain the increased cardiorespiratory morbidity and mortality associated with air pollution is the elevation of blood coagulation [9]. Inhaled particles may activate platelets and promote thrombosis rapidly and directly, the process of which may be independent from the instigation of systemic inflammatory responses [37, 60]. Growing evidence suggests that air pollution exposure, especially to PM, leads to elevated platelet activation and platelet aggregation.

One study evaluated 125 healthy young adults in Beijing before, during, and after the 2008 Beijing Olympics and found associations of changes in air pollution levels during the Beijing Olympics with acute changes in biomarkers of hemostasis [131]. Specifically, the findings show that circulatory levels of soluble CD40 ligand (sCD40L), soluble P-selectin, and von Willebrand factor (vWF) decreased significantly from the pre-Olympics to the during-Olympics period and reversed from those in the during-Olympics period in the post-Olympics period. Furthermore, these biomarkers showed significant associations with multiple ambient pollutants, including $PM_{2.5}$, SO_2, and element carbon. In the same study, findings on the different lag patterns of the associations of biomarkers for hemostasis with ultrafine particles to those with $PM_{2.5}$ suggest that the ultrafine size fraction and the fine size fraction of $PM_{2.5}$ are likely to affect PM-related pathophysiological pathways independently [38]. Another study of 40 healthy college students before and after their natural relocation from suburban campus to unurban campus with changing air pollution contents in Beijing found consistent positive associations between hemostatic biomarkers [fibrinogen, plasminogen activator inhibitor type 1(PAI-1), tissue-type plasminogen activator (t-PA), and vWF], and $PM_{2.5}$ chemical constituents, providing clues for the potential effects of $PM_{2.5}$ chemical constituents on the cardiovascular pathophysiologic changes [117]. Besides, more recent evidence found associations of size-fractionated PM with elevated levels of fibrinogen, PAI-1, and vWF in 35 type 2 diabetic mellitus patients in Shanghai [107].

4.6.4 Metabolomics Biomarkers Related with Ambient Air Pollution

It has been demonstrated that metabolomics may be sensitive in detecting subtle changes in response to environmental challenges. One study evaluated metabolite marker of 35 male welders and 16 male office workers with well-characterized exposure to welding fume at a Taiwanese shipyard and found differences of metabolite markers between welders and office workers by analysis of urinary metabolomics profiles of the participants [108]. Specifically, it is reported that higher levels of glycine, taurine, betaine/TMAO, serine, S-sulfocysteine, hippurate, gluconate, creatinine, and acetone and lower levels of creatine were observed among welders. These metabolites are involved in multiple metabolic processes, including carbohydrate metabolism, amino acid metabolism, oxidation/reduction pathways, and urea metabolism. It has been demonstrated that welding fumes, containing complex toxicants including heavy metals and ozone, can be viewed as oxidative pollutants and can produce adverse health effects on welders through increased oxidative stress [44]. Moreover, metabolomics analyses of urine samples collected from 566 healthy volunteers in Shanxi Province demonstrated that PAH exposure was correlated to its metabolic outcomes and that metabolites pertinent to amino acid, purine, lipid, and glucuronic acid metabolism were significantly changed with chronic environmental PAH exposure [110]. Specifically, in this study, dodecadienylcarnitine was found to be the only metabolite that showed significant linear dose-response relationship with urinary PAH metabolites among the subjects, hinting that urinary dodecadienylcarnitine may be developed as a novel and sensitive biomarker for monitoring PAH-induced oxidative stress in the non-occupationally exposed population.

Additionally, several studies have investigated the adverse cardio-metabolic effects of air pollution. A recent study of 371 residents with different residential distance to major roads in Shanghai found that the residents living within 50 m to the major road compared with those living more than 200 m away have 1.95 times higher fasting insulin, 1.30 times higher homeostasis model assessment of insulin resistance (HOMA-IR), and 1.56 times higher low-density lipoprotein cholesterol, though the clear exposure-response relationship of $PM_{2.5}$with the fasting glucose, cholesterol, and high-density lipoprotein was failed to be observed [58]. The findings suggest that long-term exposure to traffic-related air pollutant may impair cardio-metabolism. Moreover, another study of 65 adults with metabolic syndrome and insulin resistance in the Beijing metropolitan area demonstrated that both black carbon and fine particulate matter were significantly associated with increases in HOMA-IR [10] (Table 4.1).

Table 4.1 Air pollution-related biological effective biomarkers

Study site	Year	Study design	Pollutant	Participant, N	Biomarker	References
Inflammation biomarker						
Shanghai	2015	Randomized, double-blind, and crossover trials	$PM_{2.5}$	Healthy college students, 35	eNO, CRP, Fibrinogen, P-selectin, MCP-1, IL-1β TNF-α, IL-6	[19]
Shanghai	2016	Panel study	CO, NO_2, SO_2, PM_{10}, and $PM_{2.5}$	Healthy college students, 75	eNO	[140]
Beijing	2016	Panel study	NO_2, SO_2, PM_{10}, and $PM_{2.5}$	COPD patients, 23	eNO, eH_2S	[119]
Beijing	2016	Panel study	$PM_{2.5}$	COPD patients, 18	eNO, eH_2S	[84]
Shanghai	2015	Panel study	$PM_{2.5}$	COPD patients, 30	eNO	[18]
Beijing	2011	Panel study	CO, NO_2, SO_2, BC, and $PM_{2.5}$	Children, 36 (28 healthy, 8 asthma)	eNO	[69]
Shanxi, Henan, Zhejiang	2013	Cross-sectional study	Fuel (wood fuel, crop residues, cleaner fuels) Stoves(improved biomass, traditional biomass, cleaner-fuel stove)	Rural family, 3476	eCO	[130]
5 cities, 5 rural area	2005	Cross-sectional study	Heating (central heating, wood, coal)	Residents, 51	eCO	[86]
Beijing	2012	Panel study	CO, NO_2, SO_2, O_3, EC, OC, sulfate, and $PM_{2.5}$	Healthy adults, 125	eNO	[53]
Beijing	2014	Panel study	$PM_{2.5}$ and UFPs	Healthy adults, 125	eNO, fibrinogen, WBC	[38]
Shanghai	2015	Cross-sectional study	$PM_{2.5}$	Policemen, 107; officers,101	WBC, neutrophils, lymphocytes	[136]

(continued)

Table 4.1 (continued)

Study site	Year	Study design	Pollutant	Participant, N	Biomarker	References
Taiyuan	2009	Cross-sectional study	Traffic-related pollution areas	Children, 152	T lymphocyte subsets (CD3+, CD4+, CD8+, CD4+/CD8+)	[135]
Beijing	2010	Cross-sectional study	NO_2, SO_2, and PM_{10}	aged people, 236	WBC, CRP, IgG, IgA, PLT	[59]
Shanghai	2016	Cross-sectional study	The distance from main traffic road	Residents, 371	IL-6, IL-10	[58]
Shanghai	2014	Cross-sectional study	BC	Packing workers, 84	IL-6, IL-8, TNF-α	[132]
				Pump station workers, 104		
Shanghai	2015	Panel study	$PNC_{0.25–0.40}$	Diabetics, 35	IL-1b, IL-10, TNF- α, ICAM-1, VCAM-1	[107]
Beijing	2016	Panel study	NO, NO_2, NOx, PM_{10}, and $PM_{2.5}$	Healthy college students, 40	VCAM-1, ICAM-1	[121]
Beijing	2012	Panel study	NO, NO_2, NOx, PM_{10}, and $PM_{2.5}$	Healthy college students, 40	CRP, TNF-α	[117]
Beijing	2009	Panel study	API	Healthy taxi drivers, 14	CRP	[113]
Beijing	2011	Cross-sectional study	PM_{10} and $PM_{2.5}$	aged people, 236	CRP	[32]
Taiyuan	2014	Cross-sectional study	$PM_{2.5}$	Traffic policemen, 58	TNF-α	[73]
				CDC officers, 29		
Oxidative stress biomarker						
Beijing	2012	Panel study	CO, NO_2, SO_2, O_3, EC, OC, sulfate, and $PM_{2.5}$	Healthy adults, 125	EBC nitrite, EBC nitrate, EBC pH	[53]
					EBC 8-OHdG and urinary 8-OHdG	

Beijing	2014	Panel study	$PM_{2.5}$ and UFPs	Healthy adults, 125	EBC nitrite, EBC pH, EBC MDA	[38]
					Urinary 8-OHdG, urinary MDA	
Shanghai	2016	Cross-sectional study	The distance from main traffic road	Residents, 371	HDL, LDL-C, T-SOD	[58]
Beijing	2016	Panel study	$PM_{2.5}$	Healthy college students, 40	EC-SOD and GPX1	[120]
Some city	2005	Cross-sectional study		Traffic policemen, 170	SOD activity	[128]
				Healthy people, 108		
Nanning	2014	Cross-sectional study	Benzene, toluene, ethyl benzene	Gas station staff, 252	T-SOD	[98]
Taiyuan	2016	Randomized, double-blind, and crossover trials	$PM_{2.5}$	Atopic subjects, 17	Protein carbonyl	[133]
Beijing	2015	Panel study	$PM_{2.5}$	Healthy college students, 40	Ox-LDL	[122]
Genetic biomarker						
DNA damage						
Shanxi Province	2010	Case-control study	PAHs	Parents of children born with neural tube defect (cases)	DNA adducts (^{32}P-postlabeling)	[78]
				Parents of children born with no visible birth defect (controls)		

(continued)

Table 4.1 (continued)

Study site	Year	Study design	Pollutant	Participant, N	Biomarker	References
Yunnan Province	1997	Case-control study	Coal combustion	Lung cancer in Xuanwei County (cases);	DNA adducts (^{32}P-postlabeling)	[127]
				Healthy adults in Kunming City (controls)		
Zhuhai, Wuhan, Tianjin	2015	Cross-sectional study	$PM_{2.5}$	Healthy residents, 307	DNA damage (comet assay)	[22]
Shenyang City	2015	Cohort study	$PM_{2.5}$	Male field traffic policemen (exposure group)	DNA damage (comet assay)	[72]
				Male non-field traffic policemen (control group)		
Taipei City	2012	Cohort study	PAHs	Traffic conductors, 91;	DNA strand breaks (comet assay)	[52]
				Indoor office workers, 53		
–	2016	Cross-sectional study	Diesel engine exhaust	Male diesel engine testing workers, 101	DNA strand breaks (comet assay)	[33]
				Male pumps operating workers, 106	Oxidative DNA damage (FPG-comet assay)	
Zhuhai, Wuhan, Tianjin	2015	Cross-sectional study	$PM_{2.5}$	Healthy residents, 307	DNA damage (comet assay)	[100]
Beijing	2010	Repeated-measure study	PAHs and anthraquinone	Nonsmoking security guards, 2	8-OHdG	[111]

Guangzhou City	2012	Cohort study	PAHs	Children from elementary school situated near a heavy traffic road, 39	8-OHdG	[35]
				Children from a kindergarten situated in a corner of the main campus of a big university, 35		
Wuhan City	2015	Cross-sectional study	Heavy metals	Male coke-oven workers, 1,333	8-OHdG	[109]
Taiwan	2013	Longitudinal study	Cooking oil fumes	Male military cooks, 61	8-OHdG	[64]
				Office soldiers, 37		
Taipei City	2012	Cohort study	PAHs	Traffic conductors, 91	8-OHdG	[52]
				Indoor office workers, 53		
Chromosomal damage						
Hebi City, Henan Province	1999	Cohort study	Traffic-related air pollution	Traffic policemen (exposure group),45	Chromosome aberration	[16]
				Indoor office workers in Police Station (control group), 30		
Jiaole Township, Guizhou Province	2007	Cohort study	Coal combustion (As)	Villagers from Jiaole (exposure group), 184	Chromosomal aberrations	[129]
				Villagers from a town 12 km from Jiaole (control group), 53		

(continued)

Table 4.1 (continued)

Study site	Year	Study design	Pollutant	Participant, N	Biomarker	References
Lanzhou City	1998	Cohort study	Automobile exhaust	Traffic policemen working in the field (exposure group),78;	Micronuclei	[138]
				Household register policemen (control group), 57		
Shenyang City	2006	Cohort study	PAHs	Female office workers in the industrial area, 66	Micronuclei	[56]
				Female office workers in the rural area, 63		
Guangzhou City	2005	Cohort study	Cooking oil fumes	Cooks (exposure group), 65	Micronuclei	[126]
				Waiters (control group), 49		
Lanzhou City	1998	Cohort study	Automobile exhaust	Traffic policemen working in the field (exposure group), 78;	Sister-chromatid exchanges	[138]
				Household register policemen (control group), 57		
Jiaole Township, Guizhou Province	2007	Cohort study	Coal combustion (As)	Villagers from Jiaole (exposure group), 184	Sister chromatid exchanges	[129]
				Villagers from a town 12 km from Jiaole (control group), 53		

Others						
Beijing	2012	Cohort study	$PM_{2.5}$, PM_{10}, EC	Truck drivers, 120	Telomere length	[48]
				Office workers, 120		
Sichuan	2014	Cross-sectional study	Biomass smoke	Female residents, 25	Telomere length	[93]
Shanghai	2015	Longitudinal panel study	Size-fractionated PM, gaseous pollutants	Type 2 diabetes patients, 35	Telomere length	[123]
Beijing	2013	Repeated-measure study	EC, $PM_{2.5}$, PM_{10}	Truck Driver, 60	MtDNAcn	[49]
				Office worker, 60		
Shanghai	2015	Longitudinal panel study	Size-fractionated PM, gaseous pollutants	Type 2 diabetes patients, 35	MtDNAcn	[123]
Epigenetic biomarker						
Beijing	2013	Cohort study	EC	Truck drivers (low exposure), 20	DNA methylation	(Byun HM et al., [12])
				Truck drivers (high exposure), 20		
Beijing	2014	Repeated-measure study	$PM_{2.5}$, PM_{10}, EC	Truck drivers, 60	DNA Methylation	[43]
				Office workers, 60		
Beijing	2014	Repeated-measure study	$PM_{2.5}$ components	Truck drivers, 60	DNA Methylation	[50]
				Office workers, 60		
Beijing	2015	Repeated-measure study	PM_{10}, $PM_{2.5}$, $PM_{2.5}$ EC	Truck drivers, 60	Blood global 5mC	[91]
				Office workers, 60	Blood global 5hmC	
Beijing		Cohort study	EC	Truck drivers, 40	mtDNA Methylation	Byun et al. [13]
Beijing	2016	Repeated-measure study	$PM_{2.5}$, PM_{10}, EC	Truck drivers, 60	miRNA	[47]
				Office workers, 60		

(continued)

Table 4.1 (continued)

Study site	Year	Study design	Pollutant	Participant, N	Biomarker	References
Beijing	2015		$PM_{2.5}$	Healthy subjects, 4	Histone modification H3K27ac	[71]
Cardiac autonomic function						
Beijing	2009	Panel study	$PM_{2.5}$	Young healthy taxi drivers, 11	HRV	[116]
Beijing	2013	Panel study	Temperature and traffic-related air pollutants ($PM_{2.5}$ and CO)	Healthy taxi drivers, 14	HRV	[124]
Beijing	2012	Panel study	$PM_{2.5}$	Healthy elderly subjects, 30	HRV	[57]
Shanghai	2016	Cross-sectional study	$PM_{2.5}$	Urban residents, 371	HR	[58]
Beijing	2013	Panel study	$PM_{2.5}$,SO_2, CO, NO_2, O_3	Healthy Young Adults, 125	HR	[131]
Beijing	2012	Repeated-measure study	$PM_{2.5}$, BC, NO_2, SO_2, O_3, CO	Cardiovascular disease patients, 40	HRV (SDNN, r-MSSD, LF, HF)	[54]
Shanghai	2015	Repeated-measure study	Ambient particles, black carbon, SO_2, NO_2, CO, O_3	Type 2 diabetes patients, 53	HRV (SDNN, rMSSD, LF, HF)	[101]
Beijing	2012	Randomized crossover trial	$PM_{2.5}$	Coronary heart disease patients, 98	HRV (LF, HF, LFn, HFn, HF/LF)	[65]
Beijing	2014	Repeated-measure study	UFPs, $PM_{2.5}$	Residents, 125	HR	[38]
Beijing	2016	Panel study	$PM_{2.5}$, BC	Adults with metabolic syndrome and insulin resistance, 65	HR, HRV (LF, HF, LF/HF, SDNN, rMSSD, pNN50)	[10]

Blood pressure						
Beijing	2012	Repeated-measure study	$PM_{2.5}$, BC, NO_2, SO_2, O_3, CO	Cardiovascular disease patients, 40	SBP, DBP	[54]
Beijing	2014	Panel study	PM	Healthy adults, 201	DBP, SBP, the pulse pressure	[76]
Liaoning Province	2013	Cross-sectional study	PM_{10}, SO_2, NO_2, and O_3	Adults, 24,845	DBP, SBP	[31]
Beijing	2013	Panel study	$PM_{2.5}$	University students, 39	DBP, SBP, the pulse pressure	[114]
Beijing	2014	Panel study	$PM_{2.5}$ and chemical constituents	University students, 39	DBP, SBP, the pulse pressure	[118]
Yunnan Province	2011	Repeated-measure study	Biomass fuels ($PM_{2.5}$)	Women, 280	DBP, SBP	[3]
Yunnan Province	2014	Repeated-measure study	Biomass fuels (BC)	Women, 280	DBP, SBP	[4]
Shanghai	2015	Panel study	Size-fractionated PM	Type 2 diabetes mellitus patients, 35	DBP, SBP, the pulse pressure	[137]
Beijing	2014	Repeated-measure study	UFPs, $PM_{2.5}$	Residents, 125	DBP, SBP	[38]
Beijing	2016	Panel study	$PM_{2.5}$, BC	Adults with metabolic syndrome and insulin resistance, 65	DBP, SBP	[10]
Coagulation						
Beijing	2014	Repeated-measure study	UFPs, $PM_{2.5}$	Residents, 125	vWF, sCD40L, P-selectin	[38]
Beijing	2012	Panel study	$PM_{2.5}$ chemical constituents	University students, 40	Fibrinogen, PAI-1, t-PA, vWF, sP-selectin	[117]

(continued)

Table 4.1 (continued)

Study site	Year	Study design	Pollutant	Participant, N	Biomarker	References
Beijing	2013	Panel study	$PM_{2.5}$ and its constituents	Healthy young adults, 125	CRP, fibrinogen, vWF, sCD40L, sP-selectin	[89]
Shanghai	2015	Panel study	Size-fractionated PM	Type 2 diabetic mellitus patients, 35	Fibrinogen, PAI-1, vWF, sCD40L	[107]
Metabolomics						
Taiwan	2012	Cohort study	Welding fumes (Cr, Ni, Mn)	Male welders, 35	Urine metabolomics profiles	[108]
				Male office workers, 16		
Shanxi Province	2015	Cohort study	PAHs	Children and elderly people living in an area polluted by the coking industry, 197	Urine metabolomics profiles	[110]
				Children and elderly people living in nonpolluted control area, 369		
Beijing	2016	Panel study	$PM_{2.5}$, BC	Adults with metabolic syndrome and insulin resistance, 65	Fasting glucose, fasting insulin, HOMA-IR	[10]

PM particulate matter, *CO* carbon monoxide, NO_2 nitrogen dioxide, SO_2 sulfur dioxide, *BC* black carbon, O_3 ozone, *EC* element carbon, *OC* organic carbon, *UFPs* ultrafine particles, *PNC* particulate number concentration, *NO* nitrogen monoxide, NO_x nitrogen oxides, *PAHs* polycyclic aromatic hydrocarbons, *As* arsenic, *Cr* chromium, *Ni* nickel, *Mn* manganese, *eNO* exhaled nitric oxide, *BP* blood pressure, *CRP* C-reactive protein, *IL* interleukin, *TNF-α* tumor necrosis factor α, *sCD40L* soluble CD40 ligand, *t-PA* tissue-type plasminogen activator, eH_2S exhaled hydrogen sulfide, *eCO* exhaled carbon monoxide, *EBC* exhaled breath condensate, *8-OHdG* 8-hydroxydeoxyguanosine, *vWF* von Willebrand factor, *WBC* white blood cell, *MDA* malonyldialdehyde, *PLT* platelets, *HDL* high density lipoprotein, *LDL-C* low-density lipoprotein cholesterol, *Ox-LDL* oxidized low-density lipoprotein, *SOD* superoxide dismutase, *T-SOD* total superoxide dismutase, *EC-SOD* extracellular superoxide dismutase, *ICAM-1* intercellular adhesion molecule-1, *VCAM-1* vascular cell adhesion molecule-1, *GPX1* glutathione peroxidase1, *FPG-comet assay* formamidopyrimidine-DNA glycosylase comet assay, *mtDNA* mitochondrial DNA, *mtDNAcn* mitochondrial DNA copy number, *5mC* 5-methylcytosine, *miRNA* microRNA, *HRV* heart rate variability, *HR* heart rate, *SDNN* standard deviation of normal-to-normal, *r-MSSD* square root of the mean squared difference between adjacent normal-to-normal intervals, *LF* low frequency, *HF* high frequency, *LFn* low-frequency normalized power, *HFn* high-frequency normalized power, *pNN50* proportion of N-N intervals >50 ms, *SBP* systolic blood pressure, *DBP* diastolic blood pressure, *HOMA-IR* homeostasis model assessment of insulin resistance

Acknowledgments This chapter was supported by grants from the National Natural Science Foundation of China (No. 81571130090, 91543112) and the National Key Research and Development Program of China (2016YFC0206506, 2017YFC0702701).

References

1. Albertini RJ, Anderson D, Douglas GR, Hagmar L, Hemminki K, Merlo F, Natarajan AT, Norppa H, Shuker DE, Tice R, Waters MD, Aitio A. Ipcs guidelines for the monitoring of genotoxic effects of carcinogens in humans. International programme on chemical safety. Mutat Res. 2000;463(2):111–72.
2. Attfield MD, Schleiff PL, Lubin JH, Blair A, Stewart PA, Vermeulen R, Coble JB, Silverman DT. The diesel exhaust in miners study: a cohort mortality study with emphasis on lung cancer. J Natl Cancer Inst. 2012;104(11):869–83.
3. Baumgartner J, Schauer JJ, Ezzati M, Lu L, Cheng C, Patz JA, Bautista LE. Indoor air pollution and blood pressure in adult women living in rural china. Environ Health Perspect. 2011;119(10):1390–5.
4. Baumgartner J, Zhang Y, Schauer JJ, Huang W, Wang Y, Ezzati M. Highway proximity and black carbon from cookstoves as a risk factor for higher blood pressure in rural china. Proc Natl Acad Sci U S A. 2014;111(36):13229–34.
5. Blackburn EH. Telomeres and telomerase: the means to the end (nobel lecture). Angew Chem Int Ed Engl. 2010;49(41):7405–21.
6. Bonassi S, Znaor A, Norppa H, Hagmar L. Chromosomal aberrations and risk of cancer in humans: an epidemiologic perspective. Cytogenet Genome Res. 2004;104(1–4):376–82.
7. Branco MR, Ficz G, Reik W. Uncovering the role of 5-hydroxymethylcytosine in the epigenome. Nat Rev Genet. 2012;13(1):7–13.
8. Britton LM, Gonzales-Cope M, Zee BM, Garcia BA. Breaking the histone code with quantitative mass spectrometry. Expert Rev Proteomics. 2011;8(5):631–43.
9. Brook RD, Rajagopalan S, Pope CA 3rd, Brook JR, Bhatnagar A, Diez-Roux AV, Holguin F, Hong Y, Luepker RV, Mittleman MA, Peters A, Siscovick D, Smith SC Jr, Whitsel L, Kaufman JD. American Heart Association Council on E, Prevention CotKiCD, Council on Nutrition PA, Metabolism. Particulate matter air pollution and cardiovascular disease: an update to the scientific statement from the American heart association. Circulation. 2010;121(21):2331–78.
10. Brook RD, Sun Z, Brook JR, Zhao X, Ruan Y, Yan J, Mukherjee B, Rao X, Duan F, Sun L, Liang R, Lian H, Zhang S, Fang Q, Gu D, Sun Q, Fan Z, Rajagopalan S. Extreme air pollution conditions adversely affect blood pressure and insulin resistance: the air pollution and cardiometabolic disease study. Hypertension. 2016;67(1):77–85.
11. Buschini A, Cassoni F, Anceschi E, Pasini L, Poli P, Rossi C. Urban airborne particulate: genotoxicity evaluation of different size fractions by mutagenesis tests on microorganisms and comet assay. Chemosphere. 2001;44(8):1723–36.
12. Byun HM, Motta V, Panni T, Bertazzi PA, Apostoli P, Hou L, Aa B. Evolutionary age of repetitive element subfamilies and sensitivity of DNA methylation to airborne pollutants. Part Fibre Toxicol. 2013a;10:28.
13. Byun HM, Panni T, Motta V, Hou L, Nordio F, Apostoli P, Bertazzi PA, Baccarelli AA. Effects of airborne pollutants on mitochondrial DNA methylation. Part Fibre Toxicol. 2013b;10:18.
14. Cavalcanti S, Severi S, Chiari L, Avanzolini G, Enzmann G, Bianco G, Panzetta G. Autonomic nervous function during haemodialysis assessed by spectral analysis of heart-rate variability. Clin Sci (Lond). 1997;92(4):351–9.
15. Chapman RS, Mumford JL, Harris DB, He ZZ, Jiang WZ, Yang RD. The epidemiology of lung cancer in xuan wei, china: current progress, issues, and research strategies. Arch Environ Health. 1988;43(2):180–5.

16. Chen C, Lv Y, Zhang K, Han L, Fu B, Fu B, Fu Y. The analysis of chromosomal aberration of peripheral blood lymphocyte of traffic police. J Hyg Res. 1999;28(06):324–5. (in Chinese)
17. Chen H, Burnett RT, Copes R, Kwong JC, Villeneuve PJ, Goldberg MS, Brook RD, van Donkelaar A, Jerrett M, Martin RV, Brook JR, Kopp A, Tu JV. Ambient fine particulate matter and mortality among survivors of myocardial infarction: population-based cohort study. Environ Health Perspect. 2016;124(9):1421–8.
18. Chen R, Qiao L, Li H, Zhao Y, Zhang Y, Xu W, Wang C, Wang H, Zhao Z, Xu X, Hu H, Kan H. Fine particulate matter constituents, nitric oxide synthase DNA methylation and exhaled nitric oxide. Environ Sci Technol. 2015;49(19):11859–65.
19. Chen R, Zhao A, Chen H, Zhao Z, Cai J, Wang C, Yang C, Li H, Xu X, Ha S, Li T, Kan H. Cardiopulmonary benefits of reducing indoor particles of outdoor origin: a randomized, double-blind crossover trial of air purifiers. J Am Coll Cardiol. 2015;65(21):2279–87.
20. Chiang TA, Wu PF, Ko YC. Identification of carcinogens in cooking oil fumes. Environ Res. 1999;81(1):18–22.
21. Chiang TY, Yuan TH, Shie RH, Chen CF, Chan CC. Increased incidence of allergic rhinitis, bronchitis and asthma, in children living near a petrochemical complex with so2 pollution. Environ Int. 2016;96:1–7.
22. Chu M, Sun C, Chen W, Jin G, Gong J, Zhu M, Yuan J, Dai J, Wang M, Pan Y, Song Y, Ding X, Guo X, Du M, Xia Y, Kan H, Zhang Z, Hu Z, Wu T, Shen H. Personal exposure to pm2.5, genetic variants and DNA damage: a multi-center population-based study in Chinese. Toxicol Lett. 2015;235(3):172–8.
23. Chuang CY, Lee CC, Chang YK, Sung FC. Oxidative DNA damage estimated by urinary 8-hydroxydeoxyguanosine: influence of taxi driving, smoking and areca chewing. Chemosphere. 2003;52(7):1163–71.
24. Chuang KJ, Chan CC, Su TC, Lee CT, Tang CS. The effect of urban air pollution on inflammation, oxidative stress, coagulation, and autonomic dysfunction in young adults. Am J Respir Crit Care Med. 2007;176(4):370–6.
25. Collas P. The current state of chromatin immunoprecipitation. Mol Biotechnol. 2010;45(1):87–100.
26. Consortium EP. An integrated encyclopedia of DNA elements in the human genome. Nature. 2012;489(7414):57–74.
27. Creason J, Neas L, Walsh D, Williams R, Sheldon L, Liao D, Shy C. Particulate matter and heart rate variability among elderly retirees: the Baltimore 1998 pm study. J Expo Anal Environ Epidemiol. 2001;11(2):116–22.
28. Creyghton MP, Cheng AW, Welstead GG, Kooistra T, Carey BW, Steine EJ, Hanna J, Lodato MA, Frampton GM, Sharp PA, Boyer LA, Young RA, Jaenisch R. Histone h3k27ac separates active from poised enhancers and predicts developmental state. Proc Natl Acad Sci U S A. 2010;107(50):21931–6.
29. De Marini DM. Genotoxicity biomarkers associated with exposure to traffic and near-road atmospheres: a review. Mutagenesis. 2013;28(5):485–505.
30. De Vries R, Kriebel D, Sama S. Outdoor air pollution and copd-related emergency department visits, hospital admissions, and mortality: a meta-analysis. COPD. 2017;14(1):113–21.
31. Dong GH, Qian ZM, Xaverius PK, Trevathan E, Maalouf S, Parker J, Yang L, Liu MM, Wang D, Ren WH, Ma W, Wang J, Zelicoff A, Fu Q, Simckes M. Association between long-term air pollution and increased blood pressure and hypertension in china. Hypertension. 2013;61(3):578–84.
32. Dong S, Cheng Y, Xu D, Hong Y, Knag J, Zhang G, Wang Q, Li D, Jin Y. Ambient particulate pollutants and c reactive protein variability in elderly residents in beijing. J Environ Health. 2011;28(12):1035–9. (in Chinese)
33. Duan H, Jia X, Zhai Q, Ma L, Wang S, Huang C, Wang H, Niu Y, Li X, Dai Y, Yu S, Gao W, Chen W, Zheng Y. Long-term exposure to diesel engine exhaust induces primary DNA damage: a population-based study. Occup Environ Med. 2016;73(2):83–90.

34. Everson RB, Randerath E, Santella RM, Cefalo RC, Avitts TA, Randerath K. Detection of smoking-related covalent DNA adducts in human placenta. Science. 1986;231(4733):54–7.
35. Fan R, Wang D, Mao C, Ou S, Lian Z, Huang S, Lin Q, Ding R, She J. Preliminary study of children's exposure to pahs and its association with 8-hydroxy-2′-deoxyguanosine in Guangzhou, China. Environ Int. 2012;42:53–8.
36. Fenech M. The cytokinesis-block micronucleus technique and its application to genotoxicity studies in human populations. Environ Health Perspect. 1993;3:101–7.
37. Franchini M, Mannucci PM. Thrombogenicity and cardiovascular effects of ambient air pollution. Blood. 2011;118(9):2405–12.
38. Gong J, Zhu T, Kipen H, Wang G, Hu M, Guo Q, Ohman-Strickland P, Lu SE, Wang Y, Zhu P, Rich DQ, Huang W, Zhang J. Comparisons of ultrafine and fine particles in their associations with biomarkers reflecting physiological pathways. Environ Sci Technol. 2014;48(9):5264–73.
39. Gong J, Zhu T, Kipen H, Wang G, Hu M, Ohman-Strickland P, Lu SE, Zhang L, Wang Y, Zhu P, Rich DQ, Diehl SR, Huang W, Zhang JJ. Malondialdehyde in exhaled breath condensate and urine as a biomarker of air pollution induced oxidative stress. J Expo Sci Environ Epidemiol. 2013;23(3):322–7.
40. Gonzalez-Flecha B. Oxidant mechanisms in response to ambient air particles. Mol Asp Med. 2004;25(1–2):169–82.
41. Greiser KH, Kluttig A, Schumann B, Swenne CA, Kors JA, Kuss O, Haerting J, Schmidt H, Thiery J, Werdan K. Cardiovascular diseases, risk factors and short-term heart rate variability in an elderly general population: the Carla study 2002–2006. Eur J Epidemiol. 2009;24(3):123–42.
42. Grundhoff A, Sullivan CS. Virus-encoded micrornas. Virology. 2011;411(2):325–43.
43. Guo L, Byun HM, Zhong J, Motta V, Barupal J, Zheng Y, Dou C, Zhang F, McCracken JP, Diaz A, Marco SG, Colicino S, Schwartz J, Wang S, Hou L, Baccarelli AA. Effects of short-term exposure to inhalable particulate matter on DNA methylation of tandem repeats. Environ Mol Mutagen. 2014;55(4):322–35.
44. Han SG, Kim Y, Kashon ML, Pack DL, Castranova V, Vallyathan V. Correlates of oxidative stress and free-radical activity in serum from asymptomatic shipyard welders. Am J Respir Crit Care Med. 2005;172(12):1541–8.
45. Harley CB, Futcher AB, Greider CW. Telomeres shorten during ageing of human fibroblasts. Nature. 1990;345(6274):458–60.
46. He L, Hannon GJ. Micrornas: small rnas with a big role in gene regulation. Nat Rev Genet. 2004;5(7):522–31.
47. Hou L, Barupal J, Zhang W, Zheng Y, Liu L, Zhang X, Dou C, McCracken JP, Diaz A, Motta V, Sanchez-Guerra M, Wolf KR, Bertazzi PA, Schwartz JD, Wang S, Baccarelli AA. Particulate air pollution exposure and expression of viral and human micrornas in blood: the Beijing truck driver air pollution study. Environ Health Perspect. 2016;124(3):344–50.
48. Hou L, Wang S, Dou C, Zhang X, Yu Y, Zheng Y, Avula U, Hoxha M, Diaz A, McCracken J, Barretta F, Marinelli B, Bertazzi PA, Schwartz J, Baccarelli AA. Air pollution exposure and telomere length in highly exposed subjects in Beijing, China: a repeated-measure study. Environ Int. 2012;48:71–7.
49. Hou L, Zhang X, Dioni L, Barretta F, Dou C, Zheng Y, Hoxha M, Bertazzi PA, Schwartz J, Wu S, Wang S, Baccarelli AA. Inhalable particulate matter and mitochondrial DNA copy number in highly exposed individuals in Beijing, China: a repeated-measure study. Part Fibre Toxicol. 2013;10:17.
50. Hou L, Zhang X, Zheng Y, Wang S, Dou C, Guo L, Byun HM, Motta V, McCracken J, Diaz A, Kang CM, Koutrakis P, Bertazzi PA, Li J, Schwartz J, Baccarelli AA. Altered methylation in tandem repeat element and elemental component levels in inhalable air particles. Environ Mol Mutagen. 2014;55(3):256–65.

51. Huang CH, Su SL, Hsieh MC, Cheng WL, Chang CC, Wu HL, Kuo CL, Lin TT, Liu CS. Depleted leukocyte mitochondrial DNA copy number in metabolic syndrome. J Atheroscler Thromb. 2011;18(10):867–73.
52. Huang HB, Lai CH, Chen GW, Lin YY, Jaakkola JJ, Liou SH, Wang SL. Traffic-related air pollution and DNA damage: a longitudinal study in Taiwanese traffic conductors. PLoS One. 2012a;7(5):e37412.
53. Huang W, Wang G, Lu SE, Kipen H, Wang Y, Hu M, Lin W, Rich D, Ohman-Strickland P, Diehl SR, Zhu P, Tong J, Gong J, Zhu T, Zhang J. Inflammatory and oxidative stress responses of healthy young adults to changes in air quality during the Beijing Olympics. Am J Respir Crit Care Med. 2012b;186(11):1150–9.
54. Huang W, Zhu T, Pan X, Hu M, Lu SE, Lin Y, Wang T, Zhang Y, Tang X. Air pollution and autonomic and vascular dysfunction in patients with cardiovascular disease: interactions of systemic inflammation, overweight, and gender. Am J Epidemiol. 2012c;176(2):117–26.
55. Hug N, Lingner J. Telomere length homeostasis. Chromosoma. 2006;115(6):413–25.
56. Ishikawa H, Tian Y, Piao F, Duan Z, Zhang Y, Ma M, Li H, Yamamoto H, Matsumoto Y, Sakai S, Cui J, Yamauchi T, Yokoyama K. Genotoxic damage in female residents exposed to environmental air pollution in Shenyang city, China. Cancer Lett. 2006;240(1):29–35.
57. Jia X, Song X, Shima M, Tamura K, Deng F, Guo X. Effects of fine particulate on heart rate variability in Beijing: a panel study of healthy elderly subjects. Int Arch Occup Environ Health. 2012;85(1):97–107.
58. Jiang S, Bo L, Gong C, Du X, Kan H, Xie Y, Song W, Zhao J. Traffic-related air pollution is associated with cardio-metabolic biomarkers in general residents. Int Arch Occup Environ Health. 2016;89(6):911–21.
59. Jin Y, Cheng Y, Xu D, Hong Y, Dong S, Kang J, Zhang G, Wang B, Li D. Relationship between improvement of air quality and immune function in elderly people during Beijing Olympic games. J Environ Health. 2010;27(10):847–50.
60. Kilinc E, Schulz H, Kuiper GJ, Spronk HM, Ten Cate H, Upadhyay S, Ganguly K, Stoeger T, Semmler-Bhenke M, Takenaka S, Kreyling WG, Pitz M, Reitmeir P, Peters A, Eickelberg O, Wichmann HE. The procoagulant effects of fine particulate matter in vivo. Part Fibre Toxicol. 2011;8:12.
61. Kim JY, Mukherjee S, Ngo LC, Christiani DC. Urinary 8-hydroxy-2′-deoxyguanosine as a biomarker of oxidative DNA damage in workers exposed to fine particulates. Environ Health Perspect. 2004;112(6):666–71.
62. Kincaid RP, Sullivan CS. Virus-encoded micrornas: an overview and a look to the future. PLoS Pathog. 2012;8(12):e1003018.
63. Kouzarides T. Chromatin modifications and their function. Cell. 2007;128(4):693–705.
64. Lai CH, Jaakkola JJ, Chuang CY, Liou SH, Lung SC, Loh CH, Yu DS, Strickland PT. Exposure to cooking oil fumes and oxidative damages: a longitudinal study in Chinese military cooks. J Expo Sci Environ Epidemiol. 2013;23(1):94–100.
65. Langrish JP, Li X, Wang S, Lee MM, Barnes GD, Miller MR, Cassee FR, Boon NA, Donaldson K, Li J, Li L, Mills NL, Newby DE, Jiang L. Reducing personal exposure to particulate air pollution improves cardiovascular health in patients with coronary heart disease. Environ Health Perspect. 2012;120(3):367–72.
66. Latalova K, Prasko J, Diveky T, Grambal A, Kamaradova D, Velartova H, Salinger J, Opavsky J. Autonomic nervous system in euthymic patients with bipolar affective disorder. Neuro Endocrinol Lett. 2010;31(6):829–36.
67. Lee HC, Wei YH. Mitochondrial role in life and death of the cell. J Biomed Sci. 2000;7(1):2–15.
68. Lee HK, Song JH, Shin CS, Park DJ, Park KS, Lee KU, Koh CS. Decreased mitochondrial DNA content in peripheral blood precedes the development of non-insulin-dependent diabetes mellitus. Diabetes Res Clin Pract. 1998;42(3):161–7.
69. Lin W, Huang W, Zhu T, Hu M, Brunekreef B, Zhang Y, Liu X, Cheng H, Gehring U, Li C, Tang X. Acute respiratory inflammation in children and black carbon in ambient air before and during the 2008 Beijing Olympics. Environ Health Perspect. 2011;119(10):1507–12.

70. Linnane AW, Marzuki S, Ozawa T, Tanaka M. Mitochondrial DNA mutations as an important contributor to ageing and degenerative diseases. Lancet. 1989;1(8639):642–5.
71. Liu C, Xu J, Chen Y, Guo X, Zheng Y, Wang Q, Chen Y, Ni Y, Zhu Y, Joyce BT, Baccarelli A, Deng F, Zhang W, Hou L. Characterization of genome-wide h3k27ac profiles reveals a distinct pm2.5-associated histone modification signature. Environ Health. 2015;14:65.
72. Ma M, Li S, Jin H, Zhang Y, Xu J, Chen D, Kuimin C, Yuan Z, Xiao C. Characteristics and oxidative stress on rats and traffic policemen of ambient fine particulate matter from Shenyang. Sci Total Environ. 2015;526:110–5.
73. Ma X, Qiu Y, Xu J, Zhang Z. Effects of different working environment pm2.5 exposure on cytokines in human being. J Environ Health. 2014;31(12):1042–4. (in Chinese)
74. McKelvey-Martin VJ, Green MH, Schmezer P, Pool-Zobel BL, De Meo MP, Collins A. The single cell gel electrophoresis assay (comet assay): a European review. Mutat Res. 1993;288(1):47–63.
75. Meng X, Zhang Y, Yang K-Q, Yang Y-K, Zhou X-L. Potential harmful effects of pm2.5 on occurrence and progression of acute coronary syndrome: epidemiology, mechanisms, and prevention measures. Int J Environ Res Public Health. 2016;13(8):748.
76. Mu L, Deng F, Tian L, Li Y, Swanson M, Ying J, Browne RW, Rittenhouse-Olson K, Zhang JJ, Zhang ZF, Bonner MR. Peak expiratory flow, breath rate and blood pressure in adults with changes in particulate matter air pollution during the Beijing Olympics: a panel study. Environ Res. 2014;133:4–11.
77. Mumford JL, He XZ, Chapman RS, Cao SR, Harris DB, Li XM, Xian YL, Jiang WZ, Xu CW, Chuang JC, et al. Lung cancer and indoor air pollution in Xuan Wei, China. Science. 1987;235(4785):217–20.
78. Naufal Z, Zhiwen L, Zhu L, Zhou GD, McDonald T, He LY, Mitchell L, Ren A, Zhu H, Finnell R, Donnelly KC. Biomarkers of exposure to combustion by-products in a human population in Shanxi, China. J Expo Sci Environ Epidemiol. 2010;20(4):310–9.
79. Ndlovu MN, Denis H, Fuks F. Exposing the DNA methylome iceberg. Trends Biochem Sci. 2011;36(7):381–7.
80. Norppa H. Cytogenetic biomarkers. IARC Sci Publ. 2004;157:179–205.
81. Norppa H, Luomahaara S, Heikanen H, Roth S, Sorsa M, Renzi L, Lindholm C. Micronucleus assay in lymphocytes as a tool to biomonitor human exposure to aneuploidogens and clastogens. Environ Health Perspect. 1993;3:139–43.
82. Oglesby IK, McElvaney NG, Greene CM. Micrornas in inflammatory lung disease – master regulators or target practice? Respir Res. 2010;11:148.
83. Ostling O, Johanson KJ. Microelectrophoretic study of radiation-induced DNA damages in individual mammalian cells. Biochem Biophys Res Commun. 1984;123(1):291–8.
84. Pan L, Ni Y, Xu J, Dong W, Yang D, Liu Y, Zhu Y, Shan J, Yang X, Chen Y, Guo X, Deng F. Effects of outdoor and indoor pm2.5 on exhaled biomarkers of patients with chronic obstructive pulmonary disease. J Environ Health. 2016;33(1):4. (in Chinese)
85. Pastor WA, Pape UJ, Huang Y, Henderson HR, Lister R, Ko M, McLoughlin EM, Brudno Y, Mahapatra S, Kapranov P, Tahiliani M, Daley GQ, Liu XS, Ecker JR, Milos PM, Agarwal S, Rao A. Genome-wide mapping of 5-hydroxymethylcytosine in embryonic stem cells. Nature. 2011;473(7347):394–7.
86. Peabody JW, Riddell TJ, Smith KR, Liu Y, Zhao Y, Gong J, Milet M, Sinton JE. Indoor air pollution in rural China: cooking fuels, stoves, and health status. Arch Environ Occup Health. 2005;60(2):86–95.
87. Perera FP, Mooney LA, Stampfer M, Phillips DH, Bell DA, Rundle A, Cho S, Tsai WY, Ma J, Blackwood A, Tang D, Physicians' Health Cohort S. Associations between carcinogen-DNA damage, glutathione s-transferase genotypes, and risk of lung cancer in the prospective physicians' health cohort study. Carcinogenesis. 2002;23(10):1641–6.
88. Podmore K, Farmer PB, Herbert KE, Jones GD, Martin EA. 32p-postlabelling approaches for the detection of 8-oxo-2′-deoxyguanosine-3′-monophosphate in DNA. Mutat Res. 1997;378(1–2):139–49.

89. Rich DQ, Kipen HM, Huang W, Wang G, Wang Y, Zhu P, Ohman-Strickland P, Hu M, Philipp C, Diehl SR, Lu SE, Tong J, Gong J, Thomas D, Zhu T, Zhang JJ. Association between changes in air pollution levels during the Beijing Olympics and biomarkers of inflammation and thrombosis in healthy young adults. JAMA. 2012;307(19):2068–78.
90. Risom L, Moller P, Loft S. Oxidative stress-induced DNA damage by particulate air pollution. Mutat Res. 2005;592(1–2):119–37.
91. Sanchez-Guerra M, Zheng Y, Osorio-Yanez C, Zhong J, Chervona Y, Wang S, Chang D, McCracken JP, Diaz A, Bertazzi PA, Koutrakis P, Kang CM, Zhang X, Zhang W, Byun HM, Schwartz J, Hou L, Baccarelli AA. Effects of particulate matter exposure on blood 5-hydroxymethylation: results from the Beijing truck driver air pollution study. Epigenetics. 2015;10(7):633–42.
92. Schwartz J, Litonjua A, Suh H, Verrier M, Zanobetti A, Syring M, Nearing B, Verrier R, Stone P, MacCallum G, Speizer FE, Gold DR. Traffic related pollution and heart rate variability in a panel of elderly subjects. Thorax. 2005;60(6):455–61.
93. Shan M, Yang X, Ezzati M, Chaturvedi N, Coady E, Hughes A, Shi Y, Yang M, Zhang Y, Baumgartner J. A feasibility study of the association of exposure to biomass smoke with vascular function, inflammation, and cellular aging. Environ Res. 2014;135:165–72.
94. Shinar D, Yoffe O, Shani M, Yaffe D. Regulated expression of muscle-specific genes introduced into mouse embryonal stem cells: inverse correlation with DNA methylation. Differentiation. 1989;41(2):116–26.
95. Silverman DT, Samanic CM, Lubin JH, Blair AE, Stewart PA, Vermeulen R, Coble JB, Rothman N, Schleiff PL, Travis WD, Ziegler RG, Wacholder S, Attfield MD. The diesel exhaust in miners study: a nested case-control study of lung cancer and diesel exhaust. J Natl Cancer Inst. 2012;104(11):855–68.
96. Singh SK, Pal Bhadra M, Girschick HJ, Bhadra U. Micrornas – micro in size but macro in function. FEBS J. 2008;275(20):4929–44.
97. Small EM, Olson EN. Pervasive roles of micrornas in cardiovascular biology. Nature. 2011;469(7330):336–42.
98. Su J, Li Q, Liang G, Zhang L, Qing L, Liang L, Yang Y, Tan L, Zou Y, Geng W. Effects of occupational exposure to low concentration of benzene series on serum oxidative stress levels in gas station workers. J Environ Health. 2014;31(03):242–4. (in Chinese)
99. Suganuma T, Workman JL. Crosstalk among histone modifications. Cell. 2008;135(4):604–7.
100. Sun C, Chu M, Chen W, Jin G, Gong J, Zhu M, Yuan J, Dai J, Wang M, Pan Y, Song Y, Ding X, Du M, Dong J, Zhang Z, Hu Z, Wu T, Shen H. Genetic variants of h2ax gene were associated with pm2.5-modulated DNA damage levels in Chinese Han populations. Mutat Res. 2015;778:41–5.
101. Sun Y, Song X, Han Y, Ji Y, Gao S, Shang Y, Lu SE, Zhu T, Huang W. Size-fractioned ultrafine particles and black carbon associated with autonomic dysfunction in subjects with diabetes or impaired glucose tolerance in Shanghai, China. Part Fibre Toxicol. 2015;12:8.
102. Sun Z, Mukherjee B, Brook RD, Gatts GA, Yang F, Sun Q, Brook JR, Fan Z, Rajagopalan S. Air-pollution and cardiometabolic diseases (aircmd): a prospective study investigating the impact of air pollution exposure and propensity for type ii diabetes. Sci Total Environ. 2013;448:72–8.
103. Tahiliani M, Koh KP, Shen Y, Pastor WA, Bandukwala H, Brudno Y, Agarwal S, Iyer LM, Liu DR, Aravind L, Rao A. Conversion of 5-methylcytosine to 5-hydroxymethylcytosine in mammalian DNA by mll partner tet1. Science. 2009;324(5929):930–5.
104. Tang D, Phillips DH, Stampfer M, Mooney LA, Hsu Y, Cho S, Tsai WY, Ma J, Cole KJ, She MN, Perera FP. Association between carcinogen-DNA adducts in white blood cells and lung cancer risk in the physicians health study. Cancer Res. 2001;61(18):6708–12.
105. von Zglinicki T. Oxidative stress shortens telomeres. Trends Biochem Sci. 2002;27(7):339–44.
106. Wallace DC. Diseases of the mitochondrial DNA. Annu Rev Biochem. 1992;61:1175–212.

107. Wang C, Chen R, Zhao Z, Cai J, Lu J, Ha S, Xu X, Chen X, Kan H. Particulate air pollution and circulating biomarkers among type 2 diabetic mellitus patients: the roles of particle size and time windows of exposure. Environ Res. 2015;140:112–8.
108. Wang KC, Kuo CH, Tian TF, Tsai MH, Chiung YM, Hsiech CM, Tsai SJ, Wang SY, Tsai DM, Huang CC, Tseng YJ. Metabolomic characterization of laborers exposed to welding fumes. Chem Res Toxicol. 2012;25(3):676–86.
109. Wang T, Feng W, Kuang D, Deng Q, Zhang W, Wang S, He M, Zhang X, Wu T, Guo H. The effects of heavy metals and their interactions with polycyclic aromatic hydrocarbons on the oxidative stress among coke-oven workers. Environ Res. 2015;140:405–13.
110. Wang Z, Zheng Y, Zhao B, Zhang Y, Liu Z, Xu J, Chen Y, Yang Z, Wang F, Wang H, He J, Zhang R, Abliz Z. Human metabolic responses to chronic environmental polycyclic aromatic hydrocarbon exposure by a metabolomic approach. J Proteome Res. 2015;14(6):2583–93.
111. Wei Y, Han IK, Hu M, Shao M, Zhang JJ, Tang X. Personal exposure to particulate pahs and anthraquinone and oxidative DNA damages in humans. Chemosphere. 2010;81(10):1280–5.
112. Wilson DM 3rd, Thompson LH. Molecular mechanisms of sister-chromatid exchange. Mutat Res. 2007;616(1–2):11–23.
113. Wu S, Deng F, GUO X. Effect of ambient air pollution on cardiovascular related blood indicators in taxi drivers. J Environ Health. 2009;26(09):755–7. (in Chinese)
114. Wu S, Deng F, Huang J, Wang H, Shima M, Wang X, Qin Y, Zheng C, Wei H, Hao Y, Lv H, Lu X, Guo X. Blood pressure changes and chemical constituents of particulate air pollution: results from the healthy volunteer natural relocation (hvnr) study. Environ Health Perspect. 2013;121(1):66–72.
115. Wu S, Deng F, Liu Y, Shima M, Niu J, Huang Q, Guo X. Temperature, traffic-related air pollution, and heart rate variability in a panel of healthy adults. Environ Res. 2013;120:82–9.
116. Wu S, Deng F, Niu J, Huang Q, Liu Y, Guo X. Association of heart rate variability in taxi drivers with marked changes in particulate air pollution in Beijing in 2008. Environ Health Perspect. 2009b;118(1):87–91.
117. Wu S, Deng F, Wei H, Huang J, Wang H, Shima M, Wang X, Qin Y, Zheng C, Hao Y, Guo X. Chemical constituents of ambient particulate air pollution and biomarkers of inflammation, coagulation and homocysteine in healthy adults: a prospective panel study. Part Fibre Toxicol. 2012;9:49.
118. Wu S, Deng F, Wei H, Huang J, Wang X, Hao Y, Zheng C, Qin Y, Lv H, Shima M, Guo X. Association of cardiopulmonary health effects with source-appointed ambient fine particulate in Beijing, China: a combined analysis from the healthy volunteer natural relocation (hvnr) study. Environ Sci Technol. 2014;48(6):3438–48.
119. Wu S, Ni Y, Li H, Pan L, Yang D, Baccarelli AA, Deng F, Chen Y, Shima M, Guo X. Short-term exposure to high ambient air pollution increases airway inflammation and respiratory symptoms in chronic obstructive pulmonary disease patients in Beijing, China. Environ Int. 2016a;94:76–82.
120. Wu S, Wang B, Yang D, Wei H, Li H, Pan L, Huang J, Wang X, Qin Y, Zheng C, Shima M, Deng F, Guo X. Ambient particulate air pollution and circulating antioxidant enzymes: a repeated-measure study in healthy adults in Beijing, China. Environ Pollut. 2016;208:16–24.
121. Wu S, Yang D, Pan L, Shan J, Li H, Wei H, Wang B, Huang J, Baccarelli AA, Shima M, Deng F, Guo X. Chemical constituents and sources of ambient particulate air pollution and biomarkers of endothelial function in a panel of healthy adults in Beijing, China. Sci Total Environ. 2016;560–561:141–9.
122. Wu S, Yang D, Wei H, Wang B, Huang J, Li H, Shima M, Deng F, Guo X. Association of chemical constituents and pollution sources of ambient fine particulate air pollution and biomarkers of oxidative stress associated with atherosclerosis: a panel study among young adults in Beijing, China. Chemosphere. 2015;135:347–53.
123. Xia Y, Chen R, Wang C, Cai J, Wang L, Zhao Z, Qian J, Kan H. Ambient air pollution, blood mitochondrial DNA copy number and telomere length in a panel of diabetes patients. Inhal Toxicol. 2015;27(10):481–7.

124. Xie Y, Bo L, Jiang S, Tian Z, Kan H, Li Y, Song W, Zhao J. Individual pm2.5 exposure is associated with the impairment of cardiac autonomic modulation in general residents. Environ Sci Pollut Res. 2016;23(10):10255–61.
125. Xing YF, Xu YH, Shi MH, Lian YX. The impact of pm2.5 on the human respiratory system. J Thorac Dis. 2016;8(1):E69–74.
126. Xu B, Fu Y, Lai Z, Zhu H. Investigation of health of workers occupationally exposed to cooking oil fume. J Environ Health. 2005;05:366–8. (in Chinese)
127. Xu K, Li X, Hu F, Lu X, He X, Liu H, Cai D, Pu Z, Fu K, Liang S, Kong H, Ma D, Li J. A study on polycyclic aromatic hydrocarbon-DNA adduct in lung cancer patients exposed to indoor coal-burning smoke. Chin J Prev M ed. 1997;31(02):33–6. (in Chinese)
128. Yang X, Wang E, Liu A, GUO S. The effect of vehicle exhaust on the lipid peroxidation in human body. Occup Health. 2005;21(6):828–9. (in Chinese)
129. Zhang A, Feng H, Yang G, Pan X, Jiang X, Huang X, Dong X, Yang D, Xie Y, Peng L, Jun L, Hu C, Jian L, Wang X. Unventilated indoor coal-fired stoves in Guizhou province, china: cellular and genetic damage in villagers exposed to arsenic in food and air. Environ Health Perspect. 2007;115(4):653–8.
130. Zhang J, Zhu T, Kipen H, Wang G, Huang W, Rich D, Zhu P, Wang Y, Lu SE, Ohman-Strickland P, Diehl S, Hu M, Tong J, Gong J, Thomas D, Committee HEIHR. Cardiorespiratory biomarker responses in healthy young adults to drastic air quality changes surrounding the 2008 Beijing Olympics. Res Rep Health Eff Inst. 2013b;174:5–174.
131. Zhang Q, Li L, Smith M, Guo Y, Whitlock G, Bian Z, Kurmi O, Collins R, Chen J, Lv S, Pang Z, Chen C, Chen N, Xiong Y, Peto R, Chen AZ. Exhaled carbon monoxide and its associations with smoking, indoor household air pollution and chronic respiratory diseases among 512 000 Chinese adults. Int J Epidemiol. 2013;42(5):1464–75.
132. Zhang R, Dai Y, Zhang X, Niu Y, Meng T, Li Y, Duan H, Bin P, Ye M, Jia X, Shen M, Yu S, Yang X, Gao W, Zheng Y. Reduced pulmonary function and increased pro-inflammatory cytokines in nanoscale carbon black-exposed workers. Part Fibre Toxicol. 2014;11:73.
133. Zhang X, Hirota JA, Yang C, Carlsten C. Effect of gst variants on lung function following diesel exhaust and allergen co-exposure in a controlled human crossover study. Free Radic Biol Med. 2016;96:385–91.
134. Zhang Z, Laden F, Forman JP, Hart JE. Long-term exposure to particulate matter and self-reported hypertension: a prospective analysis in the nurses' health study. Environ Health Perspect. 2016b;124(9):1414–20.
135. Zhang Z, Wang X, GUO L, DOng J, Mou J, Gao J, Guo L, Ren Y, Zhang Z, Song Z, Zhao W. Impact of cellular immune function induced by indoor and outdoor air pollution in school-age children. J Environ Health. 2009;26(06):501–3. (in Chinese)
136. Zhao A, Chen R, Wang C, Zhao Z, Yang C, Lu J, Chen X, Kan H. Associations between size-fractionated particulate air pollution and blood pressure in a panel of type ii diabetes mellitus patients. Environ Int. 2015;80:19–25.
137. Zhao J, Bo L, Gong C, Cheng P, Kan H, Xie Y, Song W. Preliminary study to explore gene-pm2.5 interactive effects on respiratory system in traffic policemen. Int J Occup Med Environ Health. 2015a;28(6):971–83.
138. Zhao X, Niu J, Wang Y, Yan C, Wang X, Wang J. Genotoxicity and chronic health effects of automobile exhaust: a study on the traffic policemen in the city of lanzhou. Mutat Res. 1998;415(3):185–90.
139. Zhao Y, Qian ZM, Wang J, Vaughn MG, Liu YQ, Ren WH, Dong GH. Does obesity amplify the association between ambient air pollution and increased blood pressure and hypertension in adults? Findings from the 33 communities chinese health study. Int J Cardiol. 2013;168(5):e148–50.
140. Zhao Z, Chen R, Lin Z, Cai J, Yang Y, Yang D, Norback D, Kan H. Ambient carbon monoxide associated with alleviated respiratory inflammation in healthy young adults. Environ Pollut. 2016;208(Pt A):294–8.

Chapter 5
Air Pollution and Mortality in China

Hualiang Lin, Xiaojie Wang, Tao Liu, Xing Li, Jianpeng Xiao, Weilin Zeng, and Wenjun Ma

Abstract Ambient air pollution has been an important environmental issue in most areas of China. This chapter summarized the mortality effects of short-term and long-term exposures to various air pollutants among Chinese population. The literature review identified sufficient information to support significant short-term mortality effects of various air pollutants from time series analyses and case-crossover studies, such as PM_{10}, $PM_{2.5}$, SO_2, NO_2, and O_3. On the other hand, though the literature review suggested a positive effect of long-term exposure to air pollution on mortality and lung cancer risk, the evidence has been limited among Chinese population, indicating that more studies, especially cohort studies, are warranted.

Keywords Air pollution • Mortality • Lung cancer • China

5.1 Introduction

Ambient air pollution has been an important environmental problem in urban areas of the world. Among all air pollutants, the most commonly monitored are particulate matter (PM) pollutants, sulfur dioxide (SO_2), nitrogen dioxide (NO_2), and ozone (O_3). The World Health Organization (WHO) had reviewed the health effects of these air pollutants and recommended detailed concentration guidelines for both daily and yearly scales [56].

Both short-term and long-term exposures to ambient air pollution have been associated with increased morbidity and mortality, in particular, from cardiovascular and respiratory diseases [33, 34, 39, 41]. Compared with the short-term health effects, long-term exposure to air pollution usually poses more hazardous health impacts and has higher significance of public health [33, 40].

H. Lin • T. Liu • X. Li • J. Xiao • W. Zeng • W. Ma (✉)
Guangdong Provincial Institute of Public Health, Guangdong Provincial Center for Disease Control and Prevention, Guangzhou 511430, China
e-mail: mawj@gdiph.org.cn

X. Wang
School of Public Health, Guangdong Pharmaceutical University, Guangzhou 510310, China

G.-H. Dong (ed.), *Ambient Air Pollution and Health Impact in China*, Advances in Experimental Medicine and Biology 1017, DOI 10.1007/978-981-10-5657-4_5

Being the largest developing country, China has been facing some of the worst air quality in recent decades mainly due to its rapid economic development [27]. In January 2013, a hazardous dense haze covered 1.4 million km^2 of China and affected more than 800 million people, subsequently causing unprecedented public concern about the health impacts of air pollution [29].

A great number of epidemiologic studies have linked the short-term exposures to ambient air pollution with various adverse health outcomes [1, 5, 14]. The National Morbidity, Mortality, and Air Pollution Study (NMMAPS), the Air Pollution and Health: A European Approach (APHEA), and the Public Health and Air Pollution in Asia (PAPA) studies are major examples in the USA, Europe, and Asia [44, 45, 52]. The dose-response functions of the health effects of air pollutants have been summarized in a few countries [2, 3]. However, due to the differences in air pollution, characteristics of the population, the evidence from other countries may not be applied to Chinese population.

In China, a number of studies have been conducted in recent years [10, 11, 31], providing sufficient information to draw a whole picture of the short-term health effects of air pollution in China. While limited studies have been conducted to examine the long-term health effects of air pollution in China, such as the effects on mortality of cardiopulmonary diseases and lung cancer [8, 20], an overview to summarize the long-term effects among Chinese population is also warranted [35].

In this chapter, we examined the short-term and long-term mortality effects of ambient air pollution (including five main air pollutants: PM_{10}, $PM_{2.5}$, NO_2, SO_2, and O_3) in Chinese population including Mainland China, Hong Kong, Taiwan, and Macao.

5.2 Short-Term Mortality Effects

5.2.1 Overview of the Short-Term Mortality Effects of Various Air Pollutants

Daily time series study and cross-sectional study have been generally applied to quantify the short-term health effects of air pollution. Previous studies have reported that both methods could produce comparable effect estimates [23]. We thus overviewed the studies using both methods for this analysis.

One observation from the studies among Chinese population was that the air pollution concentrations in most of the study areas under review were higher than the WHO's Air Quality Guideline (AQG) [56], which needs to be emphasized in the air pollution controlling and management. The estimated pooled estimates from these studies were of public health significance for health impact assessment and important for formulation of environmental pollution controlling measures in order to protect the public health.

5.2.1.1 Short-Term Mortality Effects of PM_{10}

A systematic literature search was conducted to find the relevant studies. After excluding studies from the same study area, a total of 17 studies reported the short-term association between ambient PM_{10} and mortality (Table 5.1). The reported daily mean concentrations ranged from 51.6 to 156 μg/m³.

Table 5.1 Summary of studies on the short-term mortality effects of PM_{10}

Study	Study area	Study years	Mean concentration	Disease	ER (%)	95% CI
Kan (2016)	Shanghai	2001–2004	102	Total	0.25	0.14–0.37
				CVD	0.27	0.1–0.44
				Resp	0.27	−0.01–0.56
Chen (2010)	Anshan	2004–2006	110.9	Total	0.24	−0.03–0.51
				CVD	0.67	0.29–1.04
				Resp	0.21	−0.82–1.24
Yu (2012)	Guangzhou	2006–2009	59.5	Total	0.96	0.1–1.82
				CVD	1.79	1.11–2.47
				Resp	0.93	0.03–1.83
Wong (2008a)	Hong Kong	1996–2002	51.6	Total	0.53	0.26–0.81
				CVD	0.61	0.11–1.1
				Resp	0.83	0.23–1.44
Wong (2008a)	Wuhan	2001–2004	141.8	Total	0.43	0.24–0.62
				CVD	0.57	0.31–0.84
				Resp	0.87	0.34–1.41
Chen (2012)	Beijing	2007–2008	139	Total	0.18	0.09–0.27
				Resp	0.24	−0.03–0.51
				CVD	0.16	0.03–0.30
Chen (2012)	Fuzhou	2004–2006	72	Total	0.64	0.19–1.09
				Resp	0.67	−0.78–2.13
				CVD	0.75	0.06–1.44
Chen (2012)	Hangzhou	2002–2004	121	Total	0.33	0.06–0.60
				Resp	0.68	0.08–1.28
				CVD	0.11	−0.36–0.58
Chen (2012)	Lanzhou	2004–2008	156	Total	0.01	−0.10–0.12
Chen (2012)	Shenyang	2005–2008	114	Total	0.19	0.04–0.34
				Resp	0.66	0.21–1.12
				CVD	0.28	0.07–0.49
Chen (2012)	Suzhou	2005–2008	90	Total	0.38	0.17–0.59
				Resp	0.57	0.00–1.13
				CVD	0.30	−0.04–0.64
Chen (2012)	Taiyuan	2004–2008	132	Total	0.22	0.05–0.39
				Resp	0.70	0.10–1.29
				CVD	0.11	−0.18–0.39

(continued)

Table 5.1 (continued)

Study	Study area	Study years	Mean concentration	Disease	ER (%)	95% CI
Chen (2012)	Tangshan	2006–2008	98	Total	0.11	−0.28–0.50
				Resp	0.21	−0.83–1.25
				CVD	0.54	−0.06–1.14
Chen (2012)	Tianjin	2005–2008	101	Total	0.76	0.45–1.07
				Resp	0.43	−0.86–1.72
				CVD	0.92	0.52–1.32
Chen (2012)	Urumqi	2006–2007	144	Total	0.10	−0.09–0.29
				Resp	0.18	−0.34–0.69
				CVD	0.18	−0.20–0.55
Chen (2012)	Xi'an	2004–2008	132	Total	0.20	0.05–0.35
				Resp	0.47	0.20–0.73
				CVD	0.35	0.13–0.56
Ren (2007)	Hangzhou	2002–2004	113	CVD	0.60	0.30–0.90

Most of the study cities reported statistically significant associations between PM_{10} and mortality. Our meta-analysis using a random-effect model estimated that the overall excess risk (ER) for all-cause mortality for each 10 μg/m^3 increment in PM_{10} was 0.33% (95% CI: 0.26%, 0.40%); the pooled ER was 0.38% (95% CI: 0.24%, 0.52%) for cardiovascular mortality and 0.48% (95% CI: 0.37%, 0.60%) for respiratory mortality. The estimates of our meta-analysis were consistent with one large multicity study in China [11] and one previous meta-analysis among Chinese population [35].

Heterogeneity was observed across the study areas. For example, small and non-significant associations were observed in Anshan with an ER of 0.24% (95% CI: −0.03%, 0.51%) for all-cause mortality and in Urumqi with an ER of 0.10% (95% CI: −0.09%, 0.29%) [9], while a high effect was reported in Guangzhou (ER = 0.96%, 95% CI: 0.10%, 1.82%) [60] and in Tianjin (ER = 0.76%, 95% CI: 0.45%, 1.07%) [11].

5.2.1.2 Short-Term Mortality Effects of $PM_{2.5}$

Ambient $PM_{2.5}$ has been only included in the China's ambient air pollution monitoring system since early 2013 [36], before which there were limited monitoring data on daily ambient $PM_{2.5}$, and thus studies reporting the health effects of $PM_{2.5}$ were scarce [26, 51]. Our review identified a total of 14 studies for short-term mortality effects of ambient $PM_{2.5}$ after excluding the duplicated studies in the same city (Table 5.2). The daily mean concentrations of ambient $PM_{2.5}$ varied among the study areas, ranging from 35.1 to 176.7 μg/m^3.

Most of the study cities reported a statistically significant association between daily variation of $PM_{2.5}$ and mortality (Table 5.2). The pooled excess risk of all-cause

Table 5.2 Summary of studies on the short-term mortality effects of $PM_{2.5}$

Study	Study area	Study years	Mean concentration	Disease	ER (%)	95% CI
Qiu (2015)	Hong Kong	2001–2011	37.5	CVD	2.50	2.10–2.90
				Resp	0.80	0.40–1.20
Chen (2011)	Shanghai	2004–2008	55	Total	0.47	0.22–0.72
				CVD	0.41	0.00–0.81
				Resp	0.71	−0.05–1.47
Huang (2012)	Xi'an	2004–2008	176.7	CVD	0.27	0.08–0.46
				Resp	0.19	−0.20–0.59
				Total	2.29	0.83–3.76
Venners (2003)	Chongqing	1995	146.8	Total	1.00	0.93–1.07
Ma (2011)	Shenyang	2006–2008	75	Total	0.49	0.19–0.79
				CVD	0.53	0.09–0.97
				Resp	0.97	0.01–1.94
Li (2016)	Beijing	2005–2009	75	Total	0.67	0.25–0.83
				CVD	1.37	0.51–1.71
				Resp	0.67	0.25–0.83
Ge (2015)	Suzhou	2010–2013	57.4	Resp	0.28	−0.23–0.78
Qian (2016)	Ningbo	2011–2014	49	CVD	0.53	0.13–0.94
Lin (2016b)	Dongguan	2013–2015	43.82	Total	1.69	0.80–2.60
				CVD	2.76	1.55–3.99
				Resp	−0.67	−3.14–1.86
Lin (2016b)	Foshan	2013–2015	45.89	Total	2.29	1.54–3.04
				CVD	2.37	1.38–3.37
				Resp	1.82	0.26–3.40
Lin (2016b)	Guangzhou	2013–2015	47.89	Total	1.79	1.32–2.27
				CVD	2.33	1.69–2.99
				Resp	1.58	0.57–2.61
Lin (2016b)	Jiangmen	2013–2015	45.43	Total	1.74	1.20–2.29
				CVD	1.88	1.14–2.62
				Resp	2.35	0.89–3.83
Lin (2016b)	Shenzhen	2013–2015	35.14	Total	0.63	−0.65–1.93
				CVD	1.12	−0.73–3.00
				Resp	3.28	−0.40–7.10
Lin (2016b)	Zhuhai	2013–2015	35.57	Total	1.09	−0.32–2.52
				CVD	1.51	−0.62–3.69
				Resp	0.93	−3.23–5.27

mortality was 1.28% (95% CI: 0.78%, 1.79%) for all-cause mortality, 1.79% (95% CI: 1.22%, 2.36%) for cardiovascular mortality, and 0.96% (95% CI: 0.57%, 1.35%) for respiratory mortality. Our effect estimates were relatively larger compared with one previous meta-analysis among Chinese population [35], which reported an ER of 0.40% (95% CI: 0.22%, 0.59%) from all-cause mortality, 0.63% (95% CI: 0.35%, 0.91%) for cardiovascular mortality, and 0.75% (95% CI: 01.39%, 1.11%) for respiratory mortality; however, it was comparable with one recent multicity study in south China [31]. The underlying reasons for the inconsistency might be due to more studies were included in our analysis.

Varying magnitudes of the effect estimate were reported among various study areas. One of the earliest studies on mortality effects of $PM_{2.5}$ was reported in Chongqing [51], which observed a negative and nonsignificant association; the relative risk for each 100 μg/m^3 increment in daily $PM_{2.5}$ was 1.00 (95% CI: 0.93, 1.07). In contrast, one Hong Kong study reported a high effect (ER = 1.86%, 95% CI: 0.85%, 2.88%) [42]. The differences across the study areas might be due to differences in the characteristics of the environment, climate, as well as the chemical compositions of the particles among the different study areas [17].

5.2.1.3 Short-Term Mortality Effects of SO_2

Our literature search identified a total of 17 individual effect estimates for the short-term mortality effects of SO_2 (Table 5.3). Among the study area, the 24-h average concentrations varied, ranging from 4.8 to 79.3 μg/m^3.

Statistically significant associations between SO_2 and mortality were reported in most of the study cities. The pooled excess risk of all-cause mortality were 0.75% (95% CI: 0.47%, 1.02%), 0.83% (95% CI: 0.47%, 1.19%) for cardiovascular mortality and 1.25% (95% CI: 0.78%, 1.73%) for respiratory mortality. Our results were similar with one large multicity study including 17 Chinese cities [10].

Significant heterogeneity was observed across the study areas. For example, one study in Beijing reported a nonsignificant association (ER = 0.08%, 95% CI: −0.17%, 0.34%) [61]. On the other hand, a large effect was reported in Taiwan (ER = 4.30%, 95% CI: 1.80%, 9.80%) [30].

5.2.1.4 Short-Term Mortality Effects of NO_2

A total of 18 individual effect estimates were identified by our literature search for the short-term mortality effects of NO_2 (Table 5.4). Among the study area, the 24-h average concentrations of NO_2 varied, ranging from 23.0 to 70.4 μg/m^3.

Most of the study cities reported statistically significant associations between NO_2 and mortality. The pooled excess risk of all-cause mortality were 1.37% (95% CI: 0.94%, 1.79%), 1.46% (95% CI: 1.09%, 1.83%) for cardiovascular mortality, and 1.74% (95% CI: 0.99%, 2.49%) for respiratory mortality.

Table 5.3 Summary of studies on the short-term mortality effects of SO_2

Study	Study area	Study years	Mean concentration	Disease	ER (%)	95% CI
Wong (2008)	Hong Kong	1996–2002	17.8	Total	0.62	0.19–1.06
				CVD	1.03	0.21–1.85
				Resp	1.06	0.06–2.06
Chen (2012)	17 cities		NA	Total	0.75	0.47–1.02
				Resp	1.25	0.78–1.73
				CVD	0.83	0.47–1.19
Ou (2012)	Hong Kong	1998	48.14	Total	0.52	−0.08–1.11
Lu (2015)	Nanjing	2009–2013	35.4	Total	1.02	0.51–1.54
Zeng (2015)	Beijing	2007–2009	39.8	Total	0.08	−0.17–0.34
				CVD	0.19	−0.16–0.55
				Resp	−0.05	−0.71–0.60
Zeng (2015)	Xi'an	2007–2009	51.6	Total	0.33	0.02–0.65
				CVD	0.42	−0.10–0.96
				Resp	0.00	−0.72–0.74
Zeng (2015)	Guangzhou	2007–2009	45.4	Total	0.52	0.06–0.99
				CVD	0.51	−0.28–1.32
				Resp	1.13	0.09–2.18
Zeng (2015)	Wuhan	2007–2009	50.5	Resp	3.10	0.74–5.53
Zeng (2015)	Six cities	2007–2009	NA	Total	0.40	0.13–0.67
				CVD	0.48	0.11–0.85
				Resp	0.51	−0.11–1.15
Wong (2008)	Hong Kong	1996–2002	17.8	CVD	1.19	0.29–2.10
				Resp	1.28	0.19–2.39
Wong (2008)	Shanghai	2001–2004	44.7	Total	0.95	0.62–1.28
				CVD	0.91	0.42–1.41
				Resp	1.37	0.51–2.23
Liang (2009)	Taiwan	1997–1999	4.81	Total	4.30	1.80–9.80
				Resp	17.6	−0.20–38.40
				CVD	8.70	2.50–15.10
Ge (2015)	Tianjin	2008–2011	54	Total	0.46	0.05–0.86
				CVD	0.73	0.02–1.44
				Resp	1.52	0.17–2.88
Zhang (2011)	Lanzhou	2004–2007	56	Total	3.10	0.30–6.60
Zhang (2008)	Taiyuan	2004	79.34	CVD	1.20	−6.60–9.70
Ren (2007)	Hangzhou	2002–2004	46	CVD	1.70	0.70–2.80
Zhu (2016)	Wuhan	2002–2010	49.3	Total	1.22	0.77–1.67
Qian (2016)	Ningbo	2011–2014	17	CVD	1.73	0.87–2.60
Ye (2010)	Hangzhou	2002–2004	50.8	CVD	0.50	−3.30–4.40
Zhang (2015)	Wuhan	2002–2010	71	CVD	1.65	0.28–3.08

Table 5.4 Summary of studies on the short-term mortality effects of NO_2

Study	Study area	Study years	Mean concentration	Disease	ER (%)	95% CI
Lu (2015)	Nanjing	2009–2013	51.5	Total	1.15	0.67–1.63
Qiu (2015)	Hong Kong	2001–2011	NA	Total	2.80	2.10–2.90
				Resp	1.30	0.90–1.80
Zeng (2015)	Beijing	2007–2009	56.6	Total	0.32	−0.08–0.72
				CVD	0.47	−0.09–1.04
				Resp	−0.15	−1.23–0.94
Zeng (2015)	Tianjin	2007–2009	41.4	Total	0.62	−0.16–1.41
Zeng (2015)	Xi'an	2007–2009	54	Total	0.80	−0.81–2.44
				CVD	1.51	−0.78–3.86
				Resp	0.98	−3.78–5.98
Zeng (2015)	Shanghai	2007–2009	54.8	Total	0.46	0.05–0.87
				CVD	0.51	−0.17–1.20
				Resp	0.32	−0.64–1.30
Zeng (2015)	Guangzhou	2007–2009	60.1	Total	1.38	0.83–1.94
				Resp	2.66	1.42–3.90
Zeng (2015)	Wuhan	2007–2009	45.2	Total	1.99	0.79–3.21
				CVD	3.03	1.31–4.79
				Resp	1.73	−1.65–5.24
Zeng (2015)	Six cities	2007–2009	NA	Total	0.81	0.35–1.28
				CVD	1.03	0.40–1.66
				Resp	0.94	−0.15–2.06
Tao (2012)	Foshan	2006–2008	70.4	Total	1.87	1.40–2.35
				CVD	2.35	1.59–3.13
				Resp	1.60	0.60–2.61
Tao (2012)	Zhuhai	2006–2008	38.1	Total	1.39	−0.09–2.89
				CVD	2.22	−0.09–4.58
				Resp	2.46	−1.59–6.67
Tao (2012)	Zhongshan	2006–2008	48.4	Total	1.22	0.44–2.01
				CVD	1.19	0.05–2.34
				Resp	3.44	1.67–5.25
Tao (2012)	Four cities	2006–2008	NA	Total	1.95	1.62–2.29
				CVD	2.12	1.58–2.65
				Resp	3.48	2.73–4.23
Wong (2008)	Hong Kong	1996–2002	58.7	CVD	1.23	0.64–1.82
Lin (2016)	Guangzhou	2009–2011	44.1	CVD	1.98	0.54–3.44
Ge (2015)	Tianjin	2008–2011	42	CVD	1.77	0.38–3.15
				Resp	2.25	0.17–4.33
Zhang (2011)	Lanzhou	2004–2007	47	Total	7.20	2.90–11.60
Zhang (2008)	Taiyuan	2004	23.03	CVD	1.40	−3.80–6.80
Ren (2007)	Hangzhou	2002–2004	53	CVD	2.00	0.90–3.20
Qian (2016)	Ningbo	2011–2014	45	CVD	1.33	0.71–1.95
Ye (2010)	Hangzhou	2002–2008	55.76	CVD	−1.20	−5.10–2.90

Varying effect sizes were observed across the study areas. For example, one study in Beijing reported a positive but nonsignificant association (ER = 0.32%, 95% CI: −0.08%, 0.72%) [61]. On the other hand, a large effect was reported in Lanzhou (ER = 7.20%, 95% CI: 2.90%, 11.60%) [62].

5.2.1.5 Short-Term Mortality Effects of O_3

A total of ten individual effect estimates were identified by our literature search for the short-term mortality effects of O_3 (Table 5.5). Among the study area, the 24-h average concentrations of O_3 varied, ranging from 17.2 to 85.7 μg/m^3.

Most of the study cities reported statistically significant associations between O_3 and mortality. The pooled excess risk of all-cause mortality were 0.40% (95% CI: 0.28%, 0.53%) for O_3, 0.46% (95% CI: 0.20%, 0.73%) for cardiovascular mortality, and 0.41% (95% CI: 0.06%, 0.75%) for respiratory mortality.

We observed between-city heterogeneity in the short-term mortality effects of O_3. For example, studies from Taiwan reported a negative but nonsignificant association between O_3 and mortality (ER = −0.10%, 95% CI: −2.80%, 2.60%) [58]. Nonsignificant associations were also reported in Zhuhai, Wuhan, and Suzhou [48, 52, 57], while a large positive association was reported in Guangzhou (ER = 0.64%, 95% CI: 0.42%, 0.86%) [48].

Several mechanisms supported the short-term association between various air pollutants and mortality. For example, the ambient air pollution exposure has been associated with increased systemic inflammatory responses, plasma viscosity [38], changes in blood pressure [16], decreased heart rate variability [18], and increased cardiac arrhythmias [47]. When inhaled into the human body, some air pollutants have the ability to carry a large amount of toxic substance, including oxidant gas pollutants, organic compounds, and transition metals, which have been identified as generating pro-inflammatory response through the reactive oxygen species (ROS) [15].

5.2.2 *Effect Modifiers of Short-Term Mortality Effects*

Some studies have examined the effect modifiers of the short-term mortality effects of various air pollutants in order to identify the vulnerable subpopulations, such as sex, age, cause of death, etc. [31, 60].

One interesting observation for the mortality effects of particulate matter air pollution was that the mortality effects for each 10 μg/m^3 increase in daily air pollution were inversely associated with the annual average concentration of particulate pollutions; similar finding has also been reported in one previous systematic review study [35]. This phenomenon might be partly explained by the different particle compositions among different study areas; it was possible that the particles in some cities were more toxic though the annual concentration was relatively low [11, 35].

Table 5.5 Summary of studies on the short-term mortality effects of O_3

Study	Study area	Study years	Mean concentration	Disease	ER (%)	95% CI
Yang (2004)	Taipei	1994–1998	17.2	Total	−0.10	−2.80–2.60
				Resp	−0.90	−10.30–9.40
Tao (2012)	Foshan	2006–2008	70.7	Total	0.36	−0.06–0.78
				CVD	0.43	−0.25–1.12
				Resp	0.46	−0.43–1.36
Tao (2012)	Guangzhou	2006–2008	78.2	Total	0.64	0.42–0.86
				CVD	0.98	0.61–1.35
				Resp	0.89	0.38–1.41
Tao (2012)	Zhuhai	2006–2008	85.5	Total	0.22	−0.36–0.81
				CVD	−0.08	−1.00–0.85
				Resp	1.61	−0.05–3.30
Tao (2012)	Zhongshan	2006–2008	85.7	Total	0.61	0.22–1.00
				CVD	0.77	0.19–1.35
				Resp	0.61	−0.32–1.55
Wong (2008)	Hong Kong	1996–2002	36.9	Total	0.27	0.00–0.53
				CVD	0.45	−0.04–0.94
				Resp	0.11	0.48–0.72
Wong (2008)	Wuhan	2001–2004	85.7	Total	0.29	−0.05–0.63
				CVD	−0.07	−0.53–0.39
				Resp	0.12	−0.89–1.15
Wong (2008)	Shanghai	2001–2004	63.4	Total	0.31	0.04–0.58
				CVD	0.38	−0.03–0.80
				Resp	0.29	−0.04–1.03
Yang (2012)	Suzhou	2006–2008	57.7	Total	2.21	−0.18–4.59
				CVD	4.47	1.43–7.51
				Resp	−1.85	−0.69–3.22
Liang (2009)	Taiwan	1997–1999	28.6	Total	4.50	−0.20–9.50
				Resp	42.80	12.60–81.10
				CVD	−5.70	−13.80–3.10

It was also possible that in areas with higher annual concentrations, the residents might be more adaptive to the impact of the exposure to the particulate pollution.

In general, age group was reported to be an important effect modifier; there was a larger mortality effect of air pollutants among older populations than younger populations. This finding has also been reported in some previous studies [4, 6, 21, 43]. For example, Cakmak et al. reported a twofold magnitude of the mortality effects of air pollution among the older population in seven Chilean urban areas [6]. The underlying mechanisms might be that the older persons were generally at poor health condition, usually had preexisting medical illnesses, and thus had reduced capacity to respond to the acute exposure to high level of air pollution, which increased their vulnerability [25, 31].

Some studies reported a larger mortality effect in males due to higher exposure associated with more outdoor activities, but some other studies also observed a stronger effect among females [31, 60]; the underlying reasons need to be further investigated. Higher effects were also reported among those with lower socioeconomic status, such as lower education level [11, 31].

One important finding of this study was that the short-term mortality effects of $PM_{2.5}$ were relatively larger than those of PM_{10}, indicating that the particle size might be one important effect modifier of the health effects of particulate matter air pollution. This research question has been identified by the US National Research Council Committee on Research Priorities for Airborne Particulate Matter as one of the priority areas for future studies [37]. Similar findings have been reported in a few previous studies [12, 32]. Epidemiological and toxicological studies suggest that smaller particles might be more harmful to human health than larger particles [46].

5.3 Long-Term Mortality Effects

5.3.1 Overview of the Long-Term Mortality Effects of Various Air Pollutants

Our literature search identified seven eligible studies on the long-term mortality effects of air pollution [8, 13, 20, 50, 53, 64, 65]. All these studies (Table 5.6) were cohort studies in terms of study design. Five studies were conducted in mainland China, one in Hong Kong, and one in Taiwan. Five studies examined the effects of PM_{10}, two studies evaluated the effects of $PM_{2.5}$, five studies evaluated the effects of SO_2, four studies evaluated the effects of NO_2, and only one study evaluated the effects of O_3 [50].

In general, two methods were used for the long-term air pollution exposure assessment. Some studies assessed the air pollutant exposure using the annual average concentration of air pollution before the baseline survey by linking fixed-site monitoring data with the residential address of the participants [8, 13, 65]. The other approach assessed the air pollution exposure using the satellite-based estimates of air pollution concentration [54].

5.3.1.1 Long-Term Mortality Effects of Particulate Pollution

The meta-analysis pooling the long-term association between various air pollutants and mortality showed that PM_{10} was significantly associated with both cardiovascular and respiratory mortality, while the association with all-cause mortality was only marginally significant. For each 10 $\mu g/m^3$ increase in ambient PM_{10}, the overall relative risk (RR) was 1.34 (95% confidence interval (CI), 1.22, 1.46) for cardiovascular

Table 5.6 Summary of studies on the long-term mortality effects of various air pollutants

Study	Study area	Study years	Pollutants	Mean concentration	Disease	RR (%)	95% CI
Dong (2012)	Shenyang	1998–2009	PM_{10}	154	Resp	1.67	1.60–1.74
Zhang (2011)	Shenyang	1998–2009	PM_{10}	154	CVD	1.55	1.51–1.60
					All	1.53	1.50–1.56
Zhang (2014)	Tianjin, Shenyang, Taiyuan, Rizhao	1998–2009	PM_{10}	144	CVD	1.23	1.19–1.26
Chen (2016)	Tianjin, Shenyang, Taiyuan, Rizhao	1998–2009	PM_{10}	144.34	All	1.23	1.21–1.26
Zhou (2014)	National	1991–2006	PM_{10}	104	All	1.02	1.01–1.03
					CVD	1.02	1.01–1.03
					Resp	1.02	1.01–1.03
Wong (2015)	Hong Kong	1998–2001	$PM_{2.5}$	NA	All	1.14	1.07–1.22
					CVD	1.22	1.08–1.39
					Resp	1.05	0.90–1.22
Tseng (2015)	Taiwan	1989–2008	$PM_{2.5}$	22.8–32.9	All	0.92	0.72–1.17
					CVD	0.80	0.43–1.50
Dong (2012)	Shenyang	1998–2009	SO_2	63	Resp	1.04	0.97–1.12
Zhang (2011)	Shenyang	1998–2009	SO_2	63	CVD	0.96	0.92–1.01
					All	0.95	0.92–0.99
Chen (2016)	Tianjin, Shenyang, Taiyuan, Rizhao	1998–2009	SO_2	66.9	All	1.03	1.01–1.05
Tseng (2015)	Taiwan	1989–2008	SO_2	NA	CVD	0.91	0.67–1.25
Cao (2011)	National	1991–2000	SO_2	73	All	1.80	1.30–2.30
					CVD	3.20	2.30–4.00
					Resp	3.20	1.80–4.70
Dong (2012)	Shenyang	1998–2009	NO_2	46	Resp	2.97	2.69–3.27
Zhang (2011)	Shenyang	1998–2009	NO_2	46	CVD	2.46	2.31–2.63
					All	2.45	2.34–2.58
Chen (2016)	Tianjin, Shenyang, Taiyuan, Rizhao	1998–2009	NO_2	40.66	All	0.89	0.86–0.92
Tseng (2015)	Taiwan	1989–2008	NO_2	NA	CVD	0.99	0.95–1.04

mortality and 1.55 (95% CI: 1.30, 1.83) for respiratory mortality, respectively. In the individual studies, we observed large variations, a small effect was observed in one national cohort with an RR of 1.016 (95% CI: 1.007, 1.026) [65], and a large association was reported with an RR of 1.53 (95% CI: 1.50, 1.56) in Shenyang [63].

Long-term exposure to ambient $PM_{2.5}$ was significantly associated with all-cause and cardiovascular mortality, while the association with respiratory mortality was not statistically significant. Each 10 μg/m^3 increase in ambient $PM_{2.5}$ corresponded to an overall relative risk (RR) of 1.14 (95% CI, 1.09, 1.18) of all-cause mortality and 1.21 (95% CI: 1.13, 1.30) of cardiovascular mortality, respectively. Among the included studies, one study in Taiwan reported a negative but nonsignificant association between $PM_{2.5}$ and mortality (RR = 0.92, 95% CI: 0.72, 1.17) [50]. And the study from Hong Kong reported a significant association (RR = 1.14, 95% CI: 1.07, 1.22) [53].

5.3.1.2 Long-Term Mortality Effects of Gaseous Air Pollutants

This meta-analysis found that long-term exposure to ambient NO_2 was significantly associated with cardiovascular and respiratory mortality rather than all-cause mortality among Chinese population. The pooled RR for a 10 μg/m^3 increase in ambient SO_2 was 2.14 (95% CI: 1.57, 2.90) for cardiovascular mortality and 2.97 (95% CI: 2.81, 3.14) for respiratory mortality, respectively. While, we did not find any significant association between long-term exposure to ambient SO_2 and mortality from all cause, cardiovascular and respiratory diseases.

Different magnitudes were observed for the mortality effects of SO_2 and NO_2 among the included studies. For example, for the effects of SO_2, a negative association was observed in Shenyang (RR = 0.95, 95% CI: 0.92, 0.99) [63], while a large positive effect was reported in a national cohort study (RR = 1.80, 95% CI: 1.30, 2.30) [8].

The mechanisms by which long-term exposure to ambient air pollutants could increase the mortality risk were not well understood. Frampton and colleagues hypothesized that systemic inflammation and vasoconstriction with expression of leukocytes, endothelial adhesion molecules, oxidants, and interleukins can be induced by changes in vascular function due to long-term exposure to air pollutants [22]. Also, it has been suggested that air pollution induces a low-grade pulmonary inflammatory response and subsequent release of pro-inflammatory cytokines [64]. This may result in increased coagulability of the blood, triggering cardiovascular events in susceptible individuals [49].

5.3.2 *Effect Modifiers of the Long-Term Mortality Effects of Air Pollution*

Among the included studies, a few examined the effect modifiers of the long-term mortality effects of air pollution [8], which provided important information to identify the vulnerable population in order to formulate specific intervention measures.

Age and sex were believed to be potential effect modifiers and have been examined in a few studies. However, the mortality effects were not found to be significantly different between males and females [8, 63]. One Hong Kong study observed a larger effect among those aged between 65 and 71 years than those aged >71 years for the mortality effects of $PM_{2.5}$, which might be due to the more outdoor activities among the old participant aged between 65 and 71 years [53]. However, a few other studies did not find significant effect modification of age [63].

The possible effect modification by body mass index was examined in three studies, one Hong Kong study found that obesity could increase the mortality effect of $PM_{2.5}$ [53], one study found a significant effect modification with a higher effect of SO_2 among overweight participants [8], and one study reported that obesity could enhance the respiratory mortality effect of air pollution [19].

One study demonstrated that occupational pollution exposure and the lack of physical activity could enhance the mortality effect of air pollution in seven north Chinese cities [63], while the other study did not find significant effect modification of education levels [65].

5.4 Long-Term Effects on Lung Cancer Mortality

Our literature search identified four studies examining the lung cancer mortality effects of long-term exposure to air pollution (Table 5.7). All the four studies were cohort studies in terms of the study design. For each air pollution, there were one or two studies.

Two studies reported the association between PM_{10} and lung cancer mortality [13, 65]. The study from four cities in north China showed a statistically significant association with a large RR of 1.65, while the study using national middle-aged Chinese men did not find statistically significant association. Only one study examined the association between $PM_{2.5}$ and lung cancer mortality [54]; significant association was observed, especially for males (RR = 1.36, 95% CI: 1.05, 1.77); and no significant difference was observed between smokers and nonsmokers. This effect was relatively larger than that in one recent meta-analysis [24] and the estimate by the World Health Organization [55]. It was possible that the WHO has underestimated the effects of ambient $PM_{2.5}$ on lung cancer risk among Chinese population [55].

Two studies investigated the lung cancer effects of exposure to SO_2, both of which reported a significant association [8, 13]. The study using a national cohort

Table 5.7 The long-term effect of ambient air pollution on lung cancer mortality in Chinese population

Air pollutants	Study	Study area	Population	RR	95% CI
PM_{10}	Chen (2016)	Tianjin, Shenyang, Taiyuan, Rizhao	Overall	1.65	1.53, 1.79
PM_{10}	Zhou (2014)	National	Males	1.01	0.99, 1.03
$PM_{2.5}$	Wong (2016)	Hong Kong	Overall	1.14	0.96, 1.36
$PM_{2.5}$	Wong (2016)	Hong Kong	Males	1.36	1.05, 1.77
$PM_{2.5}$	Wong (2016)	Hong Kong	Females	0.99	0.74, 1.27
SO_2	Chen (2016)	Tianjin, Shenyang, Taiyuan, Rizhao	Overall	1.22	1.16, 1.29
SO_2	Cao (2011)	National	Overall	1.042	1.023, 1.062
NO_2	Chen (2016)	Tianjin, Shenyang, Taiyuan, Rizhao	Overall	0.837	0.746, 0.938

reported an RR associated with each 10 μg/m^3 increase in ambient SO_2 of 1.04 (95% CI: 1.02, 1.06) [8], while the study from four north Chinese cities showed an RR of 1.22 (95% CI: 1.16, 1.29) [13]. Only one study examined the effects of NO_2 and found a negative association between long-term exposure to NO_2 and lung cancer mortality [13].

The significant association between particulate pollution and lung cancer was within our expectation. One early cohort study showed that an increase of 10 μg/m^3 of total suspended particle corresponded to 1.1% (95% CI: 0.1%, 2.3%) increase of lung cancer mortality in China [7]. And one recent spatiotemporal analysis based on the National Cancer Registration of China from 1990 to 2009 revealed a significant association between $PM_{2.5}$ and lung cancer incidence (RR = 1.05, 95% CI: 1.04, 1.07 for men and 1.15, 95% CI: 1.12, 1.18 for women). Similar findings were also observed in the USA and Europe. For example, in the Harvard Six-City Study, one of the pioneer air pollution studies, Dockery et al. reported that each 10 ug/m^3 elevation in long-term average $PM_{2.5}$ concentration was associated with approximately 8% increase in lung cancer mortality [40]. And one large study including 17 cohort studies based on nine European countries reported that the RR for per 5 μg/m^3 increase in $PM_{2.5}$ was 1.18 (95% CI: 0.96, 1.46) and 1.22 (95% CI: 1.03, 1.45) for per 10 μg/m^3 [8].

Two main biologic mechanisms have been proposed to explain the air pollution-associated lung cancer mortality. First, air pollution exposure can induce oxidative stress causing macrophages to release reactive oxygen species (ROS) that can damage DNA, proteins, and lipids; the transition metals on the particle surface were capable of generating ROS through the Fenton reaction, or quinones in the particles that produce ROS through redox cycling. Second, an effect of inflammation is induced directly or indirectly by air pollution exposure, leading to the production of chemokines and cytokines to trigger angiogenesis and allowing epithelial invasion of metastatic tumor cells and then survival of the invading malignant cells in distant organs [28, 59].

5.5 Perspectives

Ambient air pollution has been a major public health concern in China alongside its rapid economic growth in recent decades. This chapter quantitatively estimated the overall short-term mortality effects of ambient air pollutants in Chinese population; the pooled associations were found to be statistically significant for various air pollutants and mortality categories. For the long-term exposure, this study generally supported the link between ambient air pollution exposure and the risk of mortality from all cause and cardiovascular and respiratory diseases. And we summarized the results on the long-term effects on lung cancer mortality and we observed some evidence of positive association, particularly for PM pollution and SO_2, but the finding was not inclusive; more studies are warranted to further examine this research question.

Acknowledgement This study was supported by the Guangdong Medical Foundation (A2017269), Guangdong Provincial Science and Technology Program (2016A020223008), and Key Special Project of National Key Research and Development Plan Program (2016YFC0207000).

References

1. Aga E, Samoli E, Touloumi G, Anderson H, Cadum E, Forsberg B, Goodman P, Goren A, Kotesovec F, Kriz B. Short-term effects of ambient particles on mortality in the elderly: results from 28 cities in the aphea2 project. Eur Respir J. 2003;21:28s–33s.
2. Atkinson R, Cohen A, Mehta S, Anderson H. Systematic review and meta-analysis of epidemiological time-series studies on outdoor air pollution and health in asia. Air Qual Atmos Health. 2012;5:383–91.
3. Bell ML, Dominici F, Samet JM. A meta-analysis of time-series studies of ozone and mortality with comparison to the national morbidity, mortality, and air pollution study. Epidemiology. 2005;16:436.
4. Bell ML, Zanobetti A, Dominici F. Evidence on vulnerability and susceptibility to health risks associated with short-term exposure to particulate matter: a systematic review and meta-analysis. Am J Epidemiol. 2013;178:865–76.
5. Bremner SA, Anderson H, Atkinson R, McMichael A, Strachan D, Bland J, Bower J. Short-term associations between outdoor air pollution and mortality in london 1992–4. Occup Environ Med. 1999;56:237–44.
6. Cakmak S, Dales RE, Rubio MA, Vidal CB. The risk of dying on days of higher air pollution among the socially disadvantaged elderly. Environ Res. 2011;111:388–93.
7. Cao J, Lee S, Ho K, Zhang X, Zou S, Fung K, Chow JC, Watson JG. Characteristics of carbonaceous aerosol in pearl river delta region, china during 2001 winter period. Atmos Environ. 2003;37:1451–60.
8. Cao J, Yang C, Li J, Chen R, Chen B, Gu D, Kan H. Association between long-term exposure to outdoor air pollution and mortality in china: a cohort study. J Hazard Mater. 2011;186:1594–600.
9. Chen R, Pan G, Kan H, Tan J, Song W, Wu Z, Xu X, Xu Q, Jiang C, Chen B. Ambient air pollution and daily mortality in Anshan, China: a time-stratified case-crossover analysis. Sci Total Environ. 2010;408:6086–91.

10. Chen R, Huang W, Wong CM, Wang Z, Quoc Thach T, Chen B, Kan H. Short-term exposure to sulfur dioxide and daily mortality in 17 Chinese cities: the china air pollution and health effects study (capes). Environ Res. 2012;118:101–6.
11. Chen R, Kan H, Chen B, Huang W, Bai Z, Song G, Pan G. Association of particulate air pollution with daily mortality the china air pollution and health effects study. Am J Epidemiol. 2012;175:1173–81.
12. Chen R, Zhao Z, Sun Q, Lin Z, Zhao A, Wang C, Xia Y, Xu X, Kan H. Size-fractionated particulate air pollution and circulating biomarkers of inflammation, coagulation, and vasoconstriction in a panel of young adults. Epidemiology. 2015;26:328–36.
13. Chen X, Zhang L-W, Huang J-J, Song F-J, Zhang L-P, Qian Z-M, Trevathan E, Mao H-J, Han B, Vaughn M. Long-term exposure to urban air pollution and lung cancer mortality: a 12-year cohort study in northern china. Sci Total Environ. 2016;571:855–61.
14. Chiusolo M, Cadum E, Stafoggia M, Galassi C, Berti G, Faustini A, Bisanti L, Vigotti MA, Dess矛 MP, Cernigliaro A. Short-term effects of nitrogen dioxide on mortality and susceptibility factors in 10 Italian cities: the epiair study. Environ Health Perspect. 2011;119:1233–8.
15. Chuang KJ, Chan CC, Su TC, Lee CT, Tang CS. The effect of urban air pollution on inflammation, oxidative stress, coagulation, and autonomic dysfunction in young adults. Am J Respir Crit Care Med. 2007;176:370–6.
16. Chuang KJ, Yan YH, Cheng TJ. Effect of air pollution on blood pressure, blood lipids, and blood sugar: a population-based approach. J Occup Environ Med. 2010;52:258–62.
17. Dai L, Zanobetti A, Koutrakis P, Schwartz JD. Associations of fine particulate matter species with mortality in the united states: a multicity time-series analysis. Environ Health Perspect. 2014;122:837–42.
18. de Paula SU, Braga ALF, Giorgi DMA, Pereira LAA, Grupi CJ, Lin CA, Bussacos MA, Zanetta DMT, do Nascimento Saldiva PH, Terra Filho M. Effects of air pollution on blood pressure and heart rate variability: a panel study of vehicular traffic controllers in the city of Sao Paulo, brazil. Eur Heart J. 2005;26:193–200.
19. Dong G, Qian Z, Liu M, Wang D, Ren W, Fu Q, Wang J, Simckes M, Ferguson T, Trevathan E. Obesity enhanced respiratory health effects of ambient air pollution in Chinese children: the seven northeastern cities study. Int J Obes. 2013;37:94–100.
20. Dong GH, Zhang P, Sun B, Zhang L, Chen X, Ma N, Yu F, Guo H, Huang H, Lee YL. Long-term exposure to ambient air pollution and respiratory disease mortality in Shenyang, China: a 12-year population-based retrospective cohort study. Respiration. 2011;84:360–8.
21. Forastiere F, Stafoggia M, Tasco C, Picciotto S, Agabiti N, Cesaroni G, Perucci CA. Socioeconomic status, particulate air pollution, and daily mortality: differential exposure or differential susceptibility. Am J Ind Med. 2007;50:208–16.
22. Frampton MW, Stewart JC, Oberdorster G, Morrow PE, Chalupa D, Pietropaoli AP, Frasier LM, Speers DM, Cox C, Huang LS, Utell MJ. Inhalation of ultrafine particles alters blood leukocyte expression of adhesion molecules in humans. Environ Health Perspect. 2006;114:51–8.
23. Fung KY, Krewski D, Chen Y, Burnett R, Cakmak S. Comparison of time series and case-crossover analyses of air pollution and hospital admission data. Int J Epidemiol. 2003;32:1064.
24. Hamra GB, Guha N, Cohen A, Laden F, Raaschou-Nielsen O, Samet JM, Vineis P, Forastiere F, Saldiva P, Yorifuji T. Outdoor particulate matter exposure and lung cancer: a systematic review and meta-analysis. Environ Health Perspect. 2014;122:906–11.
25. Huang Z, Lin H, Liu Y, Zhou M, Liu T, Xiao J, Zeng W, Li X, Zhang Y, Ebi KL, Tong S, Ma W, Wang L. Individual-level and community-level effect modifiers of the temperature–mortality relationship in 66 Chinese communities. BMJ Open. 2015;5:e009172.
26. Kan H, London SJ, Chen G, Zhang Y, Song G, Zhao N, Jiang L, Chen B. Differentiating the effects of fine and coarse particles on daily mortality in Shanghai, China. Environ Int. 2007;33:376–84.
27. Li F, Liu Y, Lü J, Liang L, Harmer P. Ambient air pollution in china poses a multifaceted health threat to outdoor physical activity. J Epidemiol Community Health. 2015;69:201–4.

28. Li R, Kou X, Xie L, Cheng F, Geng H. Effects of ambient pm2.5 on pathological injury, inflammation, oxidative stress, metabolic enzyme activity, and expression of c-fos and c-jun in lungs of rats. Environ Sci Pollut Res Int. 2015;22:20167–76.
29. Li TT, Cui LL, Chen C, Sun QH, Liu Y, Wang Q, Du YJ, Du ZH, Xu DQ. Air pollutant pm2.5 related excess mortality risk assessment in Beijing, January 2013. Dis Surveill. 2015;30:668–71.
30. Liang WM, Wei HY, Kuo HW. Association between daily mortality from respiratory and cardiovascular diseases and air pollution in Taiwan. Environ Res. 2009;109:51–8.
31. Lin H, Liu T, Xiao J, Zeng W, Li X, Guo L, Xu Y, Zhang Y, Vaughn MG, Nelson EJ, Qian Z, Ma W. Quantifying short-term and long-term health benefits of attaining ambient fine particulate pollution standards in Guangzhou, China. Atmos Environ. 2016;137:38–44.
32. Lin H, Liu T, Xiao J, Zeng W, Li X, Guo L, Zhang Y, Xu Y, Tao J, Xian H, Syberg KM, Qian Z, Ma W. Mortality burden of ambient fine particulate air pollution in six Chinese cities: results from the pearl river delta study. Environ Int. 2016;96:91–7.
33. Lin H, Tao J, Du Y, Liu T, Qian Z, Tian L, Di Q, Rutherford S, Guo L, Zeng W, Xiao J, Li X, He Z, Xu Y, Ma W. Particle size and chemical constituents of ambient particulate pollution associated with cardiovascular mortality in Guangzhou, China. Environ Pollut. 2016;208(Part B):758–66.
34. Lipsett MJ, Ostro BD, Reynolds P, Goldberg D, Hertz A, Jerrett M, Smith DF, Garcia C, Chang ET, Bernstein L. Long-term exposure to air pollution and cardiorespiratory disease in the California teachers study cohort. Am J Respir Crit Care Med. 2011;184:828–35.
35. Lu F, Xu D, Cheng Y, Dong S, Guo C, Jiang X, Zheng X. Systematic review and meta-analysis of the adverse health effects of ambient pm 2.5 and pm 10 pollution in the Chinese population. Environ Res. 2015;136:196–204.
36. Ministry of Environmental Protection of China. 2012. Ambient air quality standards (gb 3095–2012).
37. National Research Council. Committee on research priorities for airborne particulate matter. Research priorities for airborne particulate matter, iv: Continuing research progress. Washington, DC: National Academies Press; 2004.
38. Peters A, Döring A, Wichmann HE, Koenig W. Increased plasma viscosity during an air pollution episode: a link to mortality? Lancet. 1997;349:1582–7.
39. Pope CA, Young B, Dockery D. Health effects of fine particulate air pollution: lines that connect. J Air Waste Manage Assoc. 2006;56:709–42.
40. Pope CA III, Burnett RT, Thun MJ, Calle EE, Krewski D, Ito K, Thurston GD. Lung cancer, cardiopulmonary mortality, and long-term exposure to fine particulate air pollution. JAMA. 2002;287:1132–41.
41. Qian Z, He Q, Lin H-M, Kong L, Liao D, Dan J, Bentley CM, Wang B. Association of daily cause-specific mortality with ambient particle air pollution in Wuhan, China. Environ Res. 2007;105:380–9.
42. Qiu H, Yu IT, Wang X, Tian L, Tse LA, Wong TW. Differential effects of fine and coarse particles on daily emergency cardiovascular hospitalizations in Hong Kong. Atmos Environ. 2013;64:296–302.
43. Qiu H, Tian L, Kf H, Pun VC, Wang X, Yu ITS. Air pollution and mortality: effect modification by personal characteristics and specific cause of death in a case-only study. Environ Pollut. 2015;199:192–7.
44. Samet JM, Dominici F, Curriero FC, Coursac I, Zeger SL. Fine particulate air pollution and mortality in 20 U.S. cities, 1987–1994. N Engl J Med. 2000;343:1742–9.
45. Samoli E, Aga E, Touloumi G, Nisiotis K, Forsberg B, Lefranc A, Pekkanen J, Wojtyniak B, Schindler C, Niciu E. Short-term effects of nitrogen dioxide on mortality: an analysis within the aphea project. Eur Respir J. 2006;27:1129–38.
46. Selgrade MK. Air pollution and respiratory disease: extrapolating from animal models to human health effects. Immunopharmacology. 2000;48:319–24.

47. Sun Q, Hong X, Wold LE. Cardiovascular effects of ambient particulate air pollution exposure. Circulation. 2010;121:2755–65.
48. Tao Y, Huang W, Huang X, Zhong L, Lu SE, Li Y, Dai L, Zhang Y, Zhu T. Estimated acute effects of ambient ozone and nitrogen dioxide on mortality in the pearl river delta of southern China. Environ Health Perspect. 2012;120:393–8.
49. Townend JN. The way to dusty death? Thorax. 2005;60:441–2.
50. Tseng E, Ho WC, Lin MH, Cheng TJ, Chen PC, Lin HH. Chronic exposure to particulate matter and risk of cardiovascular mortality: cohort study from Taiwan. BMC Public Health. 2015;15:936.
51. Venners SA, Wang B, Xu Z, Schlatter Y, Wang L, Xu X. Particulate matter, sulfur dioxide, and daily mortality in Chongqing, China. Environ Health Perspect. 2003;111:562–7.
52. Wong CM, Vichit Vadakan N, Kan H, Qian Z. Public health and air pollution in Asia (papa): a multicity study of short-term effects of air pollution on mortality. Environ Health Perspect. 2008;116:1195–202.
53. Wong CM, Lai HK, Tsang H, Thach TQ, Thomas GN, Lam K, Chan KP, Yang L, Lau A, Ayres JG. Satellite-based estimates of long-term exposure to fine particles and association with mortality in elderlyHong Kong residents. Environ Health Perspect. 2015;123:1167–72.
54. Wong CM, Tsang H, Lai HK, Thomas GN, Lam KB, Chan KP, Zheng Q, Ayres JG, Lee SY, Lam TH. Cancer mortality risks from long-term exposure to ambient fine particle. Cancer Epidemiol Biomark Prev. 2016;25:839–45.
55. World Health Organization. Global health observatory (gho) data. http://www.who.int/gho/phe/outdoor_air_pollution/en/.
56. World Health Organization. Air quality guidelines: global update 2005: particulate matter, ozone, nitrogen dioxide and sulfur dioxide. 2006.
57. Yang C, Yang H, Guo S, Wang Z, Xu X, Duan X, Kan H. Alternative ozone metrics and daily mortality in Suzhou: the China air pollution and health effects study (capes). Sci Total Environ. 2012;426:83–9.
58. Yang CY, Chang CC, Chuang HY, Tsai SS, Wu TN, Ho CK. Relationship between air pollution and daily mortality in a subtropical city: Taipei, Taiwan. Environ Int. 2004;30:519–23.
59. Yang L, Hou XY, Wei Y, Thai P, Chai F. Biomarkers of the health outcomes associated with ambient particulate matter exposure. Sci Total Environ. 2017;579:1446–59.
60. Yu ITS, Zhang YH, San Tam WW, Yan QH, Xu YJ, Xun XJ, Wu W, Ma WJ, Tian LW, Tse LA, Lao XQ. Effect of ambient air pollution on daily mortality rates in Guangzhou, China. Atmos Environ. 2012;46:528–35.
61. Zeng Q, Li G, Zhang L, Tian L, Pan X. Acute effects of so_2 and no_2 on mortality in the six cities of China. Chin J Prev Med. 2015;49:1085–91.
62. Zhang B, Niu J, Cao J, Ruan Y, Zhao N, Dong F, Zhang L. Time series analysis of the association between ambient air pollution and health in Lanzhou. J Environ Hyg. 2011; 1.
63. Zhang P, Dong G, Sun B, Zhang L, Chen X, Ma N, Yu F, Guo H, Huang H, Lee YL. Long-term exposure to ambient air pollution and mortality due to cardiovascular disease and cerebrovascular disease in Shenyang, China. PLoS One. 2011;6:e20827.
64. Zhang P, Dong G, Sun B, Zhang L, Chen X, Ma N, Yu F, Guo H, Huang H, Lee YL, Tang N, Chen J. Long-term exposure to ambient air pollution and mortality due to cardiovascular disease and cerebrovascular disease in shenyang, china. PLoS One. 2011;6:e20827.
65. Zhou M, Liu Y, Wang L, Kuang X, Xu X, Kan H. Particulate air pollution and mortality in a cohort of Chinese men. Environ Pollut. 2014;186:1–6.

Chapter 6
Ambient Air Pollution and Morbidity in Chinese

Li-Wen Hu, Wayne R. Lawrence, Yimin Liu, Bo-Yi Yang, Xiao-Wen Zeng, Wen Chen, and Guang-Hui Dong

Abstract The rapid economic growth in China is coupled with a severe ambient air pollution, which poses a huge threat to human health and the sustainable development of social economy. The rapid urbanization and industrialization over the last three decades have placed China as one of countries with the greatest disease burden in world. Notably, the prevalence rate of chronic noncommunicable diseases (CND), including respiratory diseases, CVD, and stroke, in 2010 reaches 16.9%. The continuous growth of the incidence of CND urgent needs for effective regulatory action for health protection. This study aims to evaluate the impact of rapid urbanization on status of ambient air pollution and associated adverse health effects on the incidence and the burden of CND and risk assessment. Our findings would be greatly significant in the prediction of the risk of ambient air pollution on CND and for evidence-based policy making and risk management in China.

Keywords Ambient air pollution • Mortality • Chinese • Adults

---

L.-W. Hu • B.-Y. Yang • X.-W. Zeng • W. Chen • G.-H. Dong (✉)
Guangzhou Key Laboratory of Environmental Pollution and Health Risk Assessment, Department of Preventive Medicine, School of Public Health, Sun Yat-sen University, 74 Zhongshan 2nd Road, Yuexiu District, Guangzhou 510080, Guangdong, China
e-mail: donggh5@mail.sysu.edu.cn; donggh512@hotmail.com

W.R. Lawrence
Department of Epidemiology and Biostatistics, School of Public Health, State University of New York, Albany, NY 12144-3445, USA

Y. Liu
Laboratory of Occupational Environment and Health Effects, Guangzhou Key Medical Discipline of Occupational Health Guardianship, Guangzhou Prevention and Treatment Center for Occupational Diseases, Guangzhou Twelfth People's Hospital, Guangzhou 510620, China

G.-H. Dong (ed.), *Ambient Air Pollution and Health Impact in China*, Advances in Experimental Medicine and Biology 1017, DOI 10.1007/978-981-10-5657-4_6

6.1 Introduction

Numerous studies have documented that exposure to ambient air pollution leads to important adverse health effects resulting in increased morbidity and mortality and shortened life expectancy [3, 10, 11, 47, 58, 84]. The Global Burden of Diseases, Injuries, and Risk Factors Study 2015 [17] estimated the burden of disease attributable to 79 risk factors in 195 countries from 1990 to 2015 and identified air pollution as a leading cause of global disease burden, especially in low-income and middle-income countries [17]. By using newer technologies for more accurate pollution measurements, the WHO recently reported that exposure to air pollution resulted in an estimated seven million deaths each year, suggesting one in eight deaths worldwide in 2012 [28]. Furthermore, although air pollution has long been recognized as a risk factor in respiratory illness such as acute infections and chronic obstructive pulmonary disease (COPD), the WHO found heart disease and stroke were tied as the top causes of ambient pollution-related deaths, each accounting for 40% of deaths from such exposure, and small proportions of the deaths attributable to ambient air pollution were caused by COPD (11%), lung cancer (6%), and acute lower respiratory tract infections in children (3%) although the respiratory system is the direct target organ for air pollution attack. However, we should carefully evaluate these results for most of these studies were from the developed countries where ambient air pollution levels were usually below 40 $\mu g/m^3$. In the higher-pollution areas, such as in China, we have evaluated the relationship of air pollution with respiratory disease and CVD, respectively, and found that the impact of air pollution on respiratory mortality [relative risk (RR) = 1.67; 95% confidence interval (CI): 1.60–1.74] was much larger than that on CVD mortality (RR = 1.55; 95% CI: 1.51–1.60) and cerebrovascular disease (RR = 1.49; 95% CI: 1.45–1.53) [12, 85]. Therefore, compared with the results in populations from the Europe and America, there may be different characteristics of associations between ambient air pollution and health affects in Chinese. Furthermore, compared with the extreme event of "death," it may be more likely to show the health effects of air pollution by using "morbidity" as an outcome. Therefore, the aim of this chapter will mostly focus on the links between ambient air pollution, not only particulate matter (PM) but also gases such as carbon monoxide, ozone, sulfur oxide, and nitrogen oxide, and health affect in Chinese.

Historically, two equivalent observations have connected poor air quality to human disease. One observation is the understanding that substances in inspired air can pose health risks, and the other is the view that increase industrialization at the global level has contributed to worsening air quality. For instance, urban planning experts, economists, and environmental scientists have vocalized and written about the effect of urban development, industrial growth, and income in the context of an "EKC" relationship (environmental Kuznets curve; "inverted U-shaped" curve), which predicates on the fact that as human activities associated with industry and urban growth elevate, an initial and sharp worsening in air quality ensues. Subsequently, as a society's income levels inevitably increase, awareness, regulation, and increasing attitudes of environmental and social responsibility intervene and air quality standards improve [21].

Although ambient air quality has improved since the beginning of the twenty-first century, the levels of thoracic particles (less than 10 μm in diameter [PM_{10}]) and SO_2 are still much higher than the air quality guidance value from the WHO. Furthermore, China still relies heavily on coal consumption, which accounts for 64% of domestic energy consumption (worldwide average level 30%). Meanwhile, its rapid economic development has contributed to increased energy use and industrial waste associated with rising levels of air pollution (e.g., total yearly carbon dioxide emission increased from ~2000 billion kg in 1980 to ~5000 billion kg in 2000) [86], thus threatening public health [63].The rapid growth in motor vehicle ownership and utilization has been accompanied by an associated increase in traffic-related air pollution and progressively worsens problems in urban areas of China. Consequently, the changes in economic structure resulted in urban air pollution in China to change from the coal combustion type to a compound air pollution type due to the concurrence of coal smoke and motor vehicle emission type [12].

Here, we focus on air pollutants and their effects on morbidity, including chronic respiratory diseases, CVD, stroke, obesity, hypertension, and so on, highlighting targets for interventions (particularly for at-risk populations) and recommendations to improve air quality.

6.2 Ambient Air Pollution and Respiratory Diseases

Chronic respiratory diseases represent a substantial health-care burden in China. The Global Burden of Diseases, Injuries, and Risk Factors Study reported that chronic obstructive pulmonary disease (COPD) and lung cancer are, respectively, the third and fourth most common causes of mortality in China, with 910,809 deaths attributed to COPD, 546,259 to lung cancer, 82,328 to interstitial lung disease and sarcoidosis, and 25,163 to asthma in 2013 [82, 91]. Lung cancer was a leading cause of disability-adjusted life years (DALYs) in China, and ambient air pollution ranked fourth in risk factors that contributed most to DALYs. Each 10 $\mu g/m^3$ rise in fine PM, of any origin, has been linked to a 3.1% increase in risk of hospitalization and a 2.5% increase in mortality [19, 31]. Implementation of effective measures to minimize pollutant emissions is needed and will lead to an improvement in respiratory health in China.

6.2.1 Ambient Air Pollution and COPD

To investigate the effects of short-term exposure to traffic-related air pollution on airway oxidative stress and inflammation in COPD patients in Beijing city, a panel of 45 diagnosed COPD patients were recruited and followed with repeated measurements of biomarkers reflecting airway oxidative stress and inflammation in exhaled breath condensate (EBC), including nitrate and nitrite, 8-isoprostane, interleukin-8, and acidity of EBC (pH), between September 5, 2014 and May 26, 2015. The results showed that the concentration of $PM_{2.5}$, black carbon, and NO_2 and number concentration of particles with diameter less than 100 nm (PNC100) and particles in size ranges between 100 and 200 nm (PNC100–200) during the first follow-up were 156.5 ±117.7, 10.7 ± 0.7, and 165.9 ± 66.0 μg/m^3 and 397,521 ± 96,712 and 79,421 ± 44,090 per cubic meter, respectively; the concentrations were 67.9 ± 29.6, 3.4 ± 1.3, and 126.1 ± 10.9 μg/m^3 and 295,682 ± 39,430 and 24,693 ± 12,369 per cubic meter, respectively, during the second follow-up. In this COPD-patient panel, per interquartile range (IQR) increase in PNC100–200, the authors observed an increase of 65% (95% CI: 8–152%) in nitrate and nitrite in EBC reflecting airway oxidative stress. For an IQR increase in $PM_{2.5}$, black carbon, and PNC100–200, respective increases of 0.17 ng/ml (95% CI: 0.02–0.33), 0.12 ng/ml (95% CI: 0.01–0.24), and 0.13 ng/ml (95% CI: 0.02–0.24) in interleukin-8 in EBC reflecting airway inflammation were also observed. An IQR increase in ozone was also associated with a 0.24 (95% CI: 0.05–0.42) decrease in pH of EBC reflecting increased airway inflammation [6]. During 2009, Wang et al. [70] used GIS to investigate the spatial association between ambient air pollution and acute exacerbations of COPD (AECOPD) hospitalizations in Jinan City and at residence. The study found concentrations of SO_2, PM_{10}, NO_2, CO, O_3, and AECOPD hospitalization cases were statistical significant spatially clustered. The Z-score of SO_2, PM_{10}, CO, O_3, and NO_2 at residence is 15.88, 13.93, 12.60, 4.02, and 2.44, respectively. Additionally, while at the workplace, concentrations of PM_{10}, SO_2, O_3, CO, and AECOPD hospitalization cases revealed statistical significant spatially clustered. The Z-score of PM_{10}, SO_2, O_3, and CO at workplace is 11.39, 8.07, 6.10, and 5.08, respectively. After adjusting for potential confounders in the model, only PM_{10} concentrations at the workplace showed statistical significance, showing a 10 μg/m^3 increase of PM_{10} at workplace associated with a 7% (95% CI: [3.3%, 10%]) rise of hospitalizations due to AECOPD.

To assess associations between dust storms and emergency hospital admissions due to respiratory disease in Hong Kong, Tam et al. [60] collected the data on daily emergency admissions for respiratory diseases to major hospitals in Hong Kong and indices of air pollutants and meteorological variables from January 1998 to December 2002 from several government departments. Significant increases in emergency hospital admission due to COPD were found 2 days after dust storm episode, and the relative risk of PM_{10} for lag 2 days was 1.05 (95% CI: 1.01–1.09) per 10 μg/m^3. In four-pollutant models of Hong Kong, the RR for COPD admission per 10 μg/m^3 increase in ozone was 1.03 (95% CI: 1.02–1.04) and 1.01 (1.01–1.02) per 10 μg/m^3 increase in $PM_{2.5}$ [27]. Air pollution has contributed to mortality associated with COPD in Shanghai [26], despite a report documenting that each 0.4 ppm increase in carbon monoxide levels has been negatively correlated with COPD admissions (mean −1.8%; 95% CI: −3.1 to −0.4), with stronger effects following adjustment for nitrogen dioxide and $PM_{2.5}$ levels [61]. Air pollution has caused long-lasting adverse effects on COPD mortality, which warrants closer follow-up in at-risk populations.

Although the way in which air pollutants modulate the underlying biological mechanisms remains unclear, perturbation of the coagulation balance is one of the potential mechanisms [56]. Based on the GIRD COPD Biobank Project, the Asian game cohort study was conducted in 36 COPD patients to estimate whether changes in measurements of D-dimer, blood marker of coagulation, were associated with changes in pollutant concentration, comparing for 51 intervention days (November 1–December 21) in 2010 with the same calendar date of baseline years (2009 and 2011) in Guangzhou city [87]. The results showed that daily PM_{10} reduced from the baseline period of 65.86–62.63 μg/m^3 during the Asian Games period and daily NO_2 decreased from 51.33 to 42.63 μg/m^3. SO_2 and other weather variables did not substantially differ. Additionally, the authors did not observe statistically significant improvements in D-dimer levels by 9.86% from a pre-Asian game mean of 917 ng/ml to a during-Asian game mean of 1007 ng/ml, platelet number by 11.66%, PH by −0.15%, PCO_2 by −6.54%, and PO_2 by −1.16%. During post-Asian game period, when pollutant concentrations were elevated, most outcomes approximated pre-Asian game levels, and comparable effects were also demonstrated in platelet number, D-dimer, and arterial blood gas, which provide evidence for the presence of a hypercoagulative state in systemic circulation in COPD patients. As the mechanism of air pollutants effect on COPD remains unknown, future studies will be needed to unravel the underlying mechanism.

6.2.2 Ambient Air Pollution and Hospital Admissions for Respiratory Diseases

To explore the association between $PM_{2.5}$ pollution and hospital emergency room visits (ERVs) for total and cause-specific respiratory diseases in urban areas in Beijing, a total of 92,464 respiratory emergency visits were evaluated with $PM_{2.5}$ from January 1 to December 31, 2013, with the mean daily concentration of 102.1 ± 73.6 μg/m^3. Every 10 μg/m^3 increase in $PM_{2.5}$ concentration at lag 0 was associated with a rise in ERV, as follows: 0.19% for upper respiratory tract infection (URTI) (95% CI: 0.04–0.35%), 0.34% for lower respiratory tract infection (LRTI) (95% CI: 0.14–0.53%), 0.23% for total respiratory disease (95% CI: 0.11–0.34%), and 1.46% for acute exacerbation of chronic obstructive pulmonary disease (AECOPD) (95% CI: 0.13–2.79%). The strongest association was identified between AECOPD and $PM_{2.5}$ concentration at lags 0–3 (3.15%, 95% CI:

1.39–4.91%) [79]. Additionally, Wang et al. [71] explored the concentration-response relationship between ambient concentration of $PM_{2.5}$ and daily total hospital emergency room visits in Beijing during 2012 and 2013, with the annual average concentration of $PM_{2.5}$ as 90.9 μg/m^3. The results indicated that in every 10 μg/m^3 increase of concentration of ambient $PM_{2.5}$, the corresponding excess risk of daily ERV was 0.25% (95% CI: 0.07–0.43) and the odd ratios was 1.16 (95% CI: 1.09–1.22) for ERV with $PM_{2.5}$ as the primary air pollutants during heavy polluted days. Ping [46] explored the influence of hazy weather on respiratory diseases and found that air quality index (AQI) of hazy weather had significantly positive association the number of patients with five kinds of respiratory diseases (bronchitis, bronchial asthma, upper respiratory infection, COPD, pneumonia) and different pollutants had distinct influences on various respiratory diseases, in which $PM_{2.5}$ and SO_2 had the closest association with each disease, whereas PM_{10} and CO were positively related with bronchial asthma, upper respiratory tract infection, COPD, and pneumonia ($p < 0.01$) but had no connection with bronchitis. However, O_3 was in negative correlation with COPD and pneumonia. Leitte et al. [30] also analyzed the role of different particle size fractions with respect to respiratory health in Beijing, China, and found, in general, particles <50 nm were not positively associated with ERV, whereas particles 50–100 nm were adversely associated with respiratory ERV, both being fractions of ultrafine particles.

Even in the relative clean area, $PM_{2.5}$ also can significantly increase the risk of outpatient visits for RD. For example, a study conducted in Shenzhen of south China, where the annual average $PM_{2.5}$ concentration was 40.2 μg/m^3 (daily range 7.2–137.1 μg/m^3), reported that a 10 μg/m^3 increase in day-before $PM_{2.5}$ concentration was associated with a 1.81% (95% CI: 1.71–1.91%) ER of visits for respiratory disease [45]. A study from Ningbo, where the daily average concentration of SO_2 did not exceed the limit value of second class area, also found about 30% of total upper respiratory outpatients caused by SO_2 pollution [75]. Qiu et al. [50] conducted this study to estimate the health effects of coarse PM (PMc: 2.5–10 μm aerodynamic diameter) on emergency hospital admissions for respiratory diseases in Hong Kong after controlling for $PM_{2.5}$ and gaseous pollutants, and results showed that a 10.9 μg/m^3(interquartile range) increase in the 4-day moving average concentration of PMc was associated with a 1.94% (95% confidence interval: 1.24%, 2.64%) increase in emergency hospital admissions for respiratory diseases that was attenuated but still significant after controlling for $PM_{2.5}$. A recent study from Taiwan also evaluated the association between PMc and frequency of hospital admissions for respiratory diseases (RD) including COPD, asthma, and pneumonia [9]. Findings revealed increased rate of admissions for RD was significantly associated with higher coarse PM levels only on cool days (<25 °C), with a 10 μg/m^3 elevation in $PM_{2.5-10}$ concentrations associated with a 3% (95% CI = 1–5%) increase in COPD admissions, 4% (95% CI = 1–7%) rise in asthma admissions, and 3% (95% CI = 2–4%) increase in pneumonia admissions. On warm days, no significant associations were found between coarse particle levels and frequency of hospital admissions for RD. The results from the two-pollutant models showed that $PM_{2.5-10}$ levels remained significantly correlated with higher rate of admissions even when

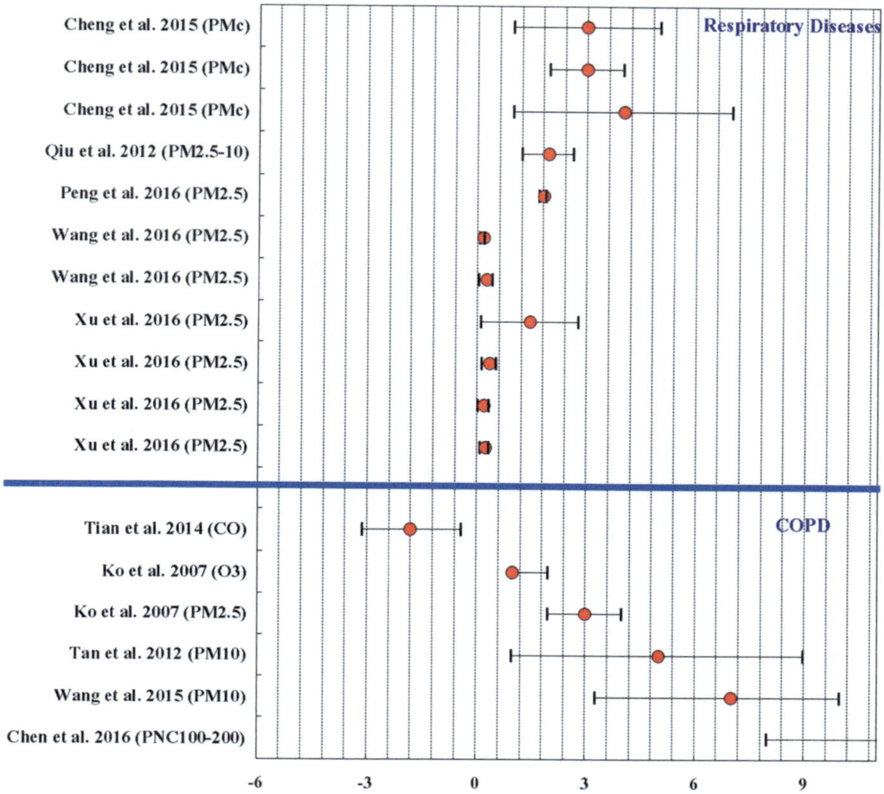

Fig. 6.1 Summary of studies of ambient air pollution and COPD and respiratory diseases

controlling for carbon monoxide, sulfur dioxide, nitrogen dioxide, or ozone on cool days. This indicated that higher levels of $PM_{2.5-10}$ enhance the risk of hospital admissions for RD on cool days. As shown in Fig. 6.1, there is a summary of some studies of ambient air pollution and COPD and respiratory diseases in China.

6.2.3 Ambient Air Pollution and Lung Function

Wu et al. [76] investigated the short-term respiratory effects (exhaled nitric oxide (FeNO) and exhaled hydrogen sulfide (FeH2S)) of ambient air pollution in a panel of 23 stable COPD patients in Beijing, China, from January to September 2014. The study found increases in interquartile range (IQR) in SO_2 (45.7 μg/m^3, 6-day), $PM_{2.5}$ (76.5 μg/m^3, 5-day), and PM_{10} (75.0 μg/m^3, 5-day) were all associated with maximum increases in FeNO of 34.2% (95% CI: 17.3%, 53.4%), 13.6% (95% CI: 4.8%, 23.2%), and 9.2% (95% CI: 2.1%, 16.8%), respectively. Additionally, the same IQR increases in SO_2 (7-day), $PM_{2.5}$ (6-day), and PM_{10} (6-day) were associated with

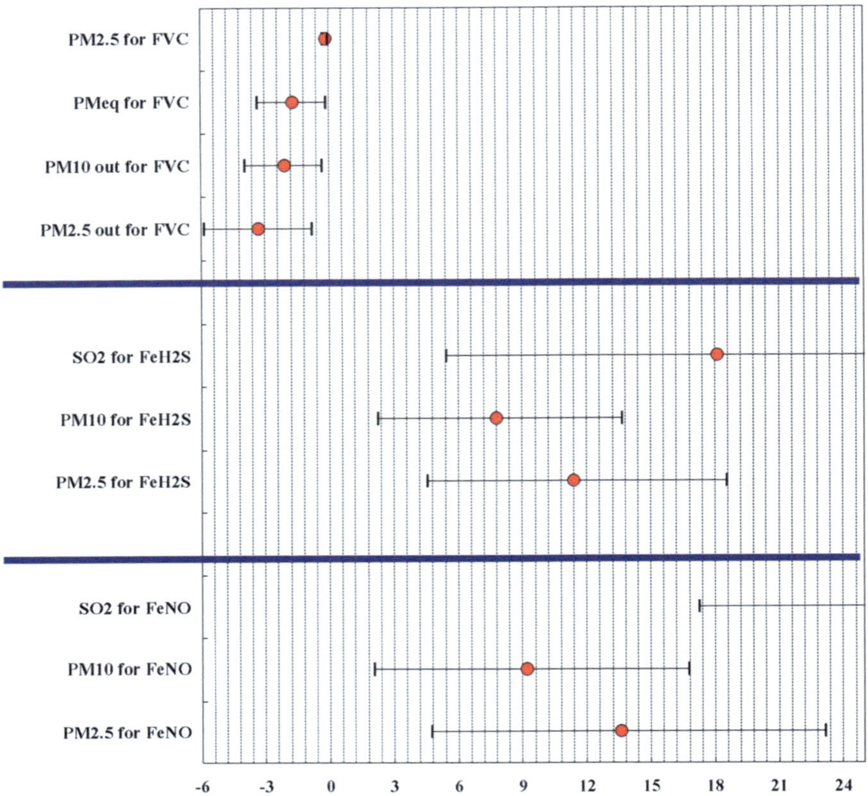

Fig. 6.2 Summary of studies of ambient air pollution and lung function

maximum increases in FeH2S of 18.1% (95% CI: 5.5%, 32.2%), 11.4% (95% CI: 4.6%, 18.6%), and 7.8% (95% CI: 2.3%, 13.7%), respectively (Fig. 6.2). Increasing levels of air pollutants were also associated with increased odds ratios for cough, sore throat, wheezing, sputum, and dyspnea. Ni et al. [43] investigated and compared the effects of short-term exposure to outdoor particulate matter (PM_{out}) and outdoor-originated equivalent personal PM (PM_{eq}) on lung function in 33 doctor-diagnosed COPD patients in 2013–2014 in Beijing. $PM_{2.5out}$ (111.0 μg/m^3, 5-day) and PM_{10out} (112.0 μg/m^3, 3-day) interquartile range increases were associated with a 3.3% (95% confidence interval [CI]: −5.8%, −0.8%) decrease and a 2.1% (95% CI: −3.9%, −0.3%) decline in forced vital capacity (FVC), respectively. Comparable results were found for forced expiratory volume in 1 s (FEV1). An interquartile range increase in $PM_{2.5}$eq (45.3 μg/m^3, 3-day), but not PM_{10}eq, remained associated with a 1.7% (95% CI: −3.3%, −0.1%) decrease in FVC (Fig. 6.2).

Transport hub is an important part of urban comprehensive transportation system. Traffic-related air pollution can reach high level because of difficulty of diffusion and increase of emission in transport hub. Huang et al. [22] performed randomized crossover study in 40 young healthy adults in Beijing. Each participant underwent 2 h exposure in a designated transport hub and, on a separate occasion, in an appointed park. Personal exposures to $PM_{2.5}$, black carbon (BC), and carbon monoxide (CO) were measured. Forced expiratory volume in 1 s (FEV_1) and peak expiratory flow (PEF) were assessed pre-, during, and post-exposure. The results showed that per 10 μg/m^3 increment in $PM_{2.5}$ was associated with −0.15 % (95 % CI: −0.28, −0.02 %) reduction in FEV_1 during exposure 2 h. However, effects became attenuate after 2 h exposure. In Guangzhou city, south China, a panel study was conducted to assess health effects of $PM_{2.5}$ on cardiopulmonary function. Results from the mixed-effect models showed that neither personal $PM_{2.5}$ exposure nor ambient air pollution exposure based on air monitoring stations was related to cardiopulmonary function, except for a marginally significant association between ambient PM_{10} and MMEF ($\beta = -0.147$; $p = 0.079$) and personal PM_{10} and MMEF ($\beta = -0.150$; $p = 0.072$).

It is well known that the health impact of PM is not only associated with concentration, size, and shape but also with the sources and components in PM [20, 52, 65]. $PM_{2.5}$ is composed of a vast number of chemical constituents, and there are increasing interests in exploring the potentially differentiated effects of various constituents [5, 20, 52, 65]. Wu et al. [68] examined the cardiopulmonary health effects of fine particles ($PM_{2.5}$) from different pollution sources in Beijing, China, among a panel of 40 healthy university students. The authors reported that seven $PM_{2.5}$ sources were identified as traffic emissions (12.0%), coal combustion (22.0%), secondary sulfate/nitrate (30.2%), metallurgical emission (0.4%), dust/soil (12.4%), industry (6.9%), and secondary organic aerosol (9.9%). $PM_{2.5}$ from dust/soil and industry were significantly associated with decreases in evening PEF as well as morning and evening FEV_1 at different exposure metrics. The significant reduction in evening PEF and morning and evening FEV_1 associated with 5-day average $PM_{2.5}$ from dust/soil was consistent over different models. Also, the declines in evening PEF and evening FEV_1 associated with $PM_{2.5}$ from industry were also consistent over various models, excluding in the source-$PM_{2.5}$ joint models. In particular, there were significant increases in morning and evening FEV_1 associated with $PM_{2.5}$ from metallurgical emission. However, these positive associations were not consistent in two-source or source-$PM_{2.5}$ joint models. The same study group also examined the associations of 32 chemical constituents of $PM_{2.5}$ with pulmonary function in a panel of 21 college students [69]. Four $PM_{2.5}$ constituents (copper, cadmium, arsenic, and stannum) were found to be most consistently associated with the reductions in these pulmonary function measures. These findings provide clues for the respiratory effects of specific particulate chemical constituents in the context of air pollution.

To investigate the acute effects of $PM_{2.5}$ constituents' fractional exhaled nitric oxide (FeNO), a well-established biomarker on respiratory inflammation, Shi et al. [55] performed a longitudinal panel study among 33 healthy young adults in Shanghai, China from January 12 to February 6, 2015. A total of 234 effective measurements of FeNO were obtained with a geometric mean of 13.1 ppb. $PM_{2.5}$-FeNO associations were strongest at lags of 0–6 h and diminished at lags longer than 12 h.

An interquartile range increase in $PM_{2.5}$ constituents $NH4^+$, $NO3^-$, K^+, $SO4^{2-}$, and elemental carbon (EC) at lags of 0–6 h was significantly associated with elevations of FeNO by 12.3%, 11.3%, 11.1%, 9.6%, and 10.7%, respectively. After controlling for $PM_{2.5}$ total mass and the collinearity, only EC remained the significant effects, indicating EC may be the primary component responsible for increased airway inflammation. Chen et al. [7] designed a time-series panel study in 28 patients to examine the effects of 10 major constituents of $PM_{2.5}$ on lung function with repeated daily measurements from December 2012 to May 2013 in Shanghai, China. The cumulative decreases in morning FEV_1, evening FEV_1, morning PEF, and evening PEF associated with an interquartile range (35.8 μg/m^3) increase in $PM_{2.5}$ concentrations were 33.49 (95% CI: 2.45, 54.53) mL, 16.80 (95% CI: 3.75, 29.86) mL, 4.48 (95% CI: 2.30, 6.66) L/min, and 1.31 (95% CI: −0.85, 3.47) L/min, respectively. The associations of elemental carbon (EC) and nitrates with morning/evening FEV_1 and the associations of EC and sulfates with morning PEF were robust after controlling for $PM_{2.5}$.

6.3 Ambient Air Pollution and Circulatory System Diseases

Cardiovascular diseases (CVDs) are actually among the main causes of death in the world [18, 38], causing more than 17 million deaths in 2008 [40]. They have a multifactorial etiology and, based on their clinical presentation, can be divided into chronic (i.e., heart failure and hypertension) and acute (i.e., acute myocardial infarction and arrhythmia) conditions. PM can cross the pulmonary epithelium into the circulation or interact with lung receptors (direct effects) to induce an acute cardiovascular response. The first evidence derives from studies of Nemmar [41, 42], which demonstrates that ultrafine particles and $PM_{2.5}$ could translocate into the pulmonary circulation and the systemic circulation. Given the continuous variability of air pollution, which is associated with the evolution of the technological progress, and individual movements through numerous microenvironments every day, it is not surprising that data from air pollution studies could be very variable. And as shown in Fig. 6.3, there is a summary of studies of ambient air pollution and CVDs in China.

6.3.1 Ambient Air Pollution, CVD, and Stroke

In order to examine whether the exposure to ambient air pollution was associated with the prevalence of stroke and CVDs among people living in a heavy industrial province of Northeast China, a large, well-characterized population-based cross-sectional study was conducted in 24,845 Chinese adults, ages 18–74 years old, from 33 communities in the 11 districts of the 3 northeastern Chinese cities during 2009 [13]. The 3-year average 24-h PM_{10} concentration was 123.1 μg/m^3 (±14.6 μg/m^3), the average 24-h SO_2 concentration was 54.4 μg/m^3 (±14.3 μg/m^3), the average 24-h

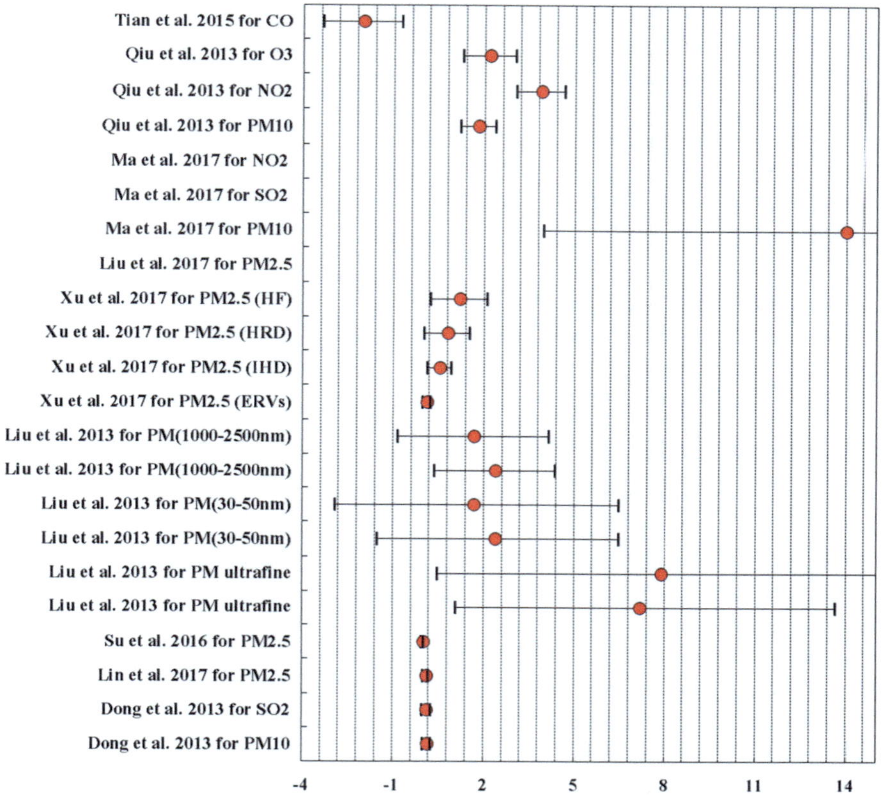

Fig. 6.3 Summary of studies of ambient air pollution and CVDs

NO_2 concentration was 35.3 μg/m^3 (±5.45 μg/m^3), and the 8-h ozone (O_3) concentration was 49.4 μg/m^3 (±14.1 μg/m^3). After adjusting for other variables, PM_{10} (adjusted OR = 1.16; 95% CI: 1.03–1.30) and SO_2 (adjusted OR = 1.14; 95% CI: 1.01–1.29) were significantly associated with stroke prevalence. Following stratification by gender, the significant association between stroke prevalence and PM_{10} (adjusted OR = 1.21; 95% CI: 1.05–1.41) and SO_2 (adjusted OR = 1.19; 95% CI: 1.01–1.40) was limited in men, but not in women. As for CVD, no significant associations were found with air pollutants in men and in women. When further stratified by BMI status, statistically significant interactions between exposure and obesity were observed for CVDs and stoke. The significant associations between annual pollutant concentrations and CVDs and stroke were greatest among obese participants (OR 1.15–1.47 for stroke, 1.33–1.59 for CVDs), less strong in overweight participants (OR 1.22–1.35 for stroke, 1.07–1.13 for CVDs), and weakest in normal-weight participants (OR ranged from 0.98 to 1.01 for stroke, 0.93 to 1.15 for CVDs). When stratified by gender, these interactions were statistically significant in women only [49].

The Study on Global Ageing and Adult Health, a prospective cohort in six low- and middle-income countries including China, has examined the effects of long-term exposure on stroke among 45,625 participants [33]. The odds of stroke were 1.13 (95% CI, 1.04–1.22) for each 10 μg/m^3 increase in $PM_{2.5}$. This effect remained after adjustment for confounding factors including age, sex, smoking, and indoor air pollution (adjusted odds ratio = 1.12; 95% confidence interval, 1.04–1.21). Further stratified analyses suggested that participants with higher levels of physical activity had greater odds of stroke, whereas those with higher consumption of fruit and vegetables had lower odds of stroke. These effects remained robust in sensitivity analyses. We further estimated that 6.55% (95% confidence interval, 1.97–12.01%) of the stroke cases could be attributable to ambient $PM_{2.5}$ in the study population.

6.3.2 *Ambient Air Pollution and Hospital Admissions for Cardiovascular Diseases*

To evaluate the association between ambient $PM_{2.5}$ and daily hospital emergency room visits (ERVs) for cardiovascular diseases in Beijing, China, Su et al. [57] obtained daily counts of cardiovascular ERV from the Peking University Third Hospital from January 2007 to December 2008 and, concurrently, data on $PM_{2.5}$, PM_{10}, nitrogen dioxide, and sulfur dioxide concentrations from monitoring networks and a fixed monitoring station. They observed adverse effects of $PM_{2.5}$ on cardiovascular ERV and IQR increase (68 μg/m^3) in $PM_{2.5}$ were associated with an overall RR of 1.022 (95% CI: 0.990–1.057) obtained from PDL model. The greatest effects of $PM_{2.5}$ on cardiovascular ERV were found at lag of 7 days; the respective estimate was 1.012 (95% CI: 1.002–1.022). The effects were strongest in spring and among females. Cerebrovascular and arrhythmia diseases revealed a greater association with $PM_{2.5}$. They also found stronger PM effects for southern and stagnant air masses, and air pollution effects were modified in the period of Olympics.

Liu et al. [36] investigated whether daily changes in number or mass concentrations of accurately size-segregated particles in the range of 3–10 μm are associated with daily cardiovascular emergency room visits in Beijing, China. The study observed delayed associations between number concentration of ultrafine particles and cardiovascular emergency room visits primarily from lag 4 to lag 10, generally contributed by 10–30 nm and 30–50 nm particles. An IQR (9040/cm^3) increase in 11-day average number concentration of ultrafine particles was associated with a 7.2% (1.1–13.7%) increase in total and a 7.9% (0.5–15.9%) rise in severe cardiovascular emergency room visits. The delayed effects of particulate mass concentration were small. In regard to immediate effects, 2-day average number concentration of Aitken mode (30–100 nm) particles had the strongest effects. An IQR (2269/cm^3) rise in 2-day average number concentration of 30–50 nm particles led to a 2.4% (−1.5 to 6.5%) increase in total and a 1.7% (−2.9 to 6.5%) increase in severe cardiovascular emergency room visits. The immediate effects of mass concentration came mainly from 1000 to 2500 nm particles. An IQR (11.7 μg/m^3) increase in 2-day average mass concentration of 1000–2500 nm particles led to approximately 2.4% (0.4–4.4%) increase in total and a 1.7% (−0.8–4.2%) increase in severe cardiovascular emergency room visits. The lagged effect curves of number and mass concentrations of 100–300 nm particles or 300–1000 nm particles were comparable indicating that using particulate number or mass concentrations seemed not to affect the cardiovascular effect (of particles within one size fraction). The effects of number concentration of ultrafine particles, sub-micrometer particles (3–1000 nm), and 10–30 nm particles were substantially higher in winter when compared with summer (reuse with permission from Liu et al. [36]).

Recently, Xu et al. [80] reported that the daily mean $PM_{2.5}$ concentration was 102.1 μg/m^3, ranging from 6.7 to 508.5 μg/m^3. Per 10 μg/m^3 increase in $PM_{2.5}$ was associated with a 0.14% (95% CI: 0.01–0.27%) increase in cardiovascular ERVs at lag 3. Cumulative delayed estimates were greatest at lag 0–5 (0.30%, 95% CI: 0.09–0.52%). The estimates of percentage change in daily ERVs per 10 μg/m^3 increase in $PM_{2.5}$ were 0.56% (95% CI: 0.16–0.95%) for ischemic heart disease (IHD) at lag 0–1, 0.81% (95% CI: 0.05–1.57%) for heart rhythm disturbances (HRD) at lag 0–1, and 1.21% (95% CI: 0.27–2.15%) for heart failure (HF) at lag 0, respectively. The effects of $PM_{2.5}$ on IHD ERVs during high-temperature days (>11.01 °C) were significantly higher than that on low-temperature days (≤11.01 °C) at lag 0, lag 0–1, lag 0–3, and lag 0–5 ($P < 0.05$). Recently, a case-crossover study examined how $PM_{2.5}$ from Asian dust storms (ADS) affects the number of emergency room (ER) admissions for cardiovascular diseases (CVDs) [37]. The data indicated that $PM_{2.5}$ concentration from ADS was highly correlated with ER visits for CVDs, with the OR increased by 2.92 (95% CI: 1.22–5.08) per 10 μg/m^3.

In western China, Ma et al. [39] investigated the short-term effects of air pollutants (PM_{10}, SO_2, and NO_2) on daily cardiovascular admissions from March 1 to May 31 during 2007 to 2011 in Lanzhou, a heavily polluted city in western China. Significant associations were found between air pollutants and hospital admissions for cardiovascular diseases both on dust event days and non-dust event days in spring. Air pollutants had lag effects on different age and gender groups. Relative risks (RRs) and their 95% confidence intervals (CIs) associated with a 10 μg/m^3 increase were 1.14 (1.04–1.26) on lag 1 for PM_{10}, 1.31 (1.21–1.51) on lag 01 for SO_2, and 1.96 (1.49–2.57) on lag 02 for NO_2 on dust days. Stronger effects of air pollutants were observed for females and the elderly (≥60 years).

Weather factors such as temperature, season, and relative humidity (RH) may modify the effects of pollution. Qiu et al. [51] conducted a study to examine the effects of air pollution on emergency ischemic heart disease (IHD) hospital admissions that varied across seasons and RH levels and to explore the possible joint modification of weather factors on pollution effects. An increase in the harmful effects of air pollution in cool season and on low humidity days was detected. During the cool and dry season, a 10 μg/m^3 increment of lag 3 exposure was associated with an increase of emergency IHD admissions by 1.82% (95% CI: 1.24–2.40%), 3.89% (95% CI: 3.08–4.70%), and 2.19% (95% CI: 1.33–3.06%) for particles with an aerodynamic diameter of less than 10 μm (PM_{10}), nitrogen dioxide

(NO_2), and ozone (O_3), respectively. The effects of pollutants declined greatly and were no longer statistically significant in warm and humid season.

In Hong Kong, during 1994 and 1995, a retrospective ecological study was performed to investigate short-term effects of concentrations of pollutants in ambient air on hospital admissions for cardiovascular and respiratory diseases [74]. Relative risk (RR) on admissions for respiratory disease for the four pollutants ranged from 1.013 (for SO_2) to 1.022 (for O_3), and for admissions for cardiovascular disease, from 1.006 (for PM_{10}) to 1.016 (for SO_2). Those aged > or = 65 years were at higher risk. Interestingly, an inversed relationship has been reported between air pollution and stroke. By using the data of daily air pollution concentrations and emergency hospital admission data from January 2004 to December 2011 in Hong Kong, a time-series study was conducted to examine the short-term association of ambient CO with emergency stroke hospitalizations [62]. Negative associations were observed between ambient CO concentrations and emergency hospital admissions for stroke. Previous 1–3 day cumulative exposure to CO was associated with a −2.0% (95% CI, −3.3% to −0.7%) reduction in stroke admissions per IQR increment in CO concentration (0.3 ppm). Similar results were found when using 1-h maximum concentration of CO as exposure indicator. The negative association was stronger compared to co-pollutant adjustment for either NO_2 or PM_{2-5}. The elderly and females appeared to be more sensitive to ambient CO exposure. Negative association tended to be greater in cool season, suggesting some acute protective effects of CO exposure against stroke onsets.

6.3.3 Ambient Air Pollution and Hypertension

As the major risk factor for cardiovascular disease (CVD) and other vascular diseases [29, 48, 73], high blood pressure (BP) and hypertension have been identified as the most important causes of disability and the leading risks for death in the world, causing an estimated 7.5 million deaths per year (13% of all deaths) [72]. To investigate the association between residential long-term exposure to air pollution and BP and hypertension, 33 communities study recruited 24,845 Chinese adults in 11 districts of 3 northeastern cities from 2009 to 2010 [14] (Fig. 6.4). High ambient pollution levels were observed in the 11 study districts. The arithmetic means of PM_{10}, SO_2, NO_2, and O_3 were 123.1, 54.4, 35.3, and 49.4 μg/m^3, respectively. These mean concentrations presented a wide variation in district gradient for PM_{10} (93–145 μg/m^3), SO_2 (36–78 μg/m^3), NO_2 (27–45 μg/m^3), and O_3 (27–71 μg/m^3). Results revealed that the odds ratio (OR) for hypertension elevated by 1.12 (95% confidence interval [CI], 1.08–1.16) per 19 μg/m^3 increase in PM_{10}, 1.11 (95% CI, 1.04–1.18) per 20 μg/m^3 increase in SO_2, and 1.13 (95% CI, 1.06–1.20) per 22 μg/m^3 increase in O_3. Estimated increases in mean systolic and diastolic BP were 0.87 mmHg (95% CI: 0.48–1.27) and 0.32 mmHg (95% CI: 0.08–0.56) per 19 μg/m^3 interquartile rise in PM_{10}, 0.80 mmHg (95% CI: 0.46–1.14) and 0.31 mmHg (95% CI: 0.10–0.51) per 20 μg/m^3 interquartile increase in SO_2, and 0.73 mmHg (95% CI: 0.35–1.11) and

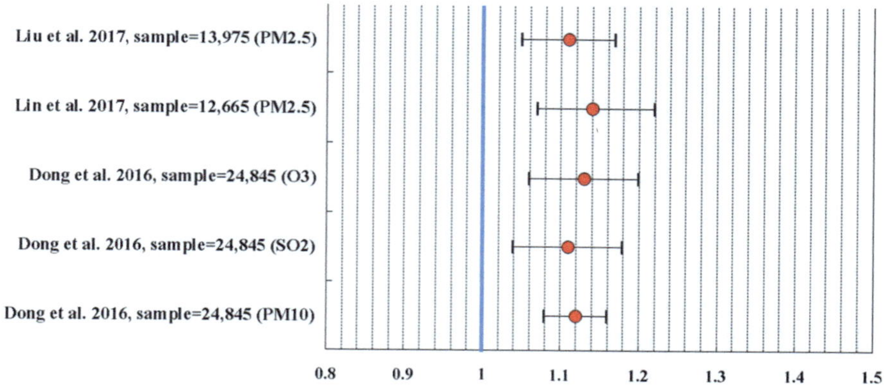

Fig. 6.4 Summary of studies of ambient air pollution and hypertension

0.37 mmHg (95% CI: 0.14–0.61) per 22 μg/m^3 interquartile increase in O_3. Additionally, these associations were statistically significant only in men. Furthermore, the authors also evaluated whether obesity modified the association between ambient air pollution and increased blood pressure and hypertension in adults [89]. For all participants, significant interactions between being overweight and obese with PM_{10} or O_3 ($P < 0.001$) were observed. However, when stratified by gender, these relationships were only obtained in males. Among men, the effects of air pollutants on hypertension were strongest in obese subjects (OR ranged from 1.27 to 1.49), less strong in overweight subjects (OR ranged from 1.09 to 1.24), and weakest in normal-weight subjects (OR ranged from 1.08 to 1.10). Among women, none of the pollutants were found to be associated with prevalence of hypertension within any subgroup. As for arterial BP, for all subjects, each of the ambient air pollutants, except for NO_2, was significantly associated with higher SBP and DBP levels and increasing BMI; this association was especially seen in SBP. When stratified by gender, these significant relationships were also limited to men [89]. Lin et al. [34] recently also examined long-term association between ambient $PM_{2.5}$ and hypertension and blood pressure by interviewing 12,665 participants aged 50 years and older and measuring their blood pressures. Annual average $PM_{2.5}$ concentrations were estimated for each community using satellite data. For each 10 μg/m^3 increase in ambient $PM_{2.5}$, the adjusted odds ratio of hypertension was 1.14 (95% confidence interval, 1.07–1.22). Stratified analyses found that overweight and obesity could enhance the association and consumption of fruit was associated with lower risk. The authors further estimated that 11.75% (95% confidence interval, 5.82–18.53%) of the hypertension cases (corresponding to 914, 95% confidence interval, 453–1442 cases) could be attributable to ambient $PM_{2.5}$ in the study population (Fig. 6.4). Findings suggest that long-term exposure to ambient $PM_{2.5}$ might be an important risk factor of hypertension and is responsible for significant hypertension burden in adults in China.

Based on a nationally representative survey (13,975 participants), a research group from Fudan University explored the associations of long-term exposure to $PM_{2.5}$ with hypertension prevalence and blood pressure (BP) in China [35] (Fig. 6.4). The long-term average exposure to $PM_{2.5}$ for all subjects during the study period (June 2011–March 2012) was estimated by a satellite-based model with a spatial resolution of 10 × 10 km. The results showed that the annual mean exposure to $PM_{2.5}$ for all participants was 72.8 μg/m^3 on average. An interquartile range increase (IQR, 41.7 μg/m^3) in $PM_{2.5}$ was associated with higher prevalence of hypertension with an odds ratio of 1.11 [95% confidence interval (CI): 1.05, 1.17]. Systolic BP increased by 0.60 mmHg (95% CI: 0.05, 1.15) per an IQR increase in $PM_{2.5}$. The effects of $PM_{2.5}$ on hypertension prevalence were stronger among middle-aged, obese, and urban participants.

The use of ambient air pollution data obtained from central air monitoring stations as surrogates for participants' exposures to outdoor air pollutants in previous studies may have introduced bias in the estimation of exposure-response associations. Individual exposure measurement, where pollution is measured using personal wearable devices (as relatively to utilizing aggregate environmental measurements from monitoring stations), can better quantify observed differences and better reflect exposure among smaller groups of people at ground level. Some consider this to be the most accurate method of pollutant exposure assessment [54, 64]. In addition, a growing number of panel studies on short-term air pollution exposures have assessed personal exposure based on ambient air concentrations measured at fixed stations within 3 km or even less than 1 km of their movement, such as community concentrations [83, 90]. Recently a panel study was conducted in 30 young adults from Guangzhou, China, to assess the effects of personal $PM_{2.5}$ exposure on blood pressure. The results showed that the 8-h mean concentration of personal $PM_{2.5}$ exposure was 65.09 ± 22.18 μg/m^3, which was 24.34 μg/m^3 higher than the ambient concentrations over the same period ($p < 0.05$); however they were strongly correlated (Spearman's $r = 0.937$, $p < 0.01$). Separate mixed-effect models for ambient and personal exposure were fitted to estimate their association with cardiopulmonary outcomes. Both models indicated $PM_{2.5}$ exposure was not related to cardiopulmonary function. However, the mean daily 8-h NO_2 exposure was significantly associated with an elevation in heart rate and blood pressure. In the personal $PM_{2.5}$ exposure model, a 10 μg/m^3 increase in NO_2 led to an increase of 6.7 beats/minute in heart rate and an increase of 6.5 mmHg and 4.4 mmHg elevation in systolic blood pressure and diastolic blood pressure, respectively.

To evaluate the association between combustion-related air pollution with ambulatory blood pressure and autonomic function during February to July 2012, personal black carbon was determined for five consecutive days using microaethalometers in patients with metabolic syndrome in Beijing, China [88]. Mean (SD) of personal black carbon and fine particulate matter during 24 h was 4.66 (2.89) and 64.2 (36.9) μg/m^3. Additionally, exposure to high levels of black carbon in the previous hours was associated significantly with adverse cardiovascular responses. One-unit increase in personal black carbon during the previous 10 h was associated with elevated systolic blood pressure of 0.53 mmHg and diastolic blood pressure of

0.37 mmHg (95% confidence interval, 0.17–0.89 and 0.10–0.65 mmHg, respectively), a percentage change in low frequency-to-high frequency ratio of 5.11, and mean interbeat interval of −0.06 (95% confidence interval, 0.62–9.60 and −0.11 to −0.01, respectively). Also, an unpublished longitudinal panel study with 6 repeated visits in 28 urban residents in Shanghai, China, was conducted from May 27 to July 5, 2014. One hundred sixty-eight pairs of effective BP measurements were obtained during the study period. An interquartile range increase of $PM_{2.5}$ (19.1 μg/m^3) was associated with increments of 1.90 mmHg (95% CI: 0.66, 3.13) in systolic BP, 0.68 mmHg (95%CI: −0.02, 1.37) in diastolic BP, and 1.23 mmHg (95% CI: 0.19, 2.29) in pulse pressure. Some constituents of $PM_{2.5}$, including organic carbon, elemental carbon, nitrate, and ammonium, were robustly associated with BP after controlling for total $PM_{2.5}$ mass and accounting for the multicollinearity.

Wu et al. [68] examined the cardiopulmonary health effects of fine particles ($PM_{2.5}$) from different pollution sources in Beijing, China, among a panel of 40 healthy university students. Significant increases in both SBP and DBP were associated with $PM_{2.5}$ from dust/soil, industry, and coal combustion at different exposure metrics, among which the associations between DBP and $PM_{2.5}$ from coal combustion were most constant. Authors additionally found significant decrease in all three BP variables associated with 1–2-day average $PM_{2.5}$ from metallurgical emission. Among these associations, the elevated DBP was associated with 1-day average $PM_{2.5}$ from coal combustion and declines in SBP and DBP associated with 1-day average $PM_{2.5}$ from metallurgical emission were consistent over different models.

Both low temperature and high air pollution have been associated with an increased risk of cardiovascular events. Wu et al. [67] evaluate the interaction between ambient temperatures with air pollution in blood pressure in healthy adults in Beijing, China, in 2010–2011. Findings revealed significant interactions between temperature and traffic-related air pollutants (particulate matter with an aerodynamic diameter ≤2.5 μm, organic carbon, elemental carbon, and nitrogen dioxide) on BP ($P < 0.05$ for all interaction tests). The estimated rises in SBP and DBP were 4.9 mmHg [95% confidence interval (CI) 2.9–6.8] and 3.7 mmHg (95% CI: 2.3–5.1) at high elemental carbon level (≥median) and were −1.3mmHg (95% CI: −6.3 to 3.6) and 0.7 mmHg (95% CI: −2.8 to 4.2) at low elemental carbon level (<median) per 10 °C decline in daily minimum temperature. This indicated low temperature and high air pollution might act synergistically to increase BP in healthy adults (Fig. 6.5).

6.3.4 Ambient Air Pollution and Prehypertension

As a new category of blood pressure (BP) classification according to the Seventh Report of the Joint National Committee, prehypertension has aroused people's great concern in recent years due to its associations with increased incidence of cardiovascular disease. The 33 communities Chinese Health Study reported that, following covariate adjustment, an increased prevalence rate of prehypertension was found

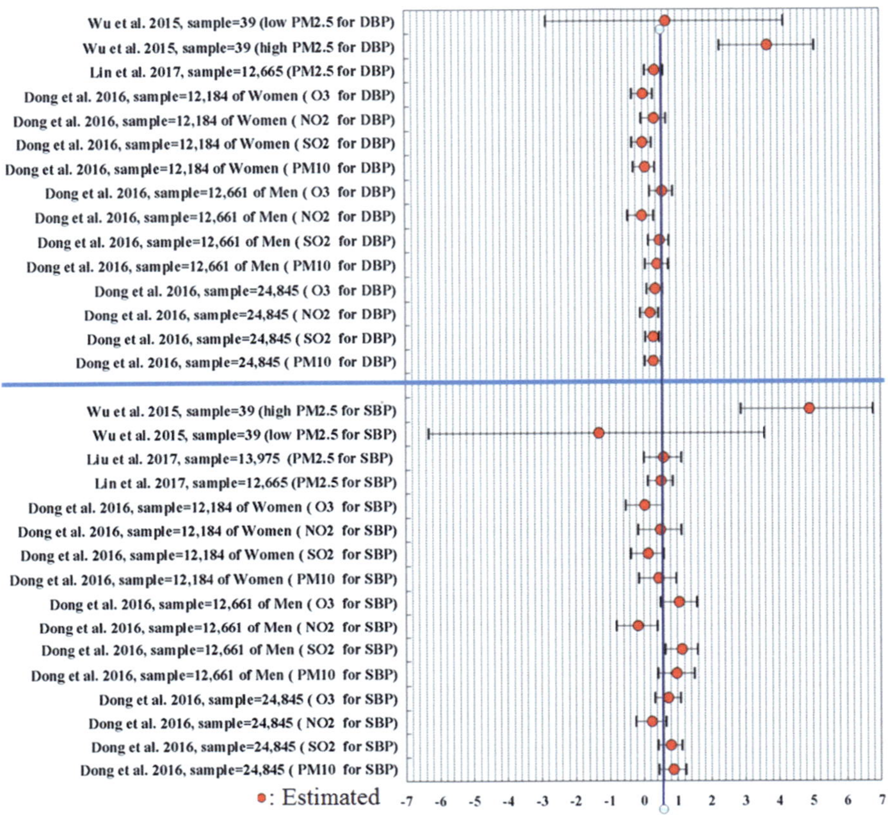

Fig. 6.5 Summary of studies of ambient air pollution and arterial blood pressure (mmHg)

to be significantly associated with an interquartile range (IQR) of PM_{10} (OR, 1.20; 95% CI, 1.12–1.28), SO_2 (OR, 1.14; 95% CI, 1.00–1.28), NO_2 (OR, 1.21; 95% CI, 1.07–1.36), and O_3 (OR, 1.16; 95% CI, 1.02–1.32). However, stratified analysis by sex showed that these relationships were only obtained in women, not in men. The corresponding *P* values for interaction of sex with PM_{10}, SO_2, NO_2, and O_3 were <0.001, 0.012, 0.217, and 0.026, respectively. When further stratified by age of the subjects, the graded relationship between air pollutants and prehypertension persisted. The ORs for prehypertension were 1.08–1.16 in subjects aged ≤35 years, 1.11–1.17 in subjects aged 35–55 years, and 1.23–1.27 in subjects aged ≥55 years. The *P* values for the interaction of age and all four air pollutants were <0.001. Subgroup analysis by BP classification was further performed, using all eligible subjects inclusive of those with hypertension [14]. The results showed that all four air pollutants were significantly associated with SBP levels in participants with prehypertension; however, only SO_2 showed a significant association in hypertensive patients. Similarly, significant associations of PM_{10}, SO_2, and O_3 with DBP were found in participants with prehypertension, but not in people with hypertension. Further sensitivity analyses by excluding the hypertensive participants on the anti-hypertensive treatment showed that the estimated changes in SBP and DBP were not materially changed. These results indicate that in comparison to hypertension, exposure to long-term ambient air pollutants may have greater effects on the development of prehypertension and that the effects on the increases in arterial BPs may be more evident among individuals with prehypertension. Authors also further analyzed the data to test whether overweight/obesity modifies the association between long-term exposure to ambient air pollution and prehypertension among Chinese adults. The associations of PM_{10}, SO_2, NO_2, and O_3 with prehypertension were significant among overweight/obese subjects (odds ratios (ORs): 1.61–1.90), but not among normal-weight subjects (ORs: 0.97–1.06). PM_{10}, SO_2, and O_3 were significantly associated with systolic blood pressure (SBP), and magnitudes of these associations were higher among overweight/obese adults (increases in SBP per interquartile range (IQR) of air pollutants: 2.24–3.48 mmHg) than those among normal-weight adults (increases in SBP per IQR of air pollutants: 0.16–0.37 mmHg). For diastolic blood pressure (DBP), significant associations were only obtained in overweight/obese participants (increases in DBP per IQR of air pollutants: 0.92–1.27 mmHg). Further stratified analyses showed that all these interactions were stronger in females and older individuals.

6.4 Ambient Air Pollution and Metabolic Disease

Experimental evidence has revealed that ambient particle ($PM_{2.5}$) exposure is capable of inducing metabolic abnormalities and obesity in nonhuman subjects (rodents) [78]. In a rodent model [66], we observed that breathing Beijing's highly polluted air resulted in weight gain and cardiorespiratory and metabolic dysfunction. Compared to rats exposed to filtered air, those exposed to unfiltered Beijing air were significantly heavier at the end of pregnancy. At 8 weeks old, the offspring prenatally and postnatally exposed to unfiltered air were heavier than those exposed to filtered air were. Among both rat dams and their offspring, following continuous exposure to unfiltered air, we observed pronounced histologic indication for both perivascular and peribronchial inflammation in the lungs, dyslipidemia, elevated tissue and systemic oxidative stress, and an enhanced proinflammatory status of epididymal fat. Findings suggest that TLR2-/TLR4-dependent inflammatory activation and lipid oxidation in the lung can spill over systemically, resulting in metabolic dysfunction and weight gain.

The 33 communities study reported that, after adjusting for confounding factors, significant associations were observed between concentrations of air pollutants and prevalence of obesity. When analysis was stratified by sex, the associations remained significant only in women. Among women, an increased prevalence of obesity was associated with an interquartile range increase in PM_{10} (19 $\mu g/m^3$; ORs = 1.18; 95%

CI: 1.06–1.32), NO_2 (9 μg/m^3; ORs = 1.24; 95% CI: 1.09–1.41), and O_3 (22 μg/m^3; ORs = 1.14; 95% CI: 1.01–1.30). Additionally, the associations were strongest among older participants [32]. Based on the same population, the authors also evaluated the ambient air pollution with diabetes prevalence, and glucose homeostasis markers including fasting and 2-h glucose and insulin levels, the homeostasis model of assessment for insulin resistance index (HOMA-IR), and the homeostasis model of assessment for beta-cell function (HOMA-B) were used as indicators of glucose homeostasis. An interquartile range increase in PM_{10}, SO_2, NO_2, and O_3 corresponds to a 1.20- (95% confidence interval (CI): 1.12–1.28), 1.12 (95% CI: 1.04–1.21), 1.22- (95% CI: 1.12–1.33), and 1.14-fold (95% CI: 1.05–1.25) increased risk of diabetes mellitus, respectively. The corresponding changes in fasting glucose, 2-h glucose, fasting insulin, 2-h insulin, HOMA-IR, and HOMA-B were 0.04–0.09 mmol/L, 0.13–0.19 mmol/L, 0.16–0.54 μU/L, 1.43–2.74 μU/L, 0.05–0.24, and −8.11 to −3.62, respectively, for an interquartile range increase of the pollutants. Stratified analyses revealed that the effects were greater in males and the young.

Sun et al. [59] used a Poisson time-series model to examine the association between 17 chemical components of PM_{10} and daily emergency hospital admissions for T2DM among residents aged 65 years or above from January 1998 to December 2007 in Hong Kong. In the single-pollutant models, PM_{10} components associated with T2DM admissions include elemental carbon, organic carbon, nitrate, and nickel. The ER% estimates per IQR increment at lag 0–1 for these four components were 3.79% (1.63, 5.95), 3.74% (0.83, 6.64), 4.58% (2.17, 6.99), and 1.91% (0.43, 3.38), respectively. In Wuhan city, Xiao et al. [77] explored the short- and long-term effects of air pollution on eight blood lipid markers among elderly hypertension patients with or without type 2 diabetes (T2D). The authors observed that in an exposure time of 7 days or longer, blood lipid markers were slightly affected by poor air quality. However, the results were unable to predict whether atherosclerosis would be promoted or inhibited by poorer air condition. Changes in blood lipid markers of hypertension among patients with or without T2D were near the same, but no blood lipid markers had an opposite trend between the two populations. Air quality index was associated with changes to blood lipid markers to some degree in a population of hypertension among patients with or without T2D.

6.5 Ambient Air Pollution and Infection Diseases

To date, most studies have focused on the effects of PM on cardiovascular and respiratory disease [1, 2]. However, studies in recent years have also reported associations between PM and infectious diseases (e.g., influenza, hemorrhagic fever with renal syndrome, and hand, foot, and mouth disease)[16, 22, 25]. For example, a study from Korea was performed to investigate the relationship between the incidence of national notifiable infectious diseases (NNIDs) and meteorological factors, air pollution levels, and hospital resources [25], and results showed that of the 34 NNIDs, malaria showed the most significant correlation with concentration of nitrogen dioxide ($r = -0.884, p < 0.01$). Inhalation transports PM deep into the lung and virus attached to particles may invade the lower part of respiratory tract directly and thus enhance the induction of infections [53].

$PM_{2.5}$ is easily absorbed into the lungs and is considered to be the primary cause of the recent hazy weather in China. Researchers have explained that $PM_{2.5}$ contains toxic chemicals and microparticles that attach to pathogenic microorganisms. When people are exposed to haze, $PM_{2.5}$ can penetrate deeply into the lungs through the respiratory tract due to its small size, thus decreasing the resistance of lung tissue to pathogenic microorganisms and thereby increasing the risk of respiratory disease [15]. Dr. Hongzhou Lu (deputy director of the Shanghai Public Health Clinical Center for H7N9 bird flu) has declared that the hazy conditions increase the risk of respiratory infection, including H7N9, but do not directly result in the proliferation of viruses [44]. Likewise, data from the Chinese Center for Disease Control and Prevention did not show severe acute respiratory syndrome or influenza virus outbreaks in the Chinese population during 2013, the worst year of haze in more than half a century in China [44]. Despite the fact that haze-associated respiratory infection is not as serious as severe acute respiratory syndrome, a wide range, long duration, and high frequency of haze occurrence, coupled with its long-term effects, will bring new challenges to the Chinese medical community in the future.

During the severe smog event in Beijing, Cao et al. [4] employed metagenomic methods to analyze the microbial composition of Beijing's PM pollutants, and the results showed that with sufficient sequencing depth, airborne microbes including bacteria, fungi, archaea, and dsDNA viruses could be identified at the species level. This suggests that most of the inhalable microorganisms were soil associated and nonpathogenic to human. However, the sequences of several respiratory microbial allergens and pathogens were identified, and their relative abundance appeared to have grown with increased concentrations of PM pollution. A recent study examined the relationships between short-term exposure to ambient particles with aerodynamic diameter ≤ 2.5 μm ($PM_{2.5}$) and measles incidence in China [8]. Data on daily numbers of new measles cases and concentrations of ambient $PM_{2.5}$ were collected from 21 cities in China during October 2013 and December 2014. Poisson regression was used to examine city-specific associations of $PM_{2.5}$ and measles, with a constrained distributed lag model, after adjusting for seasonality, day of the week, and weather conditions. A 10 μg/m^3 increase in $PM_{2.5}$ at lag 1 day, lag 2 day, and lag 3 day was significantly associated with increased measles incidence [relative risk (RR) and 95% confidence interval (CI) were 1.010 (1.003, 1.018), 1.010 (1.003, 1.016), and 1.006 (1.000, 1.012), respectively]. The cumulative relative risk of measles associated with $PM_{2.5}$ at lag 1–3 days was 1.029 (95% CI: 1.010, 1.048). Stratified analyses by meteorological factors showed that the $PM_{2.5}$ and measles associations were stronger on days with high temperature, low humidity, and high wind speed. This study provides new evidence that measles incidence is associated with exposure to ambient $PM_{2.5}$ in China.

Ye et al. [81] investigated whether the rotavirus infection rate in children is associated with temperature and air pollutants in Hangzhou, China. This study applied a distributed lag nonlinear model (DLNM) to assess the effects of daily meteorological data and air pollutants on the rotavirus-positive rate among outpatient children. The average concentrations of $PM_{2.5}$ was 53.9 μg/m^3 and PM_{10} was 80.6 μg/m^3 in Hangzhou in 2015. The highest RR of rotavirus infection occurred at lag 1–1.5 days after the increase in the concentration of these pollutants, and the RR increased gradually with the increase in concentration, and the cumulative RR caused by the cumulative effect was 2.5 and 2.2, respectively. Huang L et al. [23] performed a study in China to quantify the effects of air pollution on influenza-like illness in Nanjing, China, from January 1, 2013 to December 31, 2013. The average concentrations of $PM_{2.5}$, PM_{10}, and NO_2 were 77.37 μg/m^3, 135.20 μg/m^3, and 55.80 μg/m^3, respectively. An interquartile range increase in $PM_{2.5}$ concentration was associated with a 2.99% (95% confidence interval (CI): 1.64%, 4.36%) increase in daily influenza-like cases on the same day. However, the corresponding increase in NO_2 was associated with a 3.77% (95% CI: 2.01%, 5.56%) rise in daily cases. People ages 0–4 years were found to be significantly vulnerable to PM_{10} and NO_2; 5–14 years of age were significantly susceptible to $PM_{2.5}$ and PM_{10}; and 15–24 years of age were significantly susceptible to all the analyzed air pollutants. Air pollution effects tended to be null or negative for patients over the age of 25 years, which might be due to the small number of influenza-like cases in this age group.

Feng et al. [16] has quantified the impact of $PM_{2.5}$ on the risk of influenza-like illness (ILI) in Beijing China. Daily $PM_{2.5}$, meteorological factors, and influenza-like illness (ILI) counts during January 1, 2008, to December 31, 2014, were retrieved. Findings suggest a strong positive relationship between $PM_{2.5}$ and ILI risk during the flu season (October to April) (p-value < 0.001), after adjusting for the effects of ambient daily temperature and humidity, month, and year, whereas no significant association was identified during non-flu season (May to September) (p-value $= 0.174$). A short-term delayed effect of $PM_{2.5}$ was also identified with 2-day moving average (current day to the previous day) of $PM_{2.5}$ yielding the greatest predictive power. Moreover, $PM_{2.5}$ was strongly associated with ILI risk across all age groups (p-value < 0.001) during flu season, where the effects was most pronounced among adults (age 25–59), followed by young adults (age 15–24), school children (age 5–14), and elderly (age 60+). The effect of $PM_{2.5}$ was least pronounced for children under 5 years of age (age <5). In contrast, Huang R et al. [24] evaluated the impacts of temperature, humidity, and air pollution, particularly levels of PM_{10}, on the incidence of hand, foot, and mouth diseases (HFMD) in a city in Eastern China in the period from January 2012 to December 2014. PM_{10} level showed no relationship to the incidence of HFMD.

The increasing number of cases of respiratory infection, especially skyrocketing cases of human infection with avian influenza A H7N9 virus during January 2014, forced the Chinese Center for Disease Control and Prevention as well as medical institutions to prepare to actively meet the challenge [44]. The National Health and Family Planning Committee plans to collect firsthand data through the establishment of the National Haze Health Effects Monitoring Network to find strategies and administrate policy in the future. The Environmental Protection Department is also actively setting more stringent industry standards to control the sources of haze.

Although masks can reduce inhalation of harmful substances and the incidence of haze-related respiratory diseases, haze governance and improving air quality are effective means to reduce the attack rate of haze-related respiratory diseases.

6.6 Summary

We describe the associated health effects of ambient air pollution in Chinese. According to our knowledge, this is the first comprehensive review of the health effects of ambient air pollution in China from morbidity perspectives, and we hope this paper will provide evidence for the further development and implementation of ambient air pollution policies in China. Knowledge of the human health effects of this activity would offer a valuable rationale for policy makers to implement community-based intervention approaches to reduce personal and population exposure.

Although the exposure-response relationship has been build based on lots population-based investigation, much remains unclear regarding which pollutants account for most of the health effects and specific signaling pathways responsible for eliciting exacerbation of chronic diseases. How coarse particles from smoky coal interact with major pollutants in eliciting oncogenesis warrants investigation, as do the mechanisms underlying induction of exacerbation by components such as heavy metals, PAHs, aldehydes, and other chemicals originating from traffic and coal burning.

China is a vast geographical territory with contrasting climatic characteristics that influence the components and concentrations of air pollutants. Exploration of heterogeneous adverse effects of air pollution would benefit health-care providers in individual regions. Stricter regulations are needed to improve air quality in highly polluted regions such as Northeast China. Evaluation of health responses to air pollution in northern versus southern China, and in Han Chinese versus other ethnic groups, would unravel the mechanisms and predictors of differential adverse effects in different geographic regions and ethnic groups. Most available data have been derived from studies in developed countries where upper limits of pollution have been relatively low (50 $\mu g/m^3$ for PM_{10}) compared with Chinese levels and exclusively in white populations. Data examining dose-response relationships at high pollutant levels would be valuable for policy makers to tailor environmental protection regulations in individual regions. Well-designed studies should determine whether extreme air pollution is oncogenic or solely irritating to the airways. After all, economic development should not be at the expense of people's health.

Acknowledgments This study was funded by grants from the Major Program of National Natural Science Foundation of China (91543208), the National Natural Science Foundation of China (No. 81472936, 81673127, and 81673128), the National Key Research and Development Program of China (2016YFC0207000), the Guangdong Province Natural Science Foundation (2014A050503027), and the Fundamental Research Funds for the Central Universities (No. 16ykzd02, 17ykpy16).

References

1. Arnold C. Disease burdens associated with $PM_{2.5}$ exposure: how a new model provided global estimates. Environ Health Perspect. 2014;122(4):A111.
2. Burnett RT1, Pope CA 3rd, Ezzati M, Olives C, Lim SS, Mehta S, Shin HH, Singh G, Hubbell B, Brauer M, Anderson HR, Smith KR, Balmes JR, Bruce NG, Kan H, Laden F, Prüss-Ustün A, Turner MC, Gapstur SM, Diver WR, Cohen A. An integrated risk function for estimating the global burden of disease attributable to ambient fine particulate matter exposure. Environ Health Perspect. 2014;122(4):397–403.
3. Brauer M, Freedman G, Frostad J, van Donkelaar A, Martin RV, Dentener F, van Dingenen R, Estep K, Amini H, Apte JS, Balakrishnan K, Barregard L, Broday D, Feigin V, Ghosh S, Hopke PK, Knibbs LD, Kokubo Y, Liu Y, Ma S, Morawska L, Sangrador JL, Shaddick G, Anderson HR, Vos T, Forouzanfar MH, Burnett RT, Cohen A. Ambient air pollution exposure estimation for the global burden of disease 2013. Environ SciTechnol. 2016;50(1):79–88.
4. Cao C, Jiang W, Wang B, Fang J, Lang J, Tian G, Jiang J, Zhu TF. Inhalable microorganisms in Beijing's PM2.5 and PM10 pollutants during a severe smog event. Environ SciTechnol. 2014;48(3):1499–507.
5. Cassee FR, Héroux ME, Gerlofs-Nijland ME, Kelly FJ. Particulate matter beyond mass: recent health evidence on the role of fractions, chemical constituents and sources of emission. Inhal Toxicol. 2013;25:802–12.
6. Chen J, Zhao Q, Liu BB, Wang J, Xu HB, Zhang Y, Song XM, He B, Huang W. Airway oxidative stress and inflammation markers in chronic obstructive pulmonary diseases (COPD) patients are linked with exposure to traffic-related air pollution: a panel study. Zhonghua Yu Fang Yi XueZaZhi. 2016;50(5):411–7.
7. Chen S, Gu Y, Qiao L, Wang C, Song Y, Bai C, Sun Y, Ji H, Zhou M, Wang H, Chen R, Kan H. Fine particulate constituents and lung dysfunction: a time-series panel study. Environ SciTechnol. 2017;51(3):1687–94.
8. Chen G, Zhang W, Li S, Williams G, Liu C, Morgan GG, Jaakkola JJ, Guo Y. Is short-term exposure to ambient fine particles associated with measles incidence in China? A multi-city study. Environ Res. 2017;156:306–11.
9. Cheng MH, Chiu HF, Yang CY. Coarse particulate air pollution associated with increased risk of hospital admissions for respiratory diseases in a Tropical City, Kaohsiung, Taiwan. Int J Environ Res Public Health. 2015;12(10):13053–68.
10. Cohen AJ, Brauer M, Burnett R, Anderson HR, Frostad J, Estep K, Balakrishnan K, Brunekreef B, Dandona L, Dandona R, Feigin V, Freedman G, Hubbell B, Jobling A, Kan H, Knibbs L, Liu Y, Martin R, Morawska L, Pope CA 3rd, Shin H, Straif K, Shaddick G, Thomas M, van Dingenen R, van Donkelaar A, Vos T, Murray CJ, Forouzanfar MH. Estimates and 25-year trends of the global burden of disease attributable to ambient air pollution: an analysis of data from the Global Burden of Diseases Study 2015. Lancet. 2017; doi:10.1016/S0140-6736(17)30505-6.
11. Dockery DW, Pope CA 3rd, Xu X, Spengler JD, Ware JH, Fay ME, Ferris BG Jr, Speizer FE. An association between air pollution and mortality in six U.S. cities. N Engl J Med. 1993;329(24):1753–9.
12. Dong GH, Zhang P, Sun B, Zhang L, Chen X, Ma N, Yu F, Guo H, Huang H, Lee YL, Tang N, Chen J. Long-term exposure to ambient air pollution and respiratory disease mortality in Shenyang, China: a 12-year population-based retrospective cohort study. Respiration. 2012;84(5):360–8.
13. Dong GH, Qian Z, Wang J, Chen W, Ma W, Trevathan E, Xaverius PK, DeClue R, Wiese A, Langston M, Liu MM, Wang D, Ren WH. Association between ambient air pollution and prevalence of stroke and cardiovascular diseases in 33 Chinese communities. Atmos Environ. 2013;77:968–73.
14. Dong GH, Qian ZM, Xaverius PK, Trevathan E, Maalouf S, Parker J, Yang L, Liu MM, Wang D, Ren WH, Ma W, Wang J, Zelicoff A, Fu Q, Simckes M. Association between long-

term air pollution and increased blood pressure and hypertension in China. Hypertension. 2013;61(3):578–84.
15. Emmanuel SC. Impact to lung health of haze from forest fires: the Singapore experience. Respirology. 2000;5:175–82.
16. Feng C, Li J, Sun W, Zhang Y, Wang Q. Impact of ambient fine particulate matter (PM2.5) exposure on the risk of influenza-like-illness: a time-series analysis in Beijing, China. Environ Health. 2016;15:17.
17. GBD 2015 Risk Factors Collaborators. Global, regional, and national comparative risk assessment of 79 behavioural, environmental and occupational, and metabolic risks or clusters of risks, 1990–2015: a systematic analysis for the global burden of disease study 2015. Lancet. 2016;388:1659–724.
18. Go AS, Mozaffarian D, Roger VL, Benjamin EJ, Berry JD, Borden WB, Bravata DM, Dai S, Ford ES, Fox CS, Franco S, Fullerton HJ, Gillespie C, Hailpern SM, Heit JA, Howard VJ, Huffman MD, Kissela BM, Kittner SJ, Lackland DT, Lichtman JH, Lisabeth LD, Magid D, Marcus GM, Marelli A, Matchar DB, DK MG, Mohler ER, Moy CS, Mussolino ME, Nichol G, Paynter NP, Schreiner PJ, Sorlie PD, Stein J, Turan TN, Virani SS, Wong ND, Woo D, Turner MB. American Heart Association statistics C, stroke statistics S. Executive summary: heart disease and stroke statistics—2013 update: a report from the American Heart Association. Circulation. 2013;127(1):143–52.
19. Guan WJ, Zheng XY, Chung K, Zhong NS. Impact of air pollution on the burden of chronic respiratory diseases in China: time for urgent action. Lancet. 2016;388(10054):1939–51.
20. Harrison RM, Bousiotis D, Mohorjy AM, Alkhalaf AK, Shamy M, Alghamdi M, Khoder M, Costa M. Health risk associated with airborne particulate matter and its components in Jeddah, Saudi Arabia. Sci Total Environ. 2017;590–591:531–9.
21. Hettige HMM. Industrial pollution in economic development: the environmental Kuznets curve revisited. J Dev Econ. 2000;62(2):445–76.
22. Huang J, Deng F, Wu S, Zhao Y, Shima M, Guo B, Liu Q, Guo X. Acute effects on pulmonary function in young healthy adults exposed to traffic-related air pollution in semi-closed transport hub in Beijing. Environ Health Prev Med. 2016;21(5):312–20.
23. Huang L, Zhou L, Chen J, Chen K, Liu Y, Chen X, Tang F. Acute effects of air pollution on influenza-like illness in Nanjing, China: a population-based study. Chemosphere. 2016;147:180–7.
24. Huang R, Bian G, He T, Chen L, Xu G. Effects of meteorological parameters and PM10 on the incidence of hand, foot, and mouth disease in children in China. Int J Environ Res Public Health 2016; 13(5).
25. Jang JH, Lee JH, Je MK, Cho MJ, Bae YM, Son HS, Ahn I. Correlations between the incidence of national notifiable infectious diseases and public OpenData, including meteorological factors and medical facility resources. J Prev Med Public Health. 2015;48(4):203–15.
26. Kan H, Chen B. A case-crossover analysis for air pollution and daily mortality in Shanghai. J Occup Health. 2003;45:119–24.
27. Ko FW, Tam W, Wong TW, Chan DP, Tung AH, Lai CK, Hui DS. Temporal relationship between air pollutants and hospital admissions for chronic obstructive pulmonary disease in Hong Kong. Thorax. 2007;62:780–5.
28. Kuehn BM. WHO: more than 7 million air pollution deaths each year. JAMA. 2014;311(15):1486.
29. Lawes CM, Hoorn SV, Rodgers A. Global burden of blood-pressure-related disease, 2001. Lancet. 2008;371:1513–8.
30. Leitte AM, Schlink U, Herbarth O, Wiedensohler A, Pan XC, Hu M, Richter M, Wehner B, Tuch T, Wu Z, Yang M, Liu L, Breitner S, Cyrys J, Peters A, Wichmann HE, Franck U. Size-segregated particle number concentrations and respiratory emergency room visits in Beijing, China. Environ Health Perspect. 2011;119(4):508–13.

31. Li MH, Fan LC, Mao B, Yang JW, Choi AM, Cao WJ, Xu JF. Short-term exposure to ambient fine particulate matter increases hospitalizations and mortality in COPD: a systematic review and meta-analysis. Chest. 2016;149(2):447–58.
32. Li M, Qian Z, Vaughn M, Boutwell B, Ward P, Lu T, Lin S, Zhao Y, Zeng XW, Liu RQ, Qin XD, Zhu Y, Chen W, Dong GH. Sex-specific difference of the association between ambient air pollution and the prevalence of obesity in Chinese adults from a high pollution range area: 33 communities Chinese health study. Atmos Environ. 2015;117:227–33.
33. Lin H, Guo Y, Zheng Y, Di Q, Liu T, Xiao J, Li X, Zeng W, Cummings-Vaughn LA, Howard SW, Vaughn MG, Qian ZM, Ma W, Wu F. Long-term effects of ambient PM2.5 on hypertension and blood pressure and attributable risk among older Chinese adults. Hypertension. 2017;69(5):806–12.
34. Lin H, Guo Y, Di Q, Zheng Y, Kowal P, Xiao J, Liu T, Li X, Zeng W, Howard SW, Nelson EJ, Qian Z, Ma W, Wu F. Ambient PM2.5 and stroke: effect modifiers and population attributable risk in six low- and middle-income countries. Stroke. 2017;48(5):1191–7.
35. Liu C, Chen R, Zhao Y, Ma Z, Bi J, Liu Y, Meng X, Wang Y, Chen X, Li W, Kan H. Associations between ambient fine particulate air pollution and hypertension: a nationwide cross-sectional study in China. Sci Total Environ. 2017;584–585:869–74.
36. Liu L, Breitner S, Schneider A, Cyrys J, Brüske I, Franck U, Schlink U, Marian Leitte A, Herbarth O, Wiedensohler A, Wehner B, Pan X, Wichmann HE, Peters A. Size-fractioned particulate air pollution and cardiovascular emergency room visits in Beijing, China. Environ Res. 2013;121:52–63.
37. Liu ST, Liao CY, Kuo CY, Kuo HW. The effects of PM2.5 from Asian dust storms on emergency room visits for cardiovascular and respiratory diseases. Int J Environ Res Public Health 2017; 14(4).
38. Mathers CD, Sadana R, Salomon JA, Murray CJ, Lopez AD. Healthy life expectancy in 191 countries, 1999. Lancet. 2001;357(9269):1685–91.
39. Ma Y, Zhang H, Zhao Y, Zhou J, Yang S, Zheng X, Wang S. Short-term effects of air pollution on daily hospital admissions for cardiovascular diseases in western China. Environ SciPollut Res Int. 2017; doi:10.1007/s11356-017-8971-z.
40. Mendis S, Puska P, Norrving B. Global Atlas on cardiovascular disease prevention and control. Published by the World Health Organization in collaboration with the World Heart Federation and the World Strike Organization, 2011. ISBN 978 92 4 156437 3.
41. Nemmar A, Vanbilloen H, Hoylaerts MF, Hoet PH, Verbruggen A, Nemery B. Passage of intratracheally instilled ultrafine particles from the lung into the systemic circulation in hamster. Am J Respir Crit Care Med. 2001;164(9):1665–8.
42. Nemmar A, Hoet PH, Vanquickenborne B, Dinsdale D, Thomeer M, Hoylaerts MF, Vanbilloen H, Mortelmans L, Nemery B. Passage of inhaled particles into the blood circulation in humans. Circulation. 2002;105(4):411–4.
43. Ni Y, Wu S, Ji W, Chen Y, Zhao B, Shi S, Tu X, Li H, Pan L, Deng F, Guo X. The exposure metric choices have significant impact on the association between short-term exposure to outdoor particulate matter and changes in lung function: findings from a panel study in chronic obstructive pulmonary disease patients. Sci Total Environ. 2016;542(Pt A):264–70.
44. Pan Q, Yu Y, Tang Z, Xi M, Zang G. Haze, a hotbed of respiratory-associated infectious diseases, and a new challenge for disease control and prevention in China. Am J Infect Control. 2014;42(6):688.
45. Peng CQ, Cai JF, Yu SY, Cao ZJ, Liao YX, Liu N, He L, Zhang L, Zheng J, Shi XM, Cheng JQ. Impact of PM2.5 on daily outpatient numbers for respiratory diseases in Shenzhen, China. Zhonghua Yu Fang Yi XueZaZhi. 2016;50(10):874–9.
46. Ping J. Influence of hazy weather on patient presentation with respiratory diseases in Beijing, China. Asian Pac J Cancer Prev. 2015;16(2):607–11.
47. Pope CA 3rd, Dockery DW. Air pollution and life expectancy in China and beyond. Proc Natl Acad Sci U S A. 2013;110(32):12861–2.

48. Prabhakaran D, Roy A, Praveen PA, Ramakrishnan L, Gupta R, Amarchand R, Kondal D, Singh K, Sharma M, Shukla DK, Tandon N, Reddy KS, Krishnan A. 20-year trend of cardiovascular disease risk factors: urban and rural National Capital Region of Delhi. India Glob Heart. 2017; doi:10.1016/j.gheart.2016.11.004.
49. Qin XD, Qian Z, Vaughn MG, Trevathan E, Emo B, Paul G, Ren WH, Hao YT, Dong GH. Gender-specific differences of interaction between obesity and air pollution on stroke and cardiovascular diseases in Chinese adults from a high pollution range area: a large population based cross sectional study. Sci Total Environ. 2015;529:243–8.
50. Qiu H, Yu IT, Tian L, Wang X, Tse LA, Tam W, Wong TW. Effects of coarse particulate matter on emergency hospital admissions for respiratory diseases: a time-series analysis in Hong Kong. Environ Health Perspect. 2012;120(4):572–6.
51. Qiu H, Yu IT, Wang X, Tian L, Tse LA, Wong TW. Cool and dry weather enhances the effects of air pollution on emergency IHD hospital admissions. Int J Cardiol. 2013;168(1):500–5.
52. McDonald RA. Health effects of carbon-containing particulate matter: focus on sources and recent research program results. Crit Rev Toxicol. 2016;46(2):97–137.
53. Sedlmaier N, Hoppenheidt K, Krist H, Lehmann S, Lang H, Buttner M. Generation of avian influenza virus (AIV) contaminated fecal fine particulate matter (PM2.5): genome and infectivity detection and calculation of immission. Vet Microbiol. 2009;139(1–2):156–64.
54. Sheppard L, Burnett RT, Szpiro AA, Kim SY, Jerrett M, Pope CA, Brunekreef B. Confounding and exposure measurement error in air pollution epidemiology. Air Qual Atmos Health. 2012;5(2):203–16.
55. Shi J, Chen R, Yang C, Lin Z, Cai J, Xia Y, Wang C, Li H, Johnson N, Xu X, Zhao Z, Kan H. Association between fine particulate matter chemical constituents and airway inflammation: a panel study among healthy adults in China. Environ Res. 2016;150:264–8.
56. Stone AC, Nici L. Other systemic manifestations of chronic obstructive pulmonary disease. Clin Chest Med. 2007;28(3):553–7.
57. Su C, Breitner S, Schneider A, Liu L, Franck U, Peters A, Pan X. Short-term effects of fine particulate air pollution on cardiovascular emergency room visits: a time-series study in Beijing. China Int Arch Occup Environ Health. 2016;89(4):641–57.
58. Sun Q, Hong X, Wold LE. Cardiovascular effects of ambient particulate air pollution exposure. Circulation. 2010;121(25):2755–65.
59. Sun S, Qiu H, Ho KF, Tian L. Chemical components of respirable particulate matter associated with emergency hospital admissions for type 2 diabetes mellitus in Hong Kong. Environ Int. 2016;97:93–9.
60. Tam WW, Wong TW, Wong AH, Hui DS. Effect of dust storm events on daily emergency admissions for respiratory diseases. Respirology. 2012;17(1):143–8.
61. Tian L, Ho KF, Wang T, Qiu H, Pun VC, Chan CS, Louie PK, Yu IT. Ambient carbon monoxide and the risk of hospitalization due to chronic obstructive pulmonary disease. Am J Epidemiol. 2014;180:1159–67.
62. Tian L, Qiu H, Pun VC, Ho KF, Chan CS, Yu IT. Carbon monoxide and stroke: a time series study of ambient air pollution and emergency hospitalizations. Int J Cardiol. 2015;201:4–9.
63. Tao Y, Huang W, Huang X, Zhong L, Lu SE, Li Y, Dai L, Zhang Y, Zhu T. Estimated acute effects of ambient ozone and nitrogen dioxide on mortality in the Pearl River Delta of southern China. Environ Health Perspect. 2012;120:393–8.
64. U. S. EPA. Guidelines for exposure assessment. Washington, DC. 1992.
65. Vasilatou V, Diapouli E, Abatzoglou D, Bakeas EB, Scoullos M, Eleftheriadis K. Characterization of PM2.5 chemical composition at the Demokritos suburban station, in Athens Greece. The influence of Saharan dust. Environ SciPollut Res Int. 2017;24(12):11836–46.
66. Wei Y, Zhang JJ, Li Z, Gow A, Chung KF, Hu M, Sun Z, Zeng L, Zhu T, Jia G, Li X, Duarte M, Tang X. Chronic exposure to air pollution particles increases the risk of obesity and metabolic syndrome: findings from a natural experiment in Beijing. FASEB J. 2016;30(6):2115–22.

67. Wu S, Deng F, Huang J, Wang X, Qin Y, Zheng C, Wei H, Shima M, Guo X. Does ambient temperature interact with air pollution to alter blood pressure? A repeated-measure study in healthy adults. J Hypertens. 2015;33(12):2414–21.
68. Wu S, Deng F, Wei H, Huang J, Wang X, Hao Y, Zheng C, Qin Y, Lv H, Shima M, Guo X. Association of cardiopulmonary health effects with source-appointed ambient fine particulate in Beijing, China: a combined analysis from the Healthy Volunteer Natural Relocation (HVNR) study. Environ SciTechnol. 2014;48(6):3438–48.
69. Wu S, Deng F, Hao Y, Shima M, Wang X, Zheng C, Wei H, Lv H, Lu X, Huang J, Qin Y, Guo X. Chemical constituents of fine particulate air pollution and pulmonary function in healthy adults: the Healthy Volunteer Natural Relocation study. J Hazard Mater. 2013;260:183–91.
70. Wang W, Ying Y, Wu Q, Zhang H, Na D, Xiao W. A GIS-based spatial correlation analysis for ambient air pollution and AECOPD hospitalizations in Jinan, China. Respir Med. 2015;109(3):372–8.
71. Wang X, Li G, Jin X, Mu J, Pan J, Liang F, Tian L, Chen S, Guo Q, Dong W, Pan X. Study of relationship between atmospheric fine particulate matter concentration and one grade a tertiary hospital emergency room visits during 2012 and 2013 in Beijing. Zhonghua Yu Fang Yi Xue Za Zhi. 2016;50(1):73–8.
72. World Health Organization. Global health risks: mortality and burden of disease attributable to selected major risks. Geneva: World Health Organization; 2009. (NLM classification: WA 105). http://www.who.int/
73. Wong ND, Dede J, Chow VH, Wong KS, Franklin SS. Global cardiovascular risk associated with hypertension and extent of treatment and control according to risk group. Am J Hypertens. 2012;25:561–7.
74. Wong TW, Lau TS, Yu TS, Neller A, Wong SL, Tam W, Pang SW. Air pollution and hospital admissions for respiratory and cardiovascular diseases in Hong Kong. Occup Environ Med. 1999;56(10):679–83.
75. Wu Y, Zhao F, Qian X, Xu G, He T, Shen Y, Cai Y. Relationship beween sulfur dioxide pollution and upper respiratory outpatients in Jiangbei. Ningbo Wi Sheng Yan Jiu. 2015;44(4):565–9.
76. Wu S, Ni Y, Li H, Pan L, Yang D, Baccarelli AA, Deng F, Chen Y, Shima M, Guo X. Short-term exposure to high ambient air pollution increases airway inflammation and respiratory symptoms in chronic obstructive pulmonary disease patients in Beijing, China. Environ Int. 2016;94:76–82.
77. Xiao S, Liu R, Wei Y, Feng L, Lv X, Tang F. Air pollution and blood lipid markers levels: estimating short and long-term effects on elderly hypertension inpatients complicated with or without type 2 diabetes. Environ Pollut. 2016;215:135–40.
78. Xu X, Yavar Z, Verdin M, Ying Z, Mihai G, Kampfrath T, Wang A, Zhong M, Lippmann M, Chen LC, Rajagopalan S, Sun Q. Effect of Early Particulate Air Pollution Exposure on Obesity in Mice: Role of p47phox. Arterioscler Thromb Vasc Biol. 2010;30(12):2518-2527.
79. Xu Q, Li X, Wang S, Wang C, Huang F, Gao Q, Wu L, Tao L, Guo J, Wang W, Guo X. Fine particulate air pollution and hospital emergency room visits for respiratory disease in urban areas in Beijing, China, in 2013. PLoS One. 2016;11(4):e0153099.
80. Xu Q, Wang S, Guo Y, Wang C, Huang F, Li X, Gao Q, Wu L, Tao L, Guo J, Wang W, Guo X. Acute exposure to fine particulate matter and cardiovascular hospital emergency room visits in Beijing, China. Environ Pollut. 2017;220(Pt A):317–27.
81. Ye Q, Fu JF, Mao JH, Shen HQ, Chen XJ, Shao WX, Shang SQ, Wu YF. Haze is an important medium for the spread of rotavirus. Environ Pollut. 2016;216:324–31.
82. Yin P, Wang H, Vos T, Li Y, Liu S, Liu Y, Liu J, Wang L, Naghavi M, Murray CJ, Zhou M. A Subnational Analysis of Mortality and Prevalence of COPD in China From 1990 to 2013: Findings From the Global Burden of Disease Study 2013. Chest. 2016;150(6):1269-1280.
83. ZauliSajani S, Scotto F, Lauriola P, Galassi F, Montanari A. Urban air pollution monitoring and correlation properties between fixed-site stations. J Air Waste Manag Assoc. 2004;54:1236–41.

84. Zeng Q, Ni Y, Jiang G, Li G, Pan X. The short term burden of ambient particulate matters on non-accidental mortality and years of life lost: a ten-year multi-district study in Tianjin, China. Environ Pollut. 2017;220(Pt A):713–9.
85. Zhang P, Dong G, Sun B, Zhang L, Chen X, Ma N, Yu F, Guo H, Huang H, Lee YL, Tang N, Chen J. Long-term exposure to ambient air pollution and mortality due to cardiovascular disease and cerebrovascular disease in shenyang. China PLoS ONE. 2011;6(6):e20827.
86. Zhang J, Mauzerall DL, Zhu T, Liang S, Ezzati M, Remais JV. Environmental health in China: progress towards clean air and safe water. Lancet. 2010;375:1110–9.
87. Zhang Z, Wang J, Guo M, Xiong M, Zhou Q, Li D, Shu J, Lu W, Sun D. Air quality improvement during 2010 Asian games on blood coagulability in COPD patients. Environ SciPollut Res Int. 2016;23(7):6631–8.
88. Zhao X, Sun Z, Ruan Y, Yan J, Mukherjee B, Yang F, Duan F, Sun L, Liang R, Lian H, Zhang S, Fang Q, Gu D, Brook JR, Sun Q, Brook RD, Rajagopalan S, Fan Z. Personal black carbon exposure influences ambulatory blood pressure: air pollution and cardiometabolic disease (AIRCMD-China) study. Hypertension. 2014;63(4):871–7.
89. Zhao Y, Qian ZM, Wang J, Vaughn MG, Liu YQ, Ren WH, Dong GH. Does obesity amplify the association between ambient air pollution and increased blood pressure and hypertension in adults? Findings from the 33 communities Chinese health study. Int J Cardiol. 2013;168(5):e148–50.
90. Zhao A, Chen R, Wang C, Zhao Z, Yang C, Lu J, Chen X, Kan H. Associations between size-fractionated particulate air pollution and blood pressure in a panel of type II diabetes mellitus patients. Environ Int. 2015;80:19–25.
91. Zhou M, Wang H, Zhu J, Chen W, Wang L, Liu S, Li Y, Wang L, Liu Y, Yin P, Liu J, Yu S, Tan F, Barber RM, Coates MM, Dicker D, Fraser M, González-Medina D, Hamavid H, Hao Y, Hu G, Jiang G, Kan H, Lopez AD, Phillips MR, She J, Vos T, Wan X, Xu G, Yan LL, Yu C, Zhao Y, Zheng Y, Zou X, Naghavi M, Wang Y, Murray CJ, Yang G, Liang X. Cause-specific mortality for 240 causes in China during 1990-2013: a systematic subnational analysis for the Global Burden of Disease Study 2013. Lancet. 2016;387(10015):251-72.

Chapter 7
Air Pollution and Children's Health in Chinese

Wei-Wei Lin, Zan-Xiong Chen, Min-Li Kong, Yan-Qi Xie, and Xiao-Wen Zeng

Abstract Children can be considered as a high-risk population to environmental stress since some of the organ systems including the immune system and respiratory system are still in development during childhood. During childhood, the impact of environmental pollutants exposure may program child growth and development that have long-term consequences on later health and disease risk. Considering the rapid economic growth in recent decades in China, the impact of ambient air pollution on children health causes concerns. Therefore, we aimed to review the current epidemiological evidence on the effects of air pollution exposure on adverse health outcome, including respiratory diseases, lung function, high blood pressure, cognitive disorder, and obesity in childhood in China. We believe that the findings summarized in our review demonstrate an unequivocal relationship between air pollutants exposure and Chinese children health and these results have large public health influences.

Keywords Ambient air pollution • Children • Chinese

7.1 Introduction

Children can be considered as a high-risk population to environmental stress since some of the organ systems including the immune system and respiratory system are still in development during childhood. Their physical activity, such as spending longer time outdoors, and their closer proximity to traffic exhaust emission sources

W.-W. Lin • X.-W. Zeng (✉)
Guangzhou Key Laboratory of Environmental Pollution and Health Risk Assessment, Department of Preventive Medicine, School of Public Health, Sun Yat-sen University, 74 Zhongshan 2nd Road, Yuexiu District, Guangzhou 510080, Guangdong, China
e-mail: zxw63@mail.sysu.edu.cn; zengxw123@163.com

Z.-X. Chen • M.-L. Kong • Y.-Q. Xie
Maternal and Children's Health Hospital of Maoming City, 192, Renmin South Road, Maonan District, Maoming, Guangdong Province 525000, China

G.-H. Dong (ed.), *Ambient Air Pollution and Health Impact in China*, Advances in Experimental Medicine and Biology 1017, DOI 10.1007/978-981-10-5657-4_7

compared with adults might add to their vulnerability toward hypertensive effects of air pollution. Environmental exposures may lead to an impaired organ development resulting in long-term complications and disease in later life. There are approximately 223.2 million children aged 0–14 years in China [58]. The rapid growth of population, urbanization, and industry which encouraged an enormous increase in transportation activity and energy consumption has caused serious air pollution problem in China. Despite a growing body of literature suggesting that ambient air pollution exposure may play a role in etiology of children diseases, such as asthma and lung function disorder, information on the association of air pollution exposure and children adverse health outcome in China is still limited. Therefore, this chapter aims to summarize current epidemiological evidence on the effects of air pollution exposure on adverse health outcome, such as respiratory diseases, lung function, high blood pressure, cognitive disorder, and obesity in childhood in China.

7.2 Association Between Air Pollution and Respiratory Diseases and Symptoms in Children

Respiratory diseases, such as pneumonia, bronchitis, asthma, wheeze, runny nose, and sore throat, are the most common cause of mortality and morbidity in children all over the world, especially in young children. The state of children's respiratory health is determined by the interaction of environmental factors and genetic susceptibility. A growing body of evidence has indicated that occurrence or severity of respiratory diseases is associated with the concentration of air pollutants, one of the environmental risk factors. Not only are children with low lung function more susceptible to short-term manifestations, but also poor lung development in young children has long-term clinical symptoms with implications for lifelong respiratory problems. The aim of this section is to evaluate the epidemiological studies that have addressed the hypothesis whether ambient air pollution, one of the environmental toxicants, increases the risk of respiratory diseases in children in China in the past decade (2006–2016).

7.2.1 Lung Growth and Development

The lung is a rather complex heterogeneous structure providing the body with oxygen and removing carbon dioxide. Development of the lung includes two phrases, lung growth (structure development) and lung maturation (functional development). Lung growth can be influenced by physical, environmental, hormonal, and genetic factors. Lung functional development is primarily a biochemical process and is regulated by various hormones, growth factors, and transcriptional factors [36]. In comparison with other organs, the development of the lung plays a special role. The

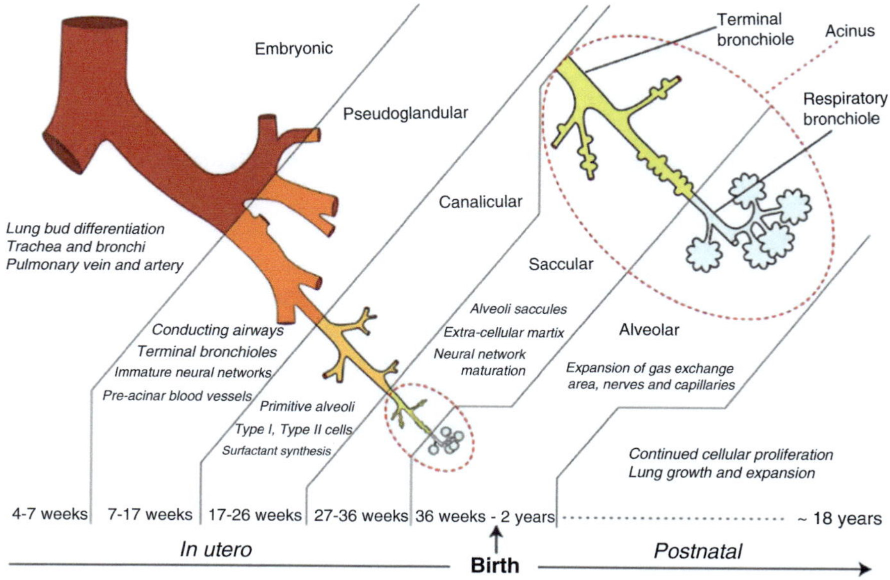

Fig. 7.1 Principal stages of lung development in humans. Diagrammatic representations of the timeline and developmental organization of the trachea, primary bronchi, intrapulmonary bronchi, and acinus in the mammalian respiratory system (Source: Kajekar [37])

milestones of development of the human lung can be divided into five stages: embryonic, pseudoglandular, canalicular, saccular, and alveolar (Fig. 7.1) [37]. During the embryonic stage, the lung begins to develop as an outgrowth of the ventral wall of the primitive foregut to form endodermal buds [37]. The respiratory diverticulum appears around 32 days after fertilization [35]. The pseudoglandular stage predominates from 7 to 17 weeks of gestation. At this stage, the entire air-conducting bronchial tree up to the terminal bronchioli are completed, but further growth and cellular differentiation continues until early adulthood [76]. Canalicular stage includes two important steps: one is the differentiation of type I and type II pneumocytes and the other is the formation of the alveolar capillary barrier [36]. The chief characteristic of this stage is the formation of the acinar structures which comprise the respiratory bronchioles, alveolar ducts, and primitive alveoli; with the detection of surfactant protein in this stage, a primary gas-exchange platform is established [36]. In the subsequent saccular stage and alveolar stage, the airways are progressive branching and finally develop to alveolar spaces capable of gas exchange, and the surfactant system reaches maturity by about 36 weeks in the last trimester. The alveolar cells remain proliferation, and their number continues to increase after birth.

The lung growth commences in the embryonic period and extends through the fetal period up to birth and even afterward. Therefore, environmental toxicants that are able to cross the placenta or interfere with the developmental program during any of the lung growth stages may influence the lung function and increase the risk of respiratory diseases in later life.

7.2.2 Air Pollutants that Influence Respiratory Health

Respiratory tract, a principal interface between air and the organism, is extremely vulnerable to the harmful effects of air pollutants. WHO estimated that air pollution kills 600,000 children under 5 years per year [77]. Not only is air pollution related to diseases that kill but also to poor health condition and high morbidity among children. Particulate matter is a mixture of solid and liquid particles suspended in air. Respiratory manifestations to PM are linked with upper airways illness and lower respiratory illness, including cough, phlegm, decreased lung function, exacerbation of asthma, increased incidence of bronchitis and respiratory infection, and increased inflammatory response in respiratory system [65, 47]. Ozone is a powerful oxidant formed by the action of sunlight on nitrogen oxides and hydrocarbons, both of which are mainly originated from industry and motor vehicles in urban areas. Once inhaled into the respiratory tract, ozone can cause peroxidation of lipids, produce reactive oxygen species and ozonation products, accelerate pulmonary inflammation, and further cause injury to the respiratory system [37]. Nitrogen dioxide is the product of the combustion of fossil fuels from motor vehicles, heating, and power generation; it is a common indoor air pollutant. Nitrogen dioxide has been linked to increase the prevalence of acute and chronic respiratory diseases, reduce lung function, and exacerbate asthma, especially in children [48].

7.2.3 Epidemiological Studies in China: Air Pollution and Respiratory Diseases

There has been an emerging literature in China indicating that air pollution including particulate matter (PM), nitrogen dioxide (NO_2), ozone (O_3), sulfur dioxide (SO_2), and black carbon has adverse effects on children's respiratory health in the past decade.

Three districts in Hong Kong with pollution levels ranging from "low" to "moderate" to "high" were studied around a 4-month period, including 2203 primary school children [24]. The authors defined the low-pollution district (LPD), moderate-pollution district (MPD), and high-pollution district (HPD) by the levels of PM_{10} over the previous 10 years in Hong Kong. The annual means for PM_{10}, NO_2, and SO_2 in the HPD in the past 12 months previous to the studied year were 8.7 μg/m^3, 12.7 μg/m^3, and 6.8 μg/m^3 higher than in the LPD, respectively, whereas the annual mean concentration of O_3 in the HPD was 10.8 μg/m^3 lower than in the LPD. Only Chinese students aged 8–10 from the elementary schools located within 1 km from the fixed air monitoring station were recruited in this cross-sectional study. Among girls, compared to those in the LPD, the odds ratios for cough at night were 1.65 (95% CI: 1.08–2.52) in the MPD and 1.81 (95% CI: 1.17–2.78) in the HPD; and the odds ratios of phlegm without colds were significantly higher in the HPD (OR = 3.84) than in the LPD. In boys, only marginal significance was reached for wheezing symptoms, phlegm without colds, and asthma in the HPD. Although this study examined chronic effects of air pollution on children's respiratory morbidities based on 10-year air pollution monitoring data, limited to the cross-sectional study design, this study could not establish the temporal order of cause and effect and was not able to separate the effects of individual exposures.

Black carbon is operationally defined as light-absorbing carbonaceous aerosol measured by an optical instrument in climate science community. In urban area, road traffic is a major source of black carbon. A time series analyses from Shanghai linked the concentration of black carbon to childhood asthma admission [30]. Daily asthma admissions from 2007 to 2012 were considered in the study. Black carbon, $PM_{2.5}$, NO_2, and SO_2 were the air pollutants evaluated. Black carbon and $PM_{2.5}$ were the main factors of interest, while NO_2 and SO_2 were treated only as potential confounders. The effects of black carbon were stronger and robust than those of $PM_{2.5}$ in both single-pollutant and multi-pollutant models. Take the single-pollutant model, for example, an inter-quartile range (2.49 $\mu g/m^3$) increase in black carbon was associated with 7% (95% CI: 5–8%) increase in asthma admission with a maximum lag of 3 days in the distributed lag model. In this study, the authors also analyzed the effects of black carbon by age. The relative risks for asthma admission in 5- to 14-year-old children were higher than those of younger children (0–4 years old). The authors attributed these differences to the differences in physiology and anatomy since asthma-like symptoms (e.g., cough and wheeze) in children under 5 eventually disappear with their ages increase. The study indicated that black carbon might serve as an additional air quality indicator in China.

A large study [44, 49] in 25 districts of 7 cities in Northeast China specifically examined the effects of outdoor and indoor air pollution on asthma and asthma-related symptoms in 23,326 6- to 13-year-old children. Outdoor air pollutants such as PM_{10}, SO_2, NO_2, and O_3 were available from municipal air pollution monitoring stations over 3 years during 2006–2008. Questionnaire-based reports of respiratory symptoms, mostly asthma and asthma-related symptoms, were collected from parents of recruited children. The study found that exposures to PM_{10}, SO_2, NO_2, and O_3 were associated with significant increases in doctor-diagnosed asthma, current asthma, persistent cough, and persistent phlegm, although current wheeze was not fully confirmed. However, the significant results for asthma and asthma-related symptoms must be considered with caution as uncontrolled confounding (e.g., temperature and relative humidity) may have occurred in this study. To examine whether gender may modify the effects of air pollution on respiratory diseases, odds ratios of respiratory illness in males and females were calculated separately, but most of the effects did not differ between genders. Nevertheless, an earlier study [15] by the same group revealed a different picture about the gender difference in the same studied population. This previous study aimed to investigate the association between exposure to PM_{10}, SO_2, NO_2, O_3, and CO and respiratory diseases and also to evaluate whether allergic predisposition and gender modified these associations. They found that boys without an allergic predisposition may be more vulnerable to air pollution, whereas girls with an allergic predisposition may be more susceptible. The authors thought the differences might be due to the gender-specific behavior and physiology and high prevalence of non-allergic type of asthma in girls.

Another study [60] in the Northeast China investigated the effects of exposures to outdoor total suspended particles (TSP), NO_2, and SO_2 on respiratory symptoms in 11,860 3- to 12-year-old children. The study considered persistent cough, persistent phlegm, current wheeze, and current asthma to be the main outcomes of interest. Six cities of Liaoning Province and three districts with different levels of air pollution within each city were selected. Outdoor levels of TSP, NO_2, and SO_2 in this region have historically been among the most polluted cities in China and often exceeded the National Standard of China and WHO guidelines during the research period [60]. Positive associations were found with the prevalence of persistent cough, persistent phlegm, and current asthma. The strength of this study is the wide concentration gradients in the air pollutants, which provided an opportunity to assess the adverse health effects of air pollution within a wide pollution range, while limitations include the cross-sectional design and unavailable data of indoor air pollution exposure.

Evaluation of the health impacts of air pollution on children requires a balanced and comprehensive risk estimate in terms of both indoor and outdoor exposures. A study [82] in the city of Taiyuan, Shanxi Province, investigated the influence of both indoor and outdoor air pollution on the asthmatic symptoms in ten elementary schools in December 2004. The outcomes, including cumulative asthma, wheeze or whistling in the chest, daytime/nocturnal attacks of breathlessness, furry pet or pollen allergy, as well as respiratory infection, were studied cross-sectionally in 1993 pupils aged 11–15. Indoor SO_2, NO_2, O_3, and formaldehyde concentrations over a 7-day continuous period were measured in selected classrooms, and a 7-day average of outdoor air pollution levels was measured by diffusion samplers. As Taiyuan city is a heavily polluted, coal-burning city, outdoor air pollution levels were two or three times higher than indoor air in the studied period. Wheeze or daytime or nocturnal attacks of breathlessness, after adjusted for several potential confounding factors, were positively associated with exposure to SO_2, NO_2, or formaldehyde. Again, it is noteworthy that the cross-sectional design has the limitation of drawing firm conclusions based on causal relationships.

In the past decade, there has been increasing evidence in China to suggest that high concentrations of air pollution have adverse effects on children's respiratory system. Children's respiratory systems are in the process of developing and growing, making them especially susceptible to high concentration of air pollution. The air pollution concentrations in those studies, especially the $PM_{2.5}$ concentration, are usually higher than WHO standards, resulting in increasing risks of chronic cough, phlegm, asthma, wheeze, and respiratory infection in children. However, most of the studies are cross-sectional or time series studies. The inherent weakness of the study design limits the possibility of drawing a convincing causal relationship.

7.3 Association Between Air Pollution and Lung Function in Children

Lung function is of interest as a marker of respiratory health in studies which examine the relationship between air pollution exposure and children's health. Spirometric measures – like forced vital capacity (FVC), peak expiratory flow (PEF), forced expiratory flow between the 25th and 75th percentile of FVC (FEF_{25-75}), and forced expiratory flow at 75% of FVC (FEF_{75}) – have been recognized as a predictor of cardiorespiratory morbidity and mortality [70]. Of additional concern, reduced lung function is characterized by significant health disparities, with low socioeconomic status children having low levels of pulmonary function [28]. A large-scale population-based cross-sectional study in Shanghai recently found strong socioeconomic disparities in lung function which widened with increasing age [22]. Although the subjects were only restricted to adult population, the study did indicate that, like many other countries, socioeconomic status was a critical risk factor for reduced lung function in China. In trying to understand the potential mechanism underlying the relationship between socioeconomic status and lung function, patterns of disparities may be informative. Environmental factors are among those patterns that contribute to disparities in lung function reduction. In 2016, based on Data Integration Model for Air Quality (DIMAQ), the World Health Organization (WHO) estimated that annual median concentration of $PM_{2.5}$ in urban area of China was 59 μg/m^3, whereas the concentration reduced to 54 μg/m^3 while including the rural areas [78]. The spatial-temporal variation of $PM_{2.5}$ was found to be correlated with socioeconomic factors; high concentrations of $PM_{2.5}$ were concentrated in regions with high populations, high gross domestic product (GDP), and rapid urban expansion areas in China [45]. People of low socioeconomic status are usually disproportionate affected by high exposures in home and neighborhood. Disparities in health outcomes from those hazardous exposures have led to extensive investigation into the relationship between environmental factors and adverse health outcomes, including lung function reduction. This section focuses on the effects of air pollution on lung function in Chinese children.

7.3.1 Identification of Evidence

In China, spirometry testing of children is widely used while evaluating the effects of ambient air pollution on children's health. In 2009, Liu and Zhang [51] collected research articles in relation to air pollution and children's lung function published from 1985 to 2006 and conducted a systematic review, focusing on FVC, FEV_1, and maximal mid-expiratory flow rate (MMEF). This review provides an interesting overview. The authors considered 11 cross-sectional studies covering 7 cities and 10,394 7- to 14-year-old children. Summary estimates of the effects of TSP, SO_2, and NOx on FVC, FEV_1, and MMEF were reported. The review indicated that FVC and EFV_1 had no effect for NOx but significant negative associations with both TSP and SO_2, whereas MMEF had a statistically significant effect only for NOx. The review suggested that TSP and SO_2 are mainly related to the decline of big airway function, while NOx is mainly related to small airway function [51]. In order to avoid an overlap with this previous review, we only considered relative studies published between 2006 and 2015 and included in Web of Science or PubMed database. Our search strategy and selection criteria included the key words "air pollution," "lung function," "children," and "China/Chinese." Evidences for the effects of air pollutants on lung function growth and decline are summarized below.

7.3.2 *Cross-Sectional Studies in Chinese Children*

Research is abundant in the children's health literature on how air pollution adversely affects lung function growth. One of the most ambitious cohort studies is the Southern California Children's Health Study (CHS) study. This study found a negative, dose-dependent association between outdoor air pollution exposure and lower levels of FEV_1 and that changes in air pollution exposure resulted in remarkable changes in lung function growth [1, 25]. As noted above, the levels of the air pollution in the CHS study were relatively low, where annual PM_{10} concentration varied from 15 to 66 μg/m^3 and annual O_3 concentration varied from 30 to 71 ppb [1]. However, children living in China have been exposed to much higher levels of air pollution. For example, annual average of PM_{10} ranged from 59 (Shenzhen) to 180 μg/m^3 (Beijing) in some mega cities of China [9]. Although there have been many important findings from western countries with low levels of air pollution, whether the impact of air pollution on lung function in children living in highly polluted areas is similar is unclear. Unresolved scientific issues dealing with the effects of higher levels of air pollution on lung function must be recognized.

Many recent cross-sectional studies have examined the relationship between long-term exposure to air pollution and lung function of children in China. Most of these studies assumed that air pollution exposure could be captured by measurements from one single station which usually monitored several pollutants. This kind of studies has several advantages: routinely collected data of multiple air pollutants and inexpensive, quick and easy to perform. As a result, the number of lung function and air pollutant parameters of interest can easily be in hundreds of thousands, leading to great statistics power to examine the association between lung function and air pollutants. Limitations of this kind of studies include difficult to identify specific causal pollutant among multiple highly correlated air pollutants, lack of personal exposure, and an assumption that air pollution concentration monitored by one single station can be generalized to all population in that area.

A cross-sectional study based on 14 communities in Taiwan [41] on children aged 12–13 years found that CO, NOx, NO_2, and NO were significantly associated with FVC and FEV_1 in children, suggesting chronic adverse effects of traffic-related air pollutants on pulmonary function. The effects of air pollutants on lung function for boys were greater than for girls. The relationship of socioeconomic status and lung function was also examined in this study – children with high parental education level were associated with better lung function [41].

A later published study [10] in Taipei city used data from a large respiratory health survey conducted in 12- to 16-year-old children (n = 2919) in five school districts. PM_{10}, SO_2, CO, NO_2, and O_3 were based on five monitoring sites in close proximity to the children's schools (within 2 km). Air pollutant concentrations of 0-day, 1-day, and 2-day before the spirometry testing were linked to the lung responses in children. The study found that short-term exposure to O_3 and PM_{10} was associated with reducing FVC and FEV_1 and CO and SO_2 exposure had a strong 1-day lag effect on FVC and FEV_1 [10]. However, the study only considered the air pollution exposure in school, and this might underestimate the real exposure level and the adverse impacts on lung function.

The effect of long-term exposure to air pollution on lung function has also been investigated in Hong Kong. Gao et al. [23] have compared FEV_1, FEF_{25-75}, and FEF_{75} of 2060 children aged 8- to 10-year-old among high, moderate, and low outdoor pollutant areas in the year 2004. The annual mean concentration of PM_{10} in the high-pollution area was nearly 8 μg/m^3 than in the other two areas. This cross-sectional study found that boys in the high-pollution district had significantly reduced lung function than those in the low- and moderate-pollution districts and the adverse effects among girls were not significant [23].

Although from these cross-sectional studies there is increasing evidence for health effects of air pollutants on children's lung growth, the age ranges of the recruited children are rather narrow, and very few studies have addressed the question of whether air pollution causes different damage to lung function of children of different ages. An important contribution to addressing this concern was made by Chen et al. [11]. They examined the association between ambient air pollution and lung function in 1494 non-asthma children of 6–15 years old in 24 communities of Taiwan. Using a pollutant-age interaction term in the regression model, the study found that sub-chronic (2 months) $PM_{2.5}$ exposure has stronger effects on lung function in smaller school children (aged 6–10 years) than those in older children (aged 11–15 years). In the full sample, sub-chronic exposure to $PM_{2.5}$ is more important than acute exposure (1-day lag) in causing adverse effects on the lung. For the health effects of O_3 exposure, acute exposure affected small airway function, while sub-chronic exposure affected lung capacity [11]. Another study including a wide range of ages of children was conducted by Zeng et al. [81] in seven northeastern cities of Liaoning Province. This study also reported significantly larger decrements in lung function among children due to long-term exposure to high concentration of air pollution. A summary of the results of cross-sectional studies on lung function is given in Table 7.1. These cross-sectional studies with community-level exposures suggest that air pollution in general affect lung capacity and lung function growth in children. However, the temporal order of cause and effect cannot be established because of the cross-sectional study design. Cohort studies with a long follow-up period are needed to provide better insight on the relationship between air pollution and lung function in children.

7.3.3 Cohort Studies in Chinese Children

Cohort studies and panel studies provide a number of advantages for examining the relationship between air pollution exposure and lung function in children. This kind of longitudinal studies with repeated measurements of children's lung function over several days, months, or years provides both acute and chronic effects of air pollution on lung function growth and offers a convincible causal relation between exposure to ambient air pollutants and lung response in children. Nevertheless, cohort studies that focused on health effects of air pollution were extremely limited. To date, only a handful of cohort studies have been conducted in Chinese children.

In four large Chinese cities of Chongqing, Guangzhou, Lanzhou, and Wuhan, Roy and colleagues assessed the relationship between ambient particulate matter exposure and lung function growth among 3273 children aged 6–12 years old [67]. The concentration range of PM in the study was 95–232 μg/m^3, that is, 2–10 times higher than other studies with similar designs which conducted in the United States [25], Mexico [66], and Austria [29]. These four Chinese cities provide a good opportunity to examine the effect of PM exposure on lung function growth at relatively high-pollution levels. The study found that long-term exposure to high levels of particulate matter was associated with deficits in lung function and decreased development of lung function and $PM_{2.5}$ had a stronger effect than PM_{10} on lung function growth [67]. Compared to children living in relatively low levels of air pollution [25, 29, 66], this study found a more modest effect of 0.7 ml/year decrease in FEV_1 growth with per 10 μg/m^3 increase in PM_{10}.

In Shenyang which is located in the northeast of China, Kasamatsu and colleagues evaluated the subacute effects of winter air pollution on lung function in children aged 8–10 years old [38]. The study included winter heating season, and the levels of TSP, PM_7, and $PM_{2.1}$ were high at the immediate end of heating season and low before the heating season. After adjusting for age and height, FVC, $FEV_{1.0}$, PEF, and FEF_{75} were significantly lower during the immediate end of heating season than those before the heating season. Over a four-time lung function measurements from 2001 to 2002, the investigators found that PM had a subacute effect on pulmonary function in children and the effects of PM_7 and $PM_{2.1}$ were stronger than TSP [38].

In Taiwan, Hwang and colleagues focused on the effects of long-term exposure to traffic-related pollutants (such as NO_2, O_3, CO, and $PM_{2.5}$) on deficits of lung function growth in children [31]. Exposure was assessed individually by relating daily averaged concentration of air pollutants to the residential address of children using the geographical information system (GIS). After adjustment for subject-specific covariates, deficits in lung function growth were significantly associated with $PM_{2.5}$ and O_3. The estimated effects were stronger for boys than girls. One of the explanations from the authors is that boys might have more extensive outdoor exposure and are more easily affected by air pollutants [31].

A summary of the results of cohort studies on lung function in Chinese children is given in Table 7.1.

Table 7.1 Studies of health effects of air pollution on lung function in Chinese children

Author/year	Location	Subject/sample size	Study design	Pollution exposure	Lung function	Main results
Kasamatsu et al. 2006 [38]	Shenyang/224 primary schools	332 primary school children, aged 8–10 year	Cohort	PM_7, $PM_{2.1}$, and TSP	FVC, $FEV_{1.0}$, PEF, and FEF_{75}	PM had a subacute effect on pulmonary function in children, and the effects of PM_7 and $PM_{2.1}$ were stronger than TSP
Lee et al. 2011 [41]	Taiwan/14 communities	3957 children aged 12–13 year	Cross-sectional	CO, NOx, NO_2, NO, O_3, SO_2, PM_{10}, and $PM_{2.5}$	FVC, FEV_1, MMEF, and PEFR	CO, NOx, NO_2, and NO had chronic adverse effects on pulmonary function in children, especially for boys
Chang et al. 2012 [10]	Taipei	2919 students, aged 12–16 year	Cross-sectional	PM_{10}, SO_2, CO, NO_2, and O_3	FVC, FEV_1/FVC, and FEV_1	Short-term exposure to O_3 and PM_{10} was associated with reducing FVC and FEV_1. CO and SO_2 exposure had a strong 1-day lag effect on FVC and FEV_1
Roy et al. 2012 [67]	Chongqing, Guangzhou, Wuhan, and Lanzhou	3273 children, aged 6–12 year, followed up from 1993 to 1996	Cohort	$PM_{2.5}$ and PM_{10}	FVC and FEV_1	Long-term exposure to high levels of particulate matter was associated with deficits in lung function and decreased development of lung function. $PM_{2.5}$ had a stronger effect than PM_{10} on lung function growth

(continued)

Table 7.1 (continued)

Author/year	Location	Subject/sample size	Study design	Pollution exposure	Lung function	Main results
Gao et al. 2013 [23]	Hong Kong/3 districts	2060 children, aged 8–10 year	Cross-sectional	PM_{10}, SO_2, NO_2, and O_3	$FEF_{25–75}$, FVC, FEV_1, and FEF_{75}	Boys in the high-pollution district had significantly reduced lung function than those in the low- and moderate-pollution districts. The adverse effects among girls were not significant
Madaniyazi et al. 2013 [53]	Baotou	107 primary school children	Panel study	$PM_{2.5}$, PM_{10}, Pb, Ni, Fe, Mn, Cr, As, Cd, and Zn in particles	PEFR	Increases of PM_{10}, $PM_{2.5}$ and Pb, Ni, Fe, Mn, and Cr in particles were associated with a decrement in the PEFR
Chen et al. 2015 [11]	Taiwan/44 schools in 24 districts	1494 non-asthma children of 6–15 year	Cross-sectional	$PM_{2.5}$, O_3, CO, SO_2, and NO_2	FVC, FEV_1, MMEF, FEV_1/FVC, and MMEF/FVC	The effects of sub-chronic exposure (2 months) to ambient $PM_{2.5}$ on lung capacity were stronger than those of acute exposure in children aged 6–15. Sub-chronic exposure to ambient $PM_{2.5}$ was associated with additional airway effect in smaller children aged 6–10

Hwang et al. 2015 [31]	Taiwan	2941 children aged 12, followed up from 2007 to 2009	Cohort	$PM_{2.5}$ and O_3	FVC, FEV_1, V_{25}, and V_{75}	Long-term exposure to $PM_{2.5}$ and O_3 may have a detrimental effect on the development of lung function in children
Zeng et al. 2016 [81]	Seven northeastern cities in Liaoning Province	6740 children aged 7–14 year	Cross-sectional	PM_{10}, O_3, O_2, and NO_2	FVC and FEV_1	Long-term exposure to high concentration of air pollution was associated with decreased lung function, especially in girls

A consistent association between long-term exposure to high level of air pollution and deficits of lung function growth and development was found in those recently published cohort studies in China. PM exposure seems to be an important pollutant surrogate in urban area, with the effects of gaseous pollutants less consistent than those of PM. However, knowledge of which PM components are responsible for lung function reduction is unclear. Future studies should implement to address the issue in order to better understand the contribution of PM components and their effects on lung function development in children.

7.4 Association Between Air Pollution and Hypertension in Children

Childhood high blood pressure (BP) is a well-established risk factor for hypertension and cardiovascular disease (CVD) in later life [40]. In the past few decades, there is mounting evidence linking air pollution and BP among adults, especially for particulate matter (PM) [4, 26]. Despite these studies among adults, little research has accrued on the linkages between air pollution and BP in children [26]. In this section, we aim to summarize evidence linking air pollution exposure (short term and long term) and BP among children in China.

7.4.1 Short-Term Exposure and Hypertension

A recent meta-analysis demonstrates that a 10 g/m^3 increase in fine particulate matter ($PM_{2.5}$, mean diameter <2.5 μm) leads to a 2.5% elevation in risk for myocardial infarctions over the following few days [57]. So far, only five published papers concerning the correlation between ambient air pollution exposure and BP in children, though limited, provide mixed results [3, 8, 64, 71]. Among these studies, only one investigation was conducted in China [80].

A cross-sectional study, the Seven Northeastern Cities (SNEC) study, was conducted from April 2012 to June 2013 [80]. Seven cities (Shenyang, Dalian, Anshan, Fushun, Benxi, Liaoyang, and Dandong) were selected based on their mean air pollution level data recorded between 2009 and 2011. A total of 9354 participants (5–17 years old) were recruited from 48 schools (24 elementary schools and 24 middle schools) in seven cities and had their BP measurements. Air pollutant level including PM_{10}, SO_2, NO_2, and O_3 at short-term exposure, 1–5 days preceding BP measurement in each district, was collected from local air monitoring station. During the study period, daily ambient PM_{10} concentrations were 108.8 μg/m^3, which exceeded the 50 μg/m^3 Grade I level set forth by the Chinese National Ambient Air Quality Standards. Results showed that there was no significant association between SO_2 and NO_2 and elevated BP in children but an association between elevated BP and an increased IQR for PM_{10} and O_3 was found. Each IQR increase in PM_{10} (ranged from 42.7 to 50.0 μg/m^3) and O_3 (ranged from 48.3 to 53.0 μg/m^3) during day 1 through day 5 of exposure was associated with an increasing of elevated BP in children, and associations were more significant for O_3 (odds ratio, ORs ranged from 2.59 to 2.85 for 1–5 day mean). Both SBP and DBP tended to increase consistently with the levels of PM_{10} averaged over the 1–5 days before BP examination. Elevated PM_{10} and O_3 were more clearly linked to increase SBP in the model. For example, in the single-pollutant model, an IQR increase in ambient PM_{10} (47.4 μg/m^3) and O_3 (51.4 μg/m^3) in the previous 5-day mean exposure predicted an increase in SBP of 2.07 mmHg (95% CI, 1.71–2.44) and 3.29 mmHg (95% CI, 2.86–3.72), respectively.

The findings from the SNEC study [80] have established consistent finding with other two studies that short-term ambient air pollutant exposure correlates with elevated BP in children [8, 71]. A cross-sectional study (n = 179) out of Pakistan found a higher SBP (115.9 mmHg, 95% CI: 114.0–117.9 mmHg) and DBP (66.4 mmHg, 95% CI: 64.4–68.4 mmHg) in 10-year-old children (8–12 years old) living in areas with heavy traffic air pollution (mean daily value of $PM_{2.5}$: 183.0 μg/m^3) than in children living in areas with less traffic air pollution (mean daily value of $PM_{2.5}$: 28.5 μg/m^3) [71]. However, conclusions from other two studies are inconsistent with results reported from Zeng et al. [80]. For example, in the PIMA study, a birth cohort consisting of 1432 children 12 years of age in the Netherlands evaluated that long-term exposure not short-term exposure may increase diastolic blood pressure in children [3]. The inconsistent findings from these five available published studies may likely be due to different methodological design in investigation, including different geographical regions (North America, Europe, and Asia), exposure concentration (low and high exposure levels), BP measurement devices (auto device and mercury sphygmomanometer), as well as population characteristics.

7.4.2 Long-Term Exposure and Hypertension

Some epidemiological evidences have conferred that higher risks occur in longer-term air pollutants exposure than short-term exposure [6]. Compared with the effect of short-term air pollution exposure on BP in children, the linkage between long-term exposure and BP in children is less unknown. In a systematic Medline search, there are only three investigations regarding the long-term effect of ambient air pollution exposure on children BP from China [16, 18], Germany [44, 49], and the Netherlands [3].

In the SNEC study, Dong et al. [16, 18] collected the level of PM_{10}, SO_2, NO_2, and O_3 from 2009 to 2012 in the air monitoring station. In the 24 districts investigated, the mean annual PM_{10} levels and SO_2 levels exceeded WHO guidelines in 100% and 95.8% of all districts, respectively. The authors observed significant associations between elevated BP and four pollutants. The OR of elevated BP was 1.68 (95% CI: 1.53–1.86), 1.33 (95% CI: 1.21–1.47), 1.12 (95% CI: 1.10–1.13), and

1.64 (95% CI: 1.48–1.82) with an IQR increasing of PM_{10} (30.6 μg/m^3), SO_2 (23.4 μg/m^3), NO_2 (13.0 μg/m^3), and O_3 (46.3 μg/m^3), respectively [16, 18].

In the two studies conducted in China [16, 18] and the Netherlands [3], results suggest that long-term exposure to PM_{10} [16, 18] and $PM_{2.5}$ [3] was associated with higher SBP in children. However, Liu et al. [44, 49] investigated associations between land-use regression modeled long-term traffic-related air pollution and BP in 2368 children aged 10 years from Germany, finding air pollution was not consistently associated with BP.

7.4.3 *Biological Mechanisms*

A pro-hypertensive response mediated by air pollution is a hypothesized pathway that links air pollution and hypertension [5]. In the most recent review, Giorgini et al. [26] proposed the possible mechanistic pathways linking air pollution and cardiovascular system. The interacting pathways include autonomic nervous system imbalance [63], endogenous proinflammatory response [2], and vascular endothelium impair [68, 69]. Mechanistic evidence supports that air pollutants can mediate increase of BP driven by increased arterial vasoconstrictor responsiveness due to endothelial dysfunction, oxidative stress, and inflammation [26]. Endothelial dysfunction may contribute to increased systemic vascular resistance, thus leading to the development of hypertension, probably due to impaired endothelium-dependent vasodilatation [13]. In addition, oxidative and inflammation play central role in the pathogenesis of endothelial dysfunction, therefore affecting hemodynamic responses [26]. Experimental studies have demonstrated that exposure to PM and O_3 promotes vascular dysfunction and oxidative stress, producing significant inflammation and injury to epithelial cells and inducing elevated atherogenesis [6, 12].

7.4.4 *Other Modifiers*

Except for the most relevant hypertension risk factors, like smoking, diabetes, BMI, and physical activity, some studies have showed that other confounding factors may modify the effect of air pollution on hypertension. Breastfeeding is associated with a lowering of later blood pressure in children [55]. In the SNEC study, Dong et al. [16, 18] showed higher ORs for hypertension in non-breastfeeding children compared with breastfeeding children, with statistically significant interactions between breastfeeding and PM_{10} ($P = 0.016$), O_3 ($P < 0.001$), CO ($P = 0.055$), and NO_2 ($P = 0.080$). If the association is causal, the wider promotion of breastfeeding is a potential component of the public health strategy to reduce the risk of childhood hypertension induced by air pollution. In the same investigated children population,

Dong et al. [19] also assessed the effect modification by obesity on the association between exposure and BP in the SNEC study. The association between air pollution exposure and hypertension was consistently larger for overweight/obese children than for children with normal weight, with ORs for hypertension ranging from 1.16 per 46.3 μg/m^3 for O_3 (95% CI, 1.12–1.20) to 2.91 per 30.6 μg/m^3 for PM_{10} (95% CI, 2.32–3.64). The estimated increases in mean systolic and diastolic BP ranged from 0.57 mmHg (95% CI, 0.36–0.78) and 0.63 mmHg (95% CI, 0.46–0.81) per 46.3 μg/m^3 for O_3 to 4.04 mmHg (95% CI, 3.00–5.09) and 2.02 mmHg (95% CI, 1.14–2.89) per 23.4 μg/m^3 for SO_2. Results indicated that obesity may be substantially associated with the susceptibility to children hypertension related to ambient air pollution.

Several studies have shown that BP tends to increase with decreasing ambient temperature in both normotensive and hypertensive populations [42]. However, the evidence of adverse impacts of ambient temperature on children's BP is rare. Li et al. [43] analyzed the association of long-term ambient temperature exposure on BP in 71,763 children from the Chinese National Survey on Students' Construction and Health (CHNSCH) in 2010. In this study, data of annual ambient temperature and air pollutants were collected in 30 cities. Results showed that low ambient temperature was negatively associated with increasing BP in children. For example, when the temperature went down from 20.4 to 9.6 °C, SBP can get the largest alteration of increasing 9.0 mmHg, suggesting that the potential adverse effect of temperature on children BP should be considered in the public health policy [43].

7.5 Association Between Air Pollution and Cognitive Function in Children

Cognitive function refers to an individual's attention, memory, language, perception, awareness, decision-making, and problem solving. It is a collective term for high-order mental action which in practice is mainly inferred from behavior [14]. Environment plays an important role in shaping cognitive development from the newborn period through adolescence. Air pollution, one of the noxious environmental factors, has been associated with increases in mortality and morbidity of respiratory and cardiovascular diseases [65].

7.5.1 Epidemiological Studies on Air Pollution and Cognitive Function

China has become known for its serious air pollution problem as a result of rapid growth of economics, industry, urbanization, population, and transportation activity, which led to a significant increase in related pollutant emissions. A key question is whether current air pollution situation in China is related with cognitive development in children.

Children rather than adults are vulnerable to toxic air pollutants because, on a body weight basis, their respiratory rate, external exposures, and time for outdoor activity are markedly higher. Studies examining cognitive function at specific points in time have assessed the effects of air pollution on neurobehavioral functions in Chinese children. Wang et al. [74] specifically sought to identify the potential effects of traffic-related air pollution in Quanzhou, Fujian Province, which is located in the southeast of China. They selected two primary schools with different levels of NO_2 and PM_{10} and traffic density: one was located in the polluted area, and the other was in the clear area. The average monthly concentrations of NO_2 and PM_{10} in the clear area were 7 μg/m^3 and 68 μg/m^3, respectively, while in the polluted area, the concentrations were 22 μg/m^3 and 80 μg/m^3. The results indicated that children living in the polluted area showed poor performance on Line Discrimination (LDT), Visual Retention (VRT), Visual Simple Reaction Time with preferred hand (VSRT-Prh), Visual Simple Reaction Time with non-preferred hand (VSRT-Nprh), Continuous Performance (CPT), Digit Erase (DET), Digit Symbol (DSB), Pursuit Aiming (PAT), and Sign Register (SRT). Of these nine outcomes, six achieved statistical significance: VSRT-Prh, VSRT-Nprh, CPT, DSB, PAT, and SRT. This study suggested that urban school-aged children with chronic low-level exposure to traffic-related air pollution might have a deficit in cognitive development. The strength of this study is the use of both computerized and manual neurobehavioral testing that made it possible to increase the sample size, while limitations include the cross-sectional design and the absence of seasonal variation of the air pollutants.

Environmental exposure factors interact frequently. Perera et al. [62] specifically studied the interaction between prenatal exposures to PAHs and environmental tobacco smoke (ETS) and their relationship with intelligence quotient (IQ) in children aged 5 in the city of Tongliang in Chongqing, China. A coal-fired power plant was the primary source of PAHs in the air. For the period of the study, PAH exposure was measured by DNA adducts in umbilical cord blood, and the IQ testing was measured by Wechsler Preschool and Primary Scale of Intelligence (Shanghai version). This study did not find significant associations of PAHs exposure with full scale and verbal IQ scores in children. However, the effect of PAHs on IQ became statistically significant when interaction with ETS was considered. Their results indicated that prenatally exposure to PAHs in combination with ETS might have adverse impact on cognitive function in children in their early life.

Findings from developed countries with relative low levels of air pollution also suggest that air pollution might be associated with cognitive function development in children [21, 50, 72]. A Spanish birth cohort of 210 children that observed nitrogen dioxide (NO_2) association with cognitive development was suggestive that traffic-related air pollution may have adverse effect on neurodevelopment on in children, even at low exposure level [21].

A cohort study [72] in the United States investigated the effects of black carbon on cognitive function assessed by the Wide Range Assessment of Memory and Learning (WRAML) and the Kaufman Brief Intelligence Test (K-BIT). There was a significant negative association between long-term exposure to black carbon and cognitive test scores. The mean annual black carbon concentration was 0.56 μg/m^3, and each 0.4 μg/m^3 increase in black carbon was associated with a small but significant decrease in the matrices (−4.0%) and composite intelligence quotient (−3.4%) scores of the K-BII and the visual subscale (−5.4%) and general index (−3.9%) of the WRAML.

Another cohort study [61] in New York City investigated the influence of prenatal exposure to airborne polycyclic aromatic hydrocarbons (PAHs) on child mental and psychomotor development. Significant difference in the mental development index at age 3 was observed between pregnant women with the less and the highly PAH exposure levels. Furthermore, the odds ratio for cognitive development delay in children with high prenatal PAHs exposure versus those with less prenatal exposure was 2.89 (95% CI, 1.33–6.25).

Given the increasing researches on neurotoxic effects of air pollution, it is important to identify the dosimetry characteristics of air pollution which determines its toxicity, for example, the physicochemical properties of particulate matter that include mass, size, number, surface area, and chemical composition, and it is also important to determine the potential contribution of air pollution to neurodevelopment in children.

7.5.2 *Biological Considerations*

The possible mechanism involved in the development of cardiopulmonary diseases has been thoroughly investigated. One mechanistic possibility of how particulate matter (PM) interacts with cardiovascular system may relate to extrapulmonary translocation. There is emerging evidence indicating that inhaled nanoparticles may translocate from respiratory tract via olfactory pathways [52, 65]. The olfactory nerve terminates in the olfactory bulb which is located within the head. That means the transport of air pollution along the olfactory not only provides information on potential mechanistic pathways that link exposure with cerebrovascular diseases but also offers a route for delivery to the brain.

There have been relevant toxicology studies using animals that have found oxidant and inflammatory effects of air pollution on neural tissue. For example, Elder et al. [20] have reported that in rat that intranasal instillation of manganese (Mn) oxide ultrafine particles (UFPs; diameters <100 nm) for 6–12 days results in significant increases of Mn in the brain, especially the olfactory bulb region. Calderon-Garciduenas et al. [7] reported that exposures to high level of air pollution exacerbated brain inflammation in dogs. Kleinman et al. [39] demonstrated that exposure to concentrated ambient particles significantly increased the levels of inflammatory markers in apolipoprotein E null mice.

Despite the suggestive evidence that exposure to traffic-related and coal-related air pollutants might have caused neurodevelopmental or neurobehavioral deficit in Chinese children, it is not easy to draw a strong conclusion because relatively few studies have been conducted. The interactions of air pollutants, the complexity of neurodevelopment process, changes in the vulnerability of children at different ages, and practical limitations to current research indicate that more studies are needed to better understand the role of the air pollutants in cognitive function in children in China.

7.6 Association Between Air Pollution and Obesity in Children

Obesity is one of the growing concerns of public health officials worldwide. Substantial epidemiological research has accrued demonstrating that obesity is a major risk factor for many chronic diseases including cardiovascular and metabolic diseases. A systematic review in The Lancet showed that prevalence of overweight and obesity increased substantially in children and adolescents in developing countries (12.9% in boys and 13.4% in girls) in 2013 [59]. The prevalence of obesity in Chinese children increased significantly from 6.1% to 13.1% during 1993–2009 [46]. While genetic and metabolic susceptibilities exist, growing evidence suggests that environmental factors, such as maternal smoking, secondhand smoking, and traffic pollution, may contribute to obesity development in childhood [32, 33, 54, 56].

7.6.1 Air Pollution Exposure and Obesity

There are few epidemiological studies that have investigated specifically whether air pollution contributes to obesity formation in childhood. In China, only three published papers concerned the association of air pollution and obesity in children which were from the Seven Northeastern Cities (SNEC) study [16, 18]. However, only one paper directly investigated the relationship between air pollution exposure and children obesity, whereas the other two publications assessed obesity as a modifier in the association between air pollution and health outcome.

On the basis of the annual air pollution level from 2006 to 2008, seven cities from Liaoning Province in Northeast China were selected, including Shenyang, Dalian, Anshan, Fushun, Benxi, Liaoyang, and Yingkou. There were 25 districts in total. One elementary school and two kindergartens were randomly selected from each district in April 2009. Every school was within 1 km from the nearest air monitoring site which provided the air pollution data. A total of 30,056 children (mean age: 8.4 ± 2.7 years old) from 25 elementary schools and 50 kindergartens were recruited and completed the questionnaire. The authors used the 3-year average concentration of the 25 districts for PM_{10} (124.2 ± 24.1 μg/m^3), SO_2 (17.6 ± 5.9 ppb), NO_2 (19.5 ± 4.0 ppb), and O_3 (27.4 ± 8.1 ppb) as air pollution level. Results showed that the overall prevalence rates of obesity and overweight were 14.1% and 12.3%, respectively. The increased prevalence of obesity was significantly associated with an IQR increasing of PM_{10} (ORs = 1.19; 95% CI: 1.11–1.26), SO_2 (ORs = 1.11; 95% CI: 1.03–1.20), NO_2 (ORs = 1.13; 95% CI: 1.04–1.22), and O_3 (ORs = 1.14; 95% CI: 1.04–1.24), respectively. Study suggests that air pollution is positively associated with an increased likelihood of obesity or overweight in children [16, 18].

In another two publications, the authors reported obesity as a modifier in the association between air pollution and health outcome, such as hypertension and respiratory health effects [17, 19]. In the first report among 30,056 children in the SNEC study, overweight and obesity were significantly more common among boys (14.4% overweight, 18.2% obese) than among girls (10.3% overweight, 9.9% obese) [17]. The authors observed that that overweight/obesity enhanced respiratory health effects of air pollution in the study children, with odds ratios (ORs) ranging from 1.17 per μg/m^3 for PM_{10} on wheeze (95% confidence interval (CI): 1.01, 1.36) to 1.50 per 10 μg/m^3 for NO_2 on phlegm (95% CI: 1.21, 1.87) and cough (95% CI: 1.24, 1.81). In another paper from the SNEC study conducted in 2013, the studied population was smaller than the previous one which consisted of 9354 children [19]. The authors investigated the effect modification by obesity on the association between exposure and blood pressure in Chinese children. The overweight and obese children accounted for 16% (n = 1460) and 17% (1571) among 9567 children, respectively [19]. Consistent with the previous study [17], higher overweight or obese rate was found in males. Results showed that obesity amplifies the association of long-term air pollution exposure with blood pressure and hypertension in Chinese children. The ORs for hypertension ranged from 1.16 per IQR increasing of O_3 (95% CI = 1.12, 1.20) to 2.91 per IQR increasing of PM_{10} (95% CI = 2.32, 3.64). The estimated increases in mean systolic and diastolic blood pressure ranged from 0.57 mmHg (95% CI = 0.36, 0.78) and 0.63 mmHg (95% CI = 0.46, 0.81) with per IQR increasing for O_3 to 4.04 mmHg (95% CI = 3.00, 5.09) and 2.02 mmHg (95% CI = 1.14, 2.89) with per IQR increasing for SO_2. In a rodent model, Wei et al. [75] found that prenatally and postnatally exposed to Beijing's highly polluted air resulted in weight gain and metabolic dysfunction in offspring, indicating that air pollution may be a contributing factor to weight gain and several related health issues.

In the limited epidemiological studies investigating association between ambient air pollution and children obesity, traffic-related air pollution is the most studied scenario. A longitudinal study assesses the obesity (using body mass index, BMI in kg/m^2) at the end of the childhood growth period in a large cohort of children (n = 3318) living in Southern California in relation to residential traffic exposure around the home [33]. The finding showed that traffic density was positively associated with attained BMI in children at age 18, suggesting traffic was a major risk factor for the development of obesity in children. In other publications, Jerrett et al. [34] investigated whether traffic-related air pollution was positively associated with growth in BMI (kg/m^2) in children aged 5–11 years (n = 4550) in Southern California in a prospective cohort. Children were enrolled while attending kindergarten and first grade and followed for 4 years. In their results, traffic pollution was positively associated with growth in BMI, with a 13.6% increase in annual BMI growth when comparing the lowest to the highest tenth percentile of air pollution exposure, which resulted in an increase of nearly 0.4 BMI units on attained BMI at age 10 [34]. Consistent with this result, McConnell et al. [56] found that near-roadway pollution exposure contributes to development of childhood obesity in a longitudinal cohort study – Southern California Children's Health Study – over an 8-year follow-up (n = 3318). Among children without a history of secondhand smoking exposure, attained BMI was 0.80 kg/m^2 higher (95% CI: 0.27–1.32) in children exposure to high near-roadway pollution exposure compared with children with near-roadway pollution exposure below the median [56].

7.6.2 Biological Mechanisms

Emerging evidence from experiments on rodent suggests that $PM_{2.5}$ exposure induces vascular insulin resistance and inflammation triggered by a mechanism involving adipose tissue inflammation and oxidative stress [73, 79]. Xu et al. [79] have evaluated the role of oxidative stress pathway in the association between early-life exposure to $PM_{2.5}$ and adiposity, inflammation, and metabolic parameters. Using genetically modified mice p47$^{phox-/-}$ deficient in the cytosolic subunit of the nicotinamide adenine dinucleotide phosphate (NADPH) oxidase, Xu et al. [79] found that NAD(P)H oxidase-derived reactive oxygen species may contribute to the obesity, insulin resistance, and inflammation respond to the early-life exposure to $PM_{2.5}$. Suppression of vascular insulin signaling by $PM_{2.5}$ may accelerate the progression to systemic insulin resistance [27]. In another experimental study, the authors also found that $PM_{2.5}$ exposure exaggerated insulin resistance and visceral inflammation/adiposity [73]. Several signaling abnormalities characteristic of insulin resistance were observed, including decreased Akt and endothelial nitric oxide synthase phosphorylation in the endothelium and increased protein kinase C expression. Consistent with the previous study, Wei et al. [75] observed pronounced histologic evidence for both perivascular and peribronchial inflammation in the lungs, increased tissue and systemic oxidative stress, dyslipidemia, and an enhanced proinflammatory status of epididymal fat. Results suggest that TLR2-/TLR4-dependent inflammatory activation and lipid oxidation in the lung can spill over systemically, leading to metabolic dysfunction and weight gain. Zou [83] proposed the hypothesized mechanism of insulin resistance and adipocyte hypertrophy in mice exposed to $PM_{2.5}$ (Fig. 7.2). Further investigation is needed to understand the mechanisms underlying the effect of air pollution on obesity.

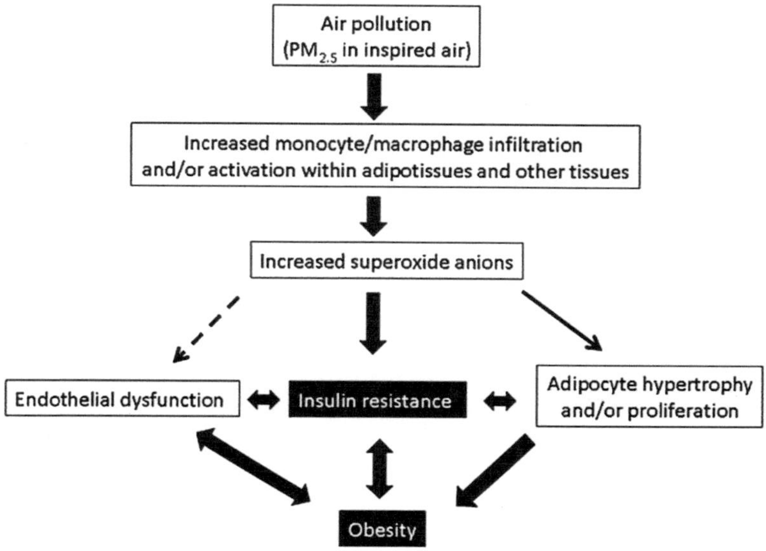

Fig. 7.2 Hypothesized mechanism of insulin resistance and adipocyte hypertrophy in mice exposed to $PM_{2.5}$. The *dashed arrow* with query is a less established mechanistic links between inspiring $PM_{2.5}$ and endothelial dysfunction. Insulin resistance and obesity, due to changes in adipocyte size and numbers, can in turn feedback to the vasculature compounding the deleterious phenotype (Quoted from Zou [83])

7.7 Summary

Children are at particulate risk for air pollution because the intrauterine, perinatal, and childhood periods provide susceptibility windows during which air pollution may influence organ development and maturation. In this chapter, we reviewed current epidemiological evidence on the effects of air pollution exposure on children health in China. Results provide broader evidence whereby air pollution can promote adverse health responses in children. We believe that the findings summarized in our review demonstrate an unequivocal relationship between air pollutants exposure and children health outcome and these results have large public health influences. Therefore, control of air pollutants and initiation of efforts to protect children from heavy air pollution exposure in China should be a priority.

Acknowledgments This study was funded by grants from the Major Program of National Natural Science Foundation of China (91543208), the National Natural Science Foundation of China (No. 81472936, 81673127, 81673128, and 41501540), the National Natural Science Foundation of China - Guangdong Joint Fund Project (No. U1301234), the National Key Research and Development Program of China (2016YFC0207000), the Guangdong Province Natural Science Foundation (2014A050503027, and 2016A030313194), and the Fundamental Research Funds for the Central Universities (No. 16ykzd02, 16ykpy18, and 17ykpy16).

References

1. Avol EL, Gauderman WJ, Tan SM, London SJ, Peters JM. Respiratory effects of relocating to areas of differing air pollution levels. Am J Resp Crit Care. 2001;164(11):2067–72.
2. Araujo JA. Particulate air pollution, systemic oxidative stress, inflammation, and atherosclerosis. Air Qual Atmos Health. 2011;4:79–93.
3. Bilenko N, Lv R, Brunekreef B, Beelen R, Eeftens M, Hoek G, Houthuijs D, de Jongste JC, Ev K, Koppelman GH, Meliefste K, Oldenwening M, Smit HA, Wijga AH, Gehring U. Traffic-related air pollution and noise and children's blood pressure: results from the PIAMA birth cohort study. Eur J Prev Cardiol. 2015;22:4–12.
4. Brook RD, Rajagopalan S. Particulate matter, air pollution, and blood pressure. J Am Soc Hypertens. 2009;3:332–50.
5. Brook RD, Urch B, Dvonch JT, Bard RL, Speck M, Keeler G, Morishita M, Marsik FJ, Kamal AS, Kaciroti N, Harkema J, Corey P, Silverman F, Gold DR, Wellenius G, Mittleman MA, Rajagopalan S, Brook JR. Insights into the mechanisms and mediators of the effects of air pollution exposure on blood pressure and vascular function in healthy humans. Hypertension.2009;54:659-67.
6. Brook RD, Rajagopalan S, Pope CA, Brook JR, Bhatnagar A, Diez-Roux AV, Holguin F, Hong Y, Luepker RV, Mittleman MA, Peters A, Siscovick D, Smith SC, Whitsel L, Kaufman JD, Epidemiology obotAHACo, Prevention CotKiCD, Council on Nutrition PA, Metabolism. Particulate matter air pollution and cardiovascular disease: an update to the scientific statement from the American Heart Association. Circulation. 2010;121:2331–78.
7. Calderon-Garciduenas L, Azzarelli B, Acuna H, Garcia R, Gambling T, Osnaya N, Monroy S, DEL Tizapanitzi M, Carson J, Villarreal-Calderon A, Rewcastle B. Air Pollution and Brain Damage. Toxicol Pathol 2002;30(3):373–89.
8. Calderón-Garcidueñas L, Vincent R, Mora-Tiscareño A, Franco-Lira M, Henríquez-Roldán C, Barragán-Mejía G, Garrido-García L, Camacho-Reyes L, Valencia-Salazar G, Paredes R, Romero L, Osnaya H, Villarreal-Calderón R, Torres-Jardón R, Hazucha MJ, Reed W. Elevated plasma endothelin-1 and pulmonary arterial pressure in children exposed to air pollution. Environ Health Perspect. 2007;115:1248–53.
9. Chan CK, Yao X. Air pollution in mega cities in China. Atmos Environ. 2008;42(1):1–42.
10. Chang YK, Wu CC, Lee LT, Lin RS, Yu YH, Chen YC. The short-term effects of air pollution on adolescent lung function in Taiwan. Chemosphere. 2012;87(1):26–30.
11. Chen CH, Chan CC, Chen BY, Cheng TJ, Guo YL. Effects of particulate air pollution and ozone on lung function in non-asthmatic children. Environ Res. 2015;137:40–8.
12. Chuang GC, Yang Z, Westbrook DG, Pompilius M, Ballinger CA, White CR, Krzywanski DM, Postlethwait EM, Ballinger SW. Pulmonary ozone exposure induces vascular dysfunction, mitochondrial damage, and atherogenesis. Am J Phys Lung Cell Mol Phys. 2009;297:L209–16.
13. Dinh QN, Drummond GR, Sobey CG, Chrissobolis S. Roles of inflammation, oxidative stress, and vascular dysfunction in hypertension. Biomed Res Int. 2014;2014:406960.
14. Dodd JW, Getov SV, Jones PW. Cognitive function in COPD. Eur Respir J. 2010;35:913–22.
15. Dong GH, Chen T, Liu MM, Wang D, Ma YN, Ren WH, Lee YL, Zhao YD, He QC. Gender differences and effect of air pollution on asthma in children with and without allergic predisposition: northeast chinese children health study. PLoS One. 2011;6:e22470.
16. Dong G-H, Qian Z, Liu M-M, Wang D, Ren W-H, Flick LH, Fu J, Wang J, Chen W, Simckes M, Trevathan E. Ambient air pollution and the prevalence of obesity in chinese children: the seven northeastern cities study. Obesity. 2014;22:795–800.
17. Dong GH, Qian Z, Liu MM, Wang D, Ren WH, Fu Q, Wang J, Simckes M, Ferguson TF, Trevathan E. Obesity enhanced respiratory health effects of ambient air pollution in Chinese children: the seven northeastern cities study. Int J Obes. 2013;37:94–100.
18. Dong G-H, Qian Z, Trevathan E, Zeng X-W, Vaughn MG, Wang J, Zhao Y, Liu Y-Q, Ren W-H, Qin X-D. Air pollution associated hypertension and increased blood pressure may be reduced

by breastfeeding in Chinese children: the seven northeastern cities chinese children's study. Int J Cardiol. 2014;176:956–61.
19. Dong G-H, Wang J, Zeng X-W, Chen L, Qin X-D, Zhou Y, Li M, Yang M, Zhao Y, Ren W-H, Hu Q-S. Interactions between air pollution and obesity on blood pressure and hypertension in Chinese children. Epidemiology. 2015;26:740–7.
20. Elder A, Gelein R, Silva V, Feikert T, Opanashuk L, Carter J, Potter R, Maynard A, Ito Y, Finkelstein J, Oberdörster G. Translocation of inhaled ultrafine manganese oxide particles to the central nervous system. Environ Health Perspect. 2006;114(8):1172–8.
21. Freire C, Ramos R, Puertas R, Lopez-Espinosa MJ, Julvez J, Aguilera I, Cruz F, Fernandez MF, Sunyer J, Olea N. Association of traffic-related air pollution with cognitive development in children. J Epidemiol Community Heath. 2010;64(3):223–8.
22. Gaffney AW, Hang JQ, Lee MS, Su L, Zhang FY, Christiani DC. Socioeconomic status is associated with reduced lung function in China: an analysis from a large cross-sectional study in Shanghai. BMC Public Health. 2016;16:96.
23. Gao Y, Chan EYY, Li LP, He QQ, Wong TW. Chronic effects of ambient air pollution on lung function among Chinese children. Arch Dis Child. 2013;98(2):128–35.
24. Gao Y, Chan EY, Li L, Lau PW, Wong TW. Chronic effects of ambient air pollution on respiratory morbidities among Chinese children: a cross-sectional study in Hong Kong. BMC Public Health. 2014;14:105.
25. Gauderman WJ, Avol E, Gilliland F, Vora H, Thomas D, Berhane K, McConnell R, Kuenzli N, Lurmann F, Rappaport E, Margolis H, Bates D, Peters J. The effect of air pollution on lung development from 10 to 18 years of age. New Engl J Med. 2004;351(11):1057–67.
26. Giorgini P, Di Giosia P, Grassi D, Rubenfire M, Brook RD, Ferri C. Air pollution exposure and blood pressure: an updated review of the literature. Curr Pharm Des. 2016;22:28–51.
27. Haberzettl P, O'Toole TE, Bhatnagar A, Conklin DJ. Exposure to fine particulate air pollution causes vascular insulin resistance by inducing pulmonary oxidative stress. Environ Health Perspect. 2016;124:1830–9.
28. Hegewald MJ, Crapo RO. Socioeconomic status and lung function. Chest. 2007;132(5):1608–14.
29. Horak F, Studnicka M, Gartner C, Spengler JD, Tauber E, Urbanek R, Veiter A, Frischer T. Particulate matter and lung function growth in children: a 3-yr follow-up study in Austrian schoolchildren. Eur Respir J. 2002;19(5):838–45.
30. Hua J, Yin Y, Peng L, Du L, Geng F, Zhu L. Acute effects of black carbon and $PM_{2.5}$ on children asthma admissions: a time-series study in a chinese city. Sci Total Environ. 2014;481:433–8.
31. Hwang BF, Chen YH, Lin YT, Wu XT, Lee YL. Relationship between exposure to fine particulates and ozone and reduced lung function in children. Environ Res. 2015;137:382–90.
32. Ino T. Maternal smoking during pregnancy and offspring obesity: meta-analysis. Pediatr Int. 2010;52:94–9.
33. Jerrett M, McConnell R, Chang CCR, Wolch J, Reynolds K, Lurmann F, Gilliland F, Berhane K. Automobile traffic around the home and attained body mass index: a longitudinal cohort study of children aged 10–18 years. Prev Med. 2010;(50, Supplement):S50–8.
34. Jerrett M, McConnell R, Wolch J, Chang R, Lam C, Dunton G, Gilliland F, Lurmann F, Islam T, Berhane K. Traffic-related air pollution and obesity formation in children: a longitudinal, multilevel analysis. Environ Health. 2014;13:49.
35. Jirásek JE. An atlas of human prenatal developmental mechanics: anatomy and staging. London: CRC Press; 2004.
36. Joshi S, Kotecha S. Lung growth and development. Early Hum Dev. 2007;83:789–94.
37. Kajekar R. Environmental factors and developmental outcomes in the lung. Pharmacol Ther. 2007;114:129–45.
38. Kasamatsu J, Shima M, Yamazaki S, Tamura K, Sun GF. Effects of winter air pollution on pulmonary function of school children in Shenyang China. Int J Hyg Environ Health. 2006;209(5):435–44.

39. Kleinman MT, Araujo JA, Nel A, Sioutas C, Campbell A, Cong PQ, Lia H, Bondy SC. Inhaled ultrafine particulate matter affects CNS inflammatory processes and may act via MAP kinase signaling pathways. Toxicol Lett. 2008;178(2):127–30.
40. Lauer RM, Burns TL, Clarke WR, Mahoney LT. Childhood predictors of future blood pressure. Hypertension. 1991;18:I74.
41. Lee YL, Wang WH, Lu CW, Lin YH, Hwang BF. Effects of ambient air pollution on pulmonary function among schoolchildren. Int J Hyg Environ Health. 2011;214(5):369–75.
42. Lewington S, Li L, Sherliker P, Guo Y, Millwood I, Bian Z, Whitlock G, Yang L, Collins R, Chen J, Wu X, Wang S, Hu Y, Jiang L, Yang L, Lacey B, Peto R, Chen Z. Seasonal variation in blood pressure and its relationship with outdoor temperature in 10 diverse regions of China: the China Kadoorie Biobank. J Hypertens. 2012;30:1383–91.
43. Li Q, Guo Y, Wei D-M, Song Y, Song J-Y, Ma J, Wang H-J. Does local ambient temperature impact children's blood pressure? A Chinese National Survey. Environ Health. 2016;15:21.
44. Liu C, Fuertes E, Tiesler CMT, Birk M, Babisch W, Bauer C-P, Koletzko S, von Berg A, Hoffmann B, Heinrich J. The associations between traffic-related air pollution and noise with blood pressure in children: results from the GINIplus and LISAplus studies. Int J Hyg Environ Health. 2014;217:499–505.
45. Lin G, Fu JY, Jiang D, Hu WS, Dong DL, Huang YH, Zhao MD. Spatio-temporal variation of PM2.5 concentrations and their relationship with geographic and socioeconomic factors in China. Int J Environ Res Public Health. 2014;11(1):173–86.
46. Liang Y-J, Xi B, Song A-Q, Liu J-X, Mi J. Trends in general and abdominal obesity among Chinese children and adolescents 1993–2009. Pediatr Obes. 2012;7:355–64.
47. Lin W, Huang W, Zhu T, Hu M, Brunekreef B, Zhang Y, Liu X, Cheng H, Gehring U, Li C, Tang X. Acute respiratory inflammation in children and black carbon in ambient air before and during the 2008 beijing olympics. Environ Health Perspect. 2011;119:1507–12.
48. Lin W, Brunekreef B, Gehring U. Meta-analysis of the effects of indoor nitrogen dioxide and gas cooking on asthma and wheeze in children. Int J Epidemiol. 2013;42:1724–37.
49. Liu F, Zhao Y, Liu YQ, Liu Y, Sun J, Huang MM, Liu Y, Dong GH. Asthma and asthma related symptoms in 23,326 chinese children in relation to indoor and outdoor environmental factors: the seven northeastern cities (snec) study. Sci Total Environ. 2014;497–498:10–7.
50. Liu JH, Lewis G. Environmental toxicity and poor cognitive outcomes in children and adults. J Environ Health. 2014;76(6):130–8.
51. Liu L, Zhang JL. Ambient air pollution and children's lung function in China. Environ Int. 2009;35:178–86.
52. Lucchini R, Dorman D, Elder A, Veronesi B. Neurological impacts from inhalation of pollutants and nose-brain connection. Neurotoxicology. 2012;33(4):838–41.
53. Madaniyazi L, Guo YM, Ye XF, Kim D, Zhang YX, Pan XC. Effects of airborne metals on lung function in inner Mongolian schoolchildren. J Occup Environ Med. 2013;55(1):92–8.
54. Mao G, Nachman R, Sun Q, Zhang X, Koehler K, Chen Z, XH, Wang G, Caruso D, Zong G, Pearson C, Ji H, Biswa S, Zuckerman B, Wills-Karp M, Wang X. Individual and joint effects of early-life ambient $PM_{2.5}$ exposure and maternal pre-pregnancy obesity on childhood overweight or obesity. Environ Health Perspect. 2016.; Sep 16. [Epub ahead of print]; doi:10.1289/EHP261.
55. Martin RM, Ness AR, Gunnell D, Emmett P, Smith GD. Does breast-feeding in infancy lower blood pressure in childhood? Avon Longitudinal Study Parents Child (ALSPAC). 2004;109:1259–66.
56. McConnell R, Shen E, Gilliland FD, Jerrett M, Wolch J, Chang C-C, Lurmann F, Berhane K. A longitudinal cohort study of body mass index and childhood exposure to secondhand tobacco smoke and air pollution: the Southern California Children's Health Study. Environ Health Perspect. 2015;123:360–6.
57. Mustafić H, Jabre P, Caussin C, Murad MH, Escolano S, Tafflet M, Périer MC, Marijon E, Vernerey D, Empana JP, Jouven X. Main air pollutants and myocardial infarction: a systematic review and meta-analysis. JAMA. 2012;307:713–21.

58. National Bureau of Statistics of China. China statistical yearbook: population by age and sex. Beijing: National Bureau of Statistics; 2014.
59. Ng M, Fleming T, Robinson M, Thomson B, Graetz N, Margono C, Mullany EC, Biryukov S, Abbafati C, Abera SF, Abraham JP, Abu-Rmeileh NME, Achoki T, AlBuhairan FS, Alemu ZA, Alfonso R, Ali MK, Ali R, Guzman NA, Ammar W, Anwari P, Banerjee A, Barquera S, Basu S, Bennett DA, Bhutta Z, Blore J, Cabral N, Nonato IC, Chang J-C, Chowdhury R, Courville KJ, Criqui MH, Cundiff DK, Dabhadkar KC, Dandona L, Davis A, Dayama A, Dharmaratne SD, Ding EL, Durrani AM, Esteghamati A, Farzadfar F, Fay DFJ, Feigin VL, Flaxman A, Forouzanfar MH, Goto A, Green MA, Gupta R, Hafezi-Nejad N, Hankey GJ, Harewood HC, Havmoeller R, Hay S, Hernandez L, Husseini A, Idrisov BT, Ikeda N, Islami F, Jahangir E, Jassal SK, Jee SH, Jeffreys M, Jonas JB, Kabagambe EK, Khalifa SEAH, Kengne AP, Khader YS, Khang Y-H, Kim D, Kimokoti RW, Kinge JM, Kokubo Y, Kosen S, Kwan G, Lai T, Leinsalu M, Li Y, Liang X, Liu S, Logroscino G, Lotufo PA, Lu Y, Ma J, Mainoo NK, Mensah GA, Merriman TR, Mokdad AH, Moschandreas J, Naghavi M, Naheed A, Nand D, Narayan KMV, Nelson EL, Neuhouser ML, Nisar MI, Ohkubo T, Oti SO, Pedroza A, Prabhakaran D, Roy N, Sampson U, Seo H, Sepanlou SG, Shibuya K, Shiri R, Shiue I, Singh GM, Singh JA, Skirbekk V, Stapelberg NJC, Sturua L, Sykes BL, Tobias M, Tran BX, Trasande L, Toyoshima H, van de Vijver S, Vasankari TJ, Veerman JL, Velasquez-Melendez G, Vlassov VV, Vollset SE, Vos T, Wang C, Wang X, Weiderpass E, Werdecker A, Wright JL, Yang YC, Yatsuya H, Yoon J, Yoon S-J, Zhao Y, Zhou M, Zhu S, Lopez AD, Murray CJL, Gakidou E. Global, regional, and national prevalence of overweight and obesity in children and adults during 1980–2013: a systematic analysis for the Global Burden of Disease Study. Lancet. 2013. 2014;384:766–81.
60. Pan G, Zhang S, Feng Y, Takahashi K, Kagawa J, Yu L, Wang P, Liu M, Liu Q, Hou S, Pan B, Li J. Air pollution and children's respiratory symptoms in six cities of Northern China. Respir Med. 2010;104:1903–11.
61. Perera FP, Rauh V, Whyatt RM, Tsai WY, Tang D, Diaz D, Hoepner L, Barr D, Tu YH, Camann D, Kinney P. Effect of prenatal exposure to airborne polycyclic aromatic hydrocarbons on neurodevelopment in the first 3 years of life among inner-city children. Environ Health Perspect. 2006;114(8):1287–92.
62. Perera F, Li TY, Lin C, Tang D. Effects of prenatal polycyclic aromatic hydrocarbon exposure and environmental tobacco smoke on child IQ in a Chinese cohort. Environ Res. 2012;114:40–6.
63. Perez CM, Hazari MS, Farraj AK. Role of autonomic reflex arcs in cardiovascular responses to air pollution exposure. Cardiovasc Toxicol. 2015;15:69–78.
64. Pieters N, Koppen G, Van Poppel M, De Prins S, Cox B, Dons E, Nelen V, Panis LI, Plusquin M, Schoeters G, Nawrot TS. Blood pressure and same-day exposure to air pollution at school: associations with nano-sized to coarse PM in children. Environ Health Perspect. 2015;123:737–42.
65. Pope CA, Dockery DW. Health effects of fine particulate air pollution: lines that connect. J Air Waste Manag Assoc. 2006;56:709–42.
66. Rojas-Martinez R, Perez-Padilla R, Olaiz-Fernandez G, Mendoza-Alvarado L, Moreno-Macias H, Fortoul T, McDonnell W, Loomis D, Romieu I. Lung function growth in children with long-term exposure to air pollutants in Mexico City. Am J Resp Crit Care. 2007;176(4):377–84.
67. Roy A, Hu W, Wei FS, Korn L, Chapman RS, Zhang JF. Ambient particulate matter and lung function growth in Chinese children. Epidemiology. 2012;23(3):464–72.
68. Schmid O, Möller W, Semmler-Behnke M, Ferron GA, Karg E, Lipka J, Schulz H, Kreyling WG, Stoeger T. Dosimetry and toxicology of inhaled ultrafine particles. Biomarkers. 2009;14:67–73.
69. Seaton A, Godden D, MacNee W, Donaldson K. Particulate air pollution and acute health effects. Lancet. 1995;345:176–8.
70. Sin DD, Wu LL, Man SFP. The relationship between reduced lung function and cardiovascular mortality – a population-based study and a systematic review of the literature. Chest. 2005;127(6):1952–9.

71. Sughis M, Nawrot TS, Ihsan-ul-Haque S, Amjad A, Nemery B. Blood pressure and particulate air pollution in schoolchildren of Lahore, Pakistan. BMC Public Health. 2012;12:1–8.
72. Suglia SF, Gryparis A, Wright RO, Schwartz J, Wright RJ. Association of black carbon with cognition among children in a prospective birth cohort study. Am J Epidemiol. 2008;167(3):280–6.
73. Sun Q, Yue P, Deiuliis JA, Lumeng CN, Kampfrath T, Mikolaj MB, Cai Y, Ostrowski MC, Lu B, Parthasarathy S, Brook RD, Moffatt-Bruce SD, Chen LC, Rajagopalan S. Ambient air pollution exaggerates adipose inflammation and insulin resistance in a mouse model of diet-induced obesity. Circulation. 2009;119(119):538–46.
74. Wang S, Zhang J, Zeng X, Zeng Y, Wang S, Chen S. Association of traffic-related air pollution with children's neurobehavioral functions in Quanzhou, China. Environ Health Perspect. 2009;117:1612–8.
75. Wei Y, Zhang J, Li Z, Gow A, Chung KF, Hu M, Sun Z, Zeng L, Zhu T, Jia G, Li X, Duarte M, Tang X. Chronic exposure to air pollution particles increases the risk of obesity and metabolic syndrome: findings from a natural experiment in Beijing. FASEB J. 2016;30:2115–22.
76. World Health Organization. Effects of air pollution on children's health and development – a review of the evidence. Copenhagen: WHO Regional Office for Europe; 2005.
77. World Health Organization. Burden of disease from household air pollution for 2012: summary of results; 2014. Accessed 20 July 2016.
78. World Health Organization. Ambient air pollution: a global assessment of exposure and burden of disease. Geneva: WHO Document Production Services; 2016.
79. Xu X, Yavar Z, Verdin M, Ying Z, Mihai G, Kampfrath T, Wang A, Zhong M, Lippmann M, Chen L-C, Rajagopalan S, Sun Q. Effect of early particulate air pollution exposure on obesity in mice: role of p47phox. Arterioscler Thromb Vasc Biol. 2010;30:2518–27.
80. Zeng X-W, Qian Z, Vaughn MG, Nelson EJ, Dharmage SC, Bowatte G, Perret J, Chen D-H, Ma H, Lin S, de Foy B, Hu L-W, Yang B-Y, Xu S-L, Zhang C, Tian Y-P, Nian M, Wang J, Xiao X, Bao W-W, Zhang Y-Z, Dong G-H. Positive association between short-term ambient air pollution exposure and children blood pressure in China–result from the Seven Northeast Cities (SNEC) study. Environ Pollut. 2017;224:698–705.
81. Zeng XW, Vivian E, Mohammed K, Jakhar S, Vaughn M, Huang J, Zelicoff A, Xaverius P, Bai ZP, Lin S, Hao YT, Paul G, Morawska L, Wang SQ, Qian ZM, Dong GH. Long-term ambient air pollution and lung function impairment in Chinese children from a high air pollution range area: the Seven Northeastern Cities (SNEC) study. Atmos Environ. 2016;138:144–51.
82. Zhao Z, Zhang Z, Wang Z, Ferm M, Liang Y, Norback D. Asthmatic symptoms among pupils in relation to winter indoor and outdoor air pollution in schools in Taiyuan, China. Environ Health Perspect. 2008;116:90–7.
83. Zou M-H. Is NAD(P)H oxidase a missing link for air pollution-enhanced obesity? Arterioscler Thromb Vasc Biol. 2010;30:2323–4.

Chapter 8
The Associations Between Air Pollution and Adverse Pregnancy Outcomes in China

Yafei Tan, Rong Yang, Jinzhu Zhao, Zhongqiang Cao, Yawen Chen, and Bin Zhang

Abstract Research on the potential impact of air pollution on the human's health has increased rapidly in recent years. Several studies have suggested that exposure to air pollutants during period of pregnancy which is a crucial time point of mother-fetus development may have long-term and serious impact on adverse pregnancy outcomes. There is lack of review to evaluate the existed epidemiologic evidence on the associations between air pollutants and adverse pregnancy outcomes in China, so we conducted a review to explore the current epidemiological evidence on the effects of air pollutants to pregnancy outcomes and possible mechanisms during the pregnancy process. We used keywords to systematically search all the English and Chinese literatures on studies that were conducted in China. Exposure to air pollutants during pregnancy had shown there were harmful effects for different birth outcomes: preterm birth, low birth weight, stillbirth, birth defects, infertility, and macrosomia fetus. Results on the effects of air pollutants on adverse pregnancy outcomes are small and inconsistent because they vary in their design and methodology. The existed available evidence is compatible with either a small negative effect of air pollutants on pregnancy outcomes or with no effect; therefore, further studies are needed to confirm and quantify the possible associations and potential biologic mechanisms between air pollutants and pregnancy outcomes.

Keywords Air pollutants • Pregnancy outcomes • Preterm birth • Stillbirth • Birth defects

---

Y. Tan • R. Yang • J. Zhao • Z. Cao • Y. Chen • B. Zhang (✉)
Wuhan Children's Hospital (Wuhan Maternal and Child Healthcare Hospital), Tongji Medical College, Huazhong University of Science & Technology, Wuhan 430016, China
e-mail: 2500070378@qq.com

G.-H. Dong (ed.), *Ambient Air Pollution and Health Impact in China*, Advances in Experimental Medicine and Biology 1017, DOI 10.1007/978-981-10-5657-4_8

8.1 Introduction

Epidemiological evidence showed that air pollution has a certain impact on people's health [99]. Previous studies have focused on the effects of air pollution on adult mortality and morbidity of various diseases, but recently there are more and more studies on the impact of air pollution on pregnancy outcomes in China [5, 12, 74]. Adverse pregnancy outcomes refer to all pathological pregnancy and childbirth complications, including preterm birth, spontaneous abortion, fetal death, low birth weight, macrosomia fetus, congenital anomalies, stillbirth, and other adverse outcomes [54]. Previous studies which showed the associations between air pollutants, such as particulate matter (PM_{10} and $PM_{2.5}$), sulfur dioxide (SO_2), nitrogen oxide (NO_x), carbon monoxide (CO), and ozone (O_3), and preterm birth, fetal death, low birth weight, macrosomia fetus, and congenital defects were mainly based on assessment of excess mortality numbers associated with air pollutant exposures, such as the monographical study on ambient air pollution and adverse pregnancy outcomes in Wuhan, China, which mainly focused on the relationships between air pollutants and adverse pregnancy outcomes, and the details would be discussed below [49, 105, 106]. Although more and more researchers have paid attention to the associations of air pollution and adverse pregnancy outcomes in recent years, the conclusions are unable to agree on which is right. In this paper, we will review studies on the relationship between main pollutants in the atmosphere and adverse pregnancy outcomes in China.

8.2 The Relationships Between Various Components of Air Pollutants and Adverse Pregnancy Outcomes

8.2.1 Air Pollution and the Risk of Preterm Birth

Preterm birth (PTB), defined as childbirth occurring before 37 completed weeks of gestation, has long-term adverse consequences for neonatal development and can lead to poor growth. According to World Health Organization (WHO) report in 2013, PTB is the main cause of newborn deaths, and among surviving infants, it is the second leading cause of death in children under 5 years (behind pneumonia) [96].Unfortunately, many local clinical and community have failed at reducing PTB, and rates of PTB are increasing in developing countries and even in some developed countries [4], but the underlying causes are still poorly understood. Multiple factors have been shown to be associated with PTB, including socioeconomic status, race, genetic influences, medical conditions, mental disorders, environmental exposures, and others [84]. Air pollution, which is one important part of environmental pollution, is recognized as one of the risk factors for numerous diseases. An increasing number of studies have estimated the association between air pollution and pregnancy outcomes; especially, a significant and relatively recent effort has been made to link environmental air pollution exposures with PTB [47]. A large number of epidemiologic studies estimated the association between air pollution and PTB, but the results were inconsistent.

A study from Beijing followed all registered pregnant women who lived in Beijing in 1988. Information for both mothers and infants and daily air pollution data were obtained independently. This study included 25,370 pregnant women. The authors reported that there was a significant dose-dependent association between gestational age and SO_2 and maternal SO_2 exposure was associated with PTB with an adjusted OR of 1.21 (95% CI, 1.01–1.46), and similar result was found in this study between PM_{10} and PTB (OR, 1.10; 95% CI, 1.01–1.20) (Table 8.1) [100]. A time series analysis from Shanghai suggested a significant effect of outdoor air pollution only with 8-week exposure before delivery [49]. A case-crossover study which included 52,488 pregnant women was conducted in Taiyuan to assess the association of air pollution and PTB with their birth surveillance system and data from environmental monitoring station; the results showed that the stronger effect of short time for PM_{10} on PTB were 5 days lag, the cumulative effect were 6 days lag, and the ORs of PTB were 1.17(95% CI, 1.06–1.29) and 1.24(95% CI, 1.12–1.37), respectively, with each 50-ug/m^3 increase; the short effect of SO_2 exposure on PTB also lagged 7 days (OR, 1.35; 95% CI, 1.12–1.62) and accumulative effect lagged 8 days (OR, 2.20; 95% CI, 1.52–3.20) [107]. A case-control study in Guangzhou was conducted to explore the relationships of air pollutants and PTB, pregnant women who delivered at 28–36 weeks during 2006–2008; the results showed that the second trimester of pregnancy for NO_2 exposure was associated with decreased risk of PTB (OR, 0.99; 95% CI, 0.99–1.00), and the third trimester for NO_2 exposure was associated with increased risk of PTB (OR, 1.01; 95% CI, 1.00–1.01), as well as the third trimester of pregnancy for SO_2 exposure (OR, 1.00; 95% CI, 1.00–1.01) [82]. Lai and his two colleagues systematically searched the studies from China including Mainland China, Hong Kong, and Taiwan on adverse health effect due to air pollution in 2012. From 1989 to 2010, the annual mean concentrations ranged 44–156 μg/m^3 for PM_{10}, 23–70 μg/m^3 for NO_2, 14–213 μg/m^3 for SO_2, and 34–86 μg/m^3 for O_3. For PTB, just one study from Shanghai reported that sub-chronic exposure (8-week average) to the PM_{10}, NO_2, SO_2, and O_3 corresponded to RR of preterm births of 1.0442 (95% CI, 1.0160–1.0725), 1.0543 (95% CI, 1.0178–1.0908), 1.1189 (95% CI, 1.0669–1.1709), and 1.0463 (95% CI, 1.0035–1.0891), respectively [55]. The other study which was conducted in Taiwan recruited all women who had first-parity singleton live births who lived in a zone along the Zhongshan Freeway in Taiwan during 1992–1997 ($N = 6251$). The zone was defined as areas 1500 m wide on either side of the freeway. The prevalence of deliveries of preterm birth infants was significantly higher among mothers who lived within 500 m of the freeway than those who resided 500–1500 m from the freeway. After controlling for several confounders, the adjusted odds ratio was 1.30 (95% CI, 1.03–1.65) for delivery of preterm infants born to mothers who lived within 500 m of the freeway. Similar to the above study results, Europe and the United States studies showed that air pollutants were associated with PTB [7, 10, 56, 65].

Table 8.1 Prenatal air pollutant exposure and preterm birth

Author/year	Location	Design	Sample size	Study period	Exposure assessment methods	Results			
						PM_{10}	SO_2	NO_2	O_3
Xu et al. 1995 [100]	Beijing	Prospective cohort	25,370	1988	Daily temperature and humidity data were obtained from the Beijing Weather Bureau	OR, 1.10; 95% CI, 1.01–1.20	OR, 1.21; 95% CI, 1.01–1.46		
Jiang et al. 2007 [49]	Shanghai	Case-control	3346	2004	Daily averaged measurements of air pollution.	RR, 1.04; 95% CI, 1.02–1.07	RR, 1.12; 95% CI, 1.07–1.17	RR, 1.05; 95% CI, 1.02–1.09	RR, 1.05; 95% CI, 1.00–1.09
Zhang et al. 2007 [107]	Taiyuan	Case-crossover	52,488	1997–2004	Daily average exposure measured by six monitoring stations from 3 months before pregnancy to 6 months after pregnancy	5 days lag: OR, 1.17; 95% CI, 1.06–1.29; 6 days lag: OR, 1.24; 95% CI, 1.12–1.37	7 days LAG: OR, 1.35; 95% CI, 1.12–1.62; 8 days lag: OR, 2.20; 95% CI, 1.52–3.20		
Ruan et al. 2010 [81]	Guangzhou	Case-control	9848	2006–2008	Average of daily measurements of pollutants from one central monitoring station		OR, 1.00; 95% CI, 1.00–1.01	OR, 1.01; 95% CI, 1.00–1.01	

A population-based study conducted recently in Wuhan, China, for a cohort of 95,911 live, single births during a 2-year period from 2011 to 2013 found that 3%, 2%, 15%, and 5% increases in risk of PTB with each 5-ug/m^3 increase in $PM_{2.5}$ and PM_{10} concentrations, 100-ug/m^3 increase in CO concentrations, and 10-ug/m^3 increase in O_3 concentrations, respectively. There was negligible evidence for associations for SO_2 and NO_2. The strongest effect for PTB was found in the second trimester for $PM_{2.5}$, PM_{10}, and CO, but for SO_2 it was in the first trimester, second month, and third month. For NO_2, it was in the first trimester and second month and for O_3, the third trimester. The first trimester and the third trimester are important for fetal development, but it is difficult to assure which gestational duration is relevant vulnerable exposure window. Most previous published literatures investigated the relationships between air pollution and PTB; some studies showed that the first trimester had greater effects, but others showed the exposure during the second trimester was associated with an increased risk of PTB, and other studies suggest that the third trimester was the role exposure window for PTB. So it is challenging to determine which pregnancy period is sensitive to air pollution exposure for PTB. Considering that the toxic influence of air pollutants may have an accumulated effect on fetal growth and development, maybe it is appropriate to study the effects of air pollutants on PTB for the exposure of the entire pregnancy.

In conclusion, heterogeneity between studies varied widely between pollutants and PTB. Three reasons can explain the inconsistent results. Firstly, methodological differences between studies can partially play a role for the heterogeneity; secondly, it is clear that PTB resulted from complicated factors including biological, physical, psychological, and social factors, and the role of these risk factors may modify the association between air pollution and PTB; it is obvious that the effect of these factors on PTB varied among different population; thirdly, air is a mixture of pollutants, the majority of gas pollutants are well correlated among themselves, and this high collinearity may make it difficult to identify which pollutant is more harmful. The study on the association between air pollution exposure and PTB may provide some reference for the control of air pollution and reduction of PTB occurrence. Given the number of studies that have found significant relationship between maternal air pollution exposures during pregnancy and risk of PTB, the epidemiological findings supported that air pollution contributes to women's risk for PTB. The effect of air pollution on PTB varied due to the specific air pollutants and different gestational age. Future studies are needed to determine an appropriate measure of time of assessment, and an accurate measurement of air pollution exposure is needed.

8.2.2 *Air Pollutants and the Risk of Low Birth Weight*

Low birth weight (LBW), which is the one of the major adverse pregnancy outcomes, is defined as a birth weight of a liveborn infant of 2499 g or less by the WHO [46]. A report by the WHO in 2015 shows that about 20 million newborn babies are born with less than 2500 g of body weight per year, accounting for 15.5% of newborns in the world, and the rate of LBW infants in developing countries (16.5%) is more than twice as high as in developed countries (7%) [56]. There are many factors associated with the occurrence of LBW, such as maternal age, education level, living habits and so on, fog and haze events which frequently occurred in recent years, and the effect of air pollutant exposure on low birth weight during the perinatal period, which have attracted more and more attention [16, 21]. However, the current study on the associations between air pollutant exposure and LBW in China is still rare except for several small-scale researches [46].

A study was conducted in Beijing based on 50,874 live births delivered between January 1, 2006, and December 31, 2010, at the Beijing Haidian Maternal and Child Health Hospital [43]. Daily average levels of air pollutants for the same period were obtained from the Beijing Municipal Environmental Monitoring Center, and the monthly average concentrations of NO_2 and PM_{10} were 58.07 ug/m^3 (23.00–83.00 ug/m^3) and 134.72 ug/m^3 (69.00–264.00 ug/m^3). Their results showed that exposure to high concentration of NO_2 in the third trimester predicted lower birth weights, with each 10-unit increment (per 10 ug/m^3) in NO_2 concentration associated with a 13.78 g (95 % CI, −21.12 to 6.43; $p < 0.001$) reduction in birth weight, but no significant association was found between PM_{10} exposure and birth weight, with or without adjustment for the concentration of other pollutants [43].The similar cohort study which covered 74,671 first-parity live births with gestational age 37–44 weeks was conducted in Beijing during 1988–1991 to explore the associations between SO_2 and total suspended particulates (TSP) and low birth weight [95]. Wang's results showed that the risk of LBW increased by 11% (95% CI, 1.06–1.16) per 100-ug/m^3 increase in $S0_2$ and 10% (95% CI, 1.05–1.14) per 100-u/m^3 increase in TSP, and the estimated reduction in birth weight was 7.3 g and 6.9 g per 100-ug/m^3 increase in SO_2 and TSP, respectively.

Similar results were obtained from the study which was conducted in Shanghai [77], the study sample was based on the official Shanghai birth registry system from 2008 to 2012, and the concentrations of air pollutants were from the records of the Shanghai Environmental Monitoring Center from 2007 to 2012. Their results revealed that the mean concentration of PM_{10} increase per 10 ug/m^3 in the first trimester would reduce 4.3 g of baby's birth weight, and the reduction in birth weight was 7.6 g and 6.0 g per 100-ug/m^3 increase in SO_2 and NO_2 [77]. Fan's study which was based on hospital records in Hainan province during 2000–2010 showed that PM_{10} exposure was significant associated with low birth weight; however, there was no significant association found between NO_2 and SO_2 exposure and low birth weight [26]. A recent meta-analysis reported that $PM_{2.5}$ was significantly associated with low birth weight [60]. However, Zhang's study [107] which was conducted in Taiyuan city found that there was no significant association between PM_{10} and SO_2 exposure during pregnancy and low birth weight. The details of studies can be seen in Table 8.2.

From above, exposure to air pollutants during pregnancy is closely related to the occurrence of LBW, and there are significant differences in exposure to different air pollutants and different stages of pregnancy in different studies; however, some studies also show that air pollutant exposure are not significantly associated with

Table 8.2 Prenatal air pollutant exposure and low birth weight

Author/Year	Location	Design	Sample size	Study period	Exposure assessment methods	Exposure period	Results			
							PM_{10}	SO_2	NO_2	TSP
Zhang et al .2007 [107]	Taiyuan	Case-crossover	52,488	1997–2004	Daily average exposure measured by six monitoring stations from 3 months before pregnancy to 6 months after pregnancy	3 months before pregnancy	OR, 0.99; 95% CI, 0.88–1.12	OR, 0.98; 95% CI, 0.89–1.07		
						First month after pregnancy	OR, 0.89; 95% CI, 0.78–1.01	OR, 1.04; 95% CI, 0.96–1.13		
						Second month after pregnancy	OR, 0.89; 95% CI, 0.78–1.02	OR, 1.01; 95% CI, 0.92–1.02		
						Third month after pregnancy	OR, 1.04; 95% CI, 0.91–1.19	OR, 0.93; 95% CI, 0.85–1.02		
						Fourth to sixth months after pregnancy	Or, 1.06; 95% CI, 0.86–1.31	OR, 0.92; 95% CI, 0.83–1.02		
						Seventh to ninth months after pregnancy	OR, 0.88; 95% CI, 0.78–0.99	OR, 1.03; 95% CI, 0.94–1.12		

(continued)

Table 8.2 (continued)

Author/Year	Location	Design	Sample size	Study period	Exposure assessment methods	Exposure period	Results			
							PM_{10}	SO_2	NO_2	TSP
Huang et al. 2015 [43]	Beijing	Case-control	50,874	2006–2010	Daily average concentration of NO_2 and PM_{10} was obtained from the Beijing Municipal Environmental Monitoring Center	First trimester	Change of birth weight, −1.18 g; 95% CI, -3.21–0.85		Change of birth weight, 5.85 g; 95% CI, −1.03–12.74	
						Second trimester	Change of birth weight, 0.43 g; 95% CI, -1.74–2.60		Change of birth weight, −3.72 g; 95% CI,-11.76–4.32	
						Third trimester	Change of birth weight, −2.55 g; 95% CI,-5.29–0.20		Change of birth weight, −13.78 g; 95% CI, −21.12–6.43	
Wang et al. 1997 [95]	Beijing	Cohort	74,671	1988–1991	The outdoor TSP and SO_2 concentrations were monitored at the World Health Organization Global Environmental Monitoring System	Third trimester	OR, 1.16; 95% CI, 1.06–1.16;			OR,1.10; 95% , 1.05–1.14

Qian et al. 2016 [77]	Shanghai	Case-crossover	964,400	2008–2012	daily average concentration of air pollutants was obtained from the environmental monitoring center in Shanghai	First trimester	Change of birth weight, −4.3 g	Change of birth weight, −7.6 g;	Change of birth weight, −6.0 g;	
Fan 2013 [26]	Hainan	Cohort	352,361	2000–2010	Daily average concentration of PM_{10}, NO_2, and SO_2 was obtained from the Hainan Environmental Monitoring Center		RR (lag1), 1.023; 95% CI, 1.010–1.036	RR (lag5), 0.899; 95% CI, 0.814–0.984	RR (lag2), 0.948; 95% CI, 0.903–0.993	

low birth weight. This may be due to pollution source, pollutant concentration in different regions, different individual exposure caused by pregnancy, as well as the research design and statistical methods; further, researches are needed to determine the potential relationships.

8.2.3 Air Pollutants and the Risk of Fetal Death and Stillbirth

Fetal death is defined as a death occurring after 20 weeks' gestation. Stillbirth, which is a kind of fetal death, is defined by the WHO as a child born with no signs of life during delivery [66]. Fetal death and stillbirth are common complications of pregnant women. There are about 2.65 million cases of fetal death and stillbirth each year in the world [6]. Although the majority of dead fetus can be delivered naturally in 2 or 3 weeks, it is possible to increase mothers' risk of disseminated intravascular coagulation, placenta adhesion, intrauterine infection, and postpartum retained placenta. Fetal death and stillbirth are not only seriously harmful to pregnant women's physical and mental health, but also bring economic burden to their families and the country, especially for low-income families. Evidence showed that exposure to air pollution has been associated with fetal death or stillbirth.

8.2.3.1 The Relationships Between Particulate Matter (PM) and Fetal Death and Stillbirth

Some researchers have conducted research on the relationship between PM and fetal death in China, and most of studies' results showed that PM exposure was associated with fetal death. Hou et al. [39, 41] (Table 8.3) preliminarily explored the relationships of air pollutants and fetal death in Tianjin city, which is located in the north of China. They collected fetal death cases from 15 comprehensive hospitals and obstetrics and gynecology hospitals which are located in six districts of Tianjin city during January 2002 to August 2007 and obtained the concentration data of air pollutants in the six districts of Tianjin city from the environmental monitoring center in the same period. The results showed that air pollution in heating period in Tianjin was more serious than that in non-heating period, and the proportion of fetal death in heating period was relatively higher than that in non-heating period, and the variations of PM_{10} and TSP pollution level were consistent with the variation of fetal death. After adjusting for pregnant women's age and pregnancy times, they found that high concentration of TSP (TSP $\geqq$ 0.996 mg/m^3) exposure may increase the risk of fetal death (OR, 1.40; 95% CI, 1.06–1.86), but they did not find that PM_{10} exposure during pregnancy was associated with fetal death [40]. Hou's another study (Hou et al. 2009) found that the concentration of BaP-DNA adduct in the peripheral blood of the mother who had fetal death was significantly higher than those who had normal pregnancy. Compared with women who had a level of BaP-DNA adduct less than 1.8 in the peripheral blood, the risk of fetal death was significantly increased

Table 8.3 Prenatal air pollutant exposure and fetal death

Author/year	Location	Design	Sample size	Study period	Exposure assessment methods	Exposure period	Results					
							TSP	PM_{10}	SO_2	NO_2	O_3	CO
Hou et al. 2012 [40]	Tianjin	Case-control	1918	2001–2006	Monthly average of pollutants at nearest of ten monitors to maternal residence	Peri pregnancy	OR, 1.40; 95% CI, 1.06–1.86					
Zhang et al. 2007 [107]	Taiyuan	Case-crossover	52,488	1997–2004	Daily average exposure measured by six monitoring stations from 3 months before pregnancy to 6 months after pregnancy	Three months before pregnancy		OR, 1.07; 95% CI, 0.74–1.55	OR, 0.88; 95% CI, 0.65–1.19			
						First month after pregnancy		OR, 1.03; 95% CI, 0.69–1.52	OR, 0.96; 95% CI, 0.74–1.25			
						Second month after pregnancy		OR, 0.94; 95% CI, 0.62–1.45	OR, 1.02; 95% CI, 0.78–1.35			
						Third month after pregnancy		OR, 1.26; 95% CI, 0.85–1.88	OR, 0.83; 95% CI, 0.61–1.11			
						Fourth to sixth months after pregnancy		OR, 1.30; 95% CI, 0.72–2.34	OR, 0.93; 95% CI, 0.69–1.26			
						Seventh to ninth months after pregnancy		OR, 1.66; 95% CI, 1.26–2.19	OR, 0.77; 95% CI, 0.59–0.99			

(continued)

Table 8.3 (continued)

Author/year	Location	Design	Sample size	Study period	Exposure assessment methods	Exposure period	Results					
							TSP	PM_{10}	SO_2	NO_2	O_3	CO
Hwang et al. 2011 [45]	Taiwan	Case-control	102,575	2001–2007	Ambient air monitoring data for SO_2, NO_2, O_3, CO, and PM_{10} are available for 72 Taiwan Environmental Protection Administration (EPA) monitoring stations	First trimester		OR, 1.03; 95% CI, 1.00–1.07	OR, 1.04; 95% CI, 1.01–1.07	OR, 1.04; 95% CI, 0.97–1.12	OR, 1.02; 95%CI, 0.95–1.09	OR, 0.99; 95% CI, 0.97–1.02
						Second trimester		OR, 0.99; 95% CI, 0.95–1.02	OR, 1.00; 95% CI, 0.94–1.07	OR, 1.00; 95% CI, 0.93–1.07	OR, 1.00; 95%CI, 0.94–1.07	OR, 0.98; 95% CI, 0.96–1.01
						Third trimester		OR, 0.97; 95% CI, 0.92–1.02	OR, 1.01; 95% CI, 0.97–1.04	OR, 0.98; 95% CI, 0.89–1.08	OR, 0.98; 95% CI, 0.93–1.08	OR, 0.98; 95% CI, 0.95–1.02
						Whole pregnancy		OR, 1.00; 95% CI, 0.96–1.05	OR, 1.03; 95% CI, 1.00–1.06	OR, 1.02; 95% CI, 0.94–1.10	OR, 1.00; 95%CI, 0.91–1.09	OR, 0.99; 95% CI, 0.96–1.01
Zou 2008 [113]	Tianjin	Case-control	1950	2001–2007	Daily average exposure measured by six monitoring stations	First month after pregnancy		OR, 1.19; 95% CI, 1.01–1.39				

Hou et al. 2014 [42]	Tianjin	Case-control	1918	2001–2006	Daily average and monthly concentrations of the gaseous pollutants	3 months prior to conception	OR, 1.80; 95% CI, 0.86–3.77	OR, 0.95; 95% CI, 0.29–4.48	OR, 2.18; 95% CI, 0.43–10.97	OR, 1.57; 95% CI, 0.03–84.01		
						2 months prior to conception	OR, 1.98; 95% CI, 0.92–4.26	OR, 0.25; 95% CI, 0.03–2.51	OR, 1.90; 95% CI, 0.36–9.90	OR, 152.05; 95% CI, 1.83–12,663.36		
						1 month prior to conception	OR, 1.28; 95% CI, 0.66–2.49	OR, 0.30; 95% CI, 0.02–4.15	OR, 3.97; 95% CI, 0.69–22.89	OR, 98.35; 95% CI, 2.96–3264.98		
						Month of conception	OR, 1.47; 95% CI, 0.65–3.32	OR, 13.99; 95% CI, 0.88–222.74	OR, 12.33; 95% CI, 2.18–69.59	OR, 3.78; 95% CI, 0.16–91.68		
						1 month after conception	OR, 2.20; 95% CI, 1.14–4.25	OR, 2.82; 95% CI, 0.44–18.22	OR, 17.46; 95% CI, 3.01–101.24	OR, 107.60; 95% CI, 2.55–4544.61		
Wang 2008 [94]	Tianjin	Case-control	1678	2002–2006	Monthly average of pollutants at nearest of ten monitors to maternal residence	Peri pregnancy		High exposure: OR, 1.88; 95% CI, 1.44–2.44; moderate exposure: OR, 1.36; 95% CI, 1.08–1.70				

when the peripheral blood adduct level was above 5.5 (OR, 47.22; 95% CI, 6.06–367.73). BaP is one of the components of $PM_{2.5}$, which means that $PM_{2.5}$ may be a risk factor for fetal death. Besides, Yang et al. [103] conducted case-control study based on hospital and found that the blood plumbum levels of pregnant women in fetal death group and control group were 87.17 ± 28.98 μg/L and 59.51 ± 13.63 μg/L, respectively, and the difference was statistically significant. Because heavy metal was an important component of $PM_{2.5}$, Yang's study suggested that $PM_{2.5}$ might be related to fetal death. Zhang et al. [107] collected the concentration of air pollutants from environment monitoring points and women's pregnancy outcome data in the same period to explore the relationship between different stages of embryonic development and air pollutant exposure in Taiyuan city. After adjusting the potential confounding factors, the results show that PM_{10} was the high-risk factor of occurrence of fetal death. In late pregnancy (seventh to ninth months), the concentration of PM_{10} in the air increased per 100 μg/m^3; the OR value of fetal death and stillbirth was 1.66 (95% CI: 1.26–2.19), which was in accordance with previous study [97], but they did not found PM_{10} in 3 months before pregnancy, and 1–6 months after pregnancy was associated with fetal death. Hwang et al. [45] conducted a population-based case-control study from 1 January 2001 to 31 December 2007 in Taiwan based on the Taiwanese Birth Registry that comprised 1,510,064 registered singleton births. They found that a 10-μg/m3 increase in PM_{10} were weekly associated with stillbirth overall during the first (adjusted OR, 1.02; 95% CI, 1.00–1.05) and second month (adjusted OR, 1.02; 95% CI, 1.00–1.04).

From above, the components of PM exposure are closely related to the occurrence of fetal death or stillbirth. Meanwhile, their results also suggest that PM exposure during pregnancy is harmful to pregnant women and fetuses. Due to the diversity and harmfulness of components of PM, we should pay much more attention to it and further prevent the occurrence of adverse pregnancy outcome.

8.2.3.2 The Relationships Between Carbon Monoxide (CO) and Fetal Death and Stillbirth

In China, there are several studies exploring the relationship between CO exposure and fetal death and stillbirth, but their results on the relationship between CO exposure and stillbirth were not consistent. Zou et al. [113] conducted a monographic study to explore the relationships of CO exposure and fetal death in Tianjin city by a retrospective case-control study. They collected fetal death cases from 15 comprehensive hospitals and obstetrics and gynecology hospitals which are located in six districts of Tianjin city during January 2001 to August 2007 and randomly selected the control group according to age group. After adjusting for confounding factors, the results showed that CO was a risk factor for the occurrence of fetal death (OR, 1.19; 95% CI, 1.01–1.39). For different CO concentration exposures, pregnant women exposed to medium concentration groups were 1.61 times (OR, 1.61; 95%

CI, 1.03–2.52) as high as that in the low exposure group in the first month of pregnancy, and pregnant women in the high concentration exposure group were 1.66 times (OR, 1.66; 95% CI, 1.06–2.61) of the low exposure group. These results above showed that pregnant women who are exposed to medium and high concentrations of CO in the first month after pregnancy would increase the possibility of the occurrence of fetal death. Moreover, Hwang et al. [45] carried out a case-control study in 2011 in Taiwan to explore the relationship between a 100-ppb increase in CO during different pregnancy and stillbirth. But the results showed that there was no significant relationship between CO exposure and stillbirth. From the results of the above case-control study, there is no unified conclusion on the causal relationship between CO exposure and stillbirths; more comprehensive research are needed to explore the causal relationships in the future.

8.2.3.3 The Relationships Between Carbon Monoxide (SO_2) and Fetal Death and Stillbirth

Hou [42] conducted a retrospective case-control's study on the relationship between air pollution and fetal death in Tianjin, and 959 fetal losses and 959 normal intrauterine pregnancies were recruited. Hou's study showed that pregnant women with SO_2 exposure during the month of pregnancy and first month after pregnancy were significantly associated with fetal death; OR values were 12.33 (95% CI, 2.18–69.59) and 17.46 (95% CI, 3.01–101.24), respectively. Wang [94] conducted the specific study to explore the relationship between different SO_2 exposure levels in periconceptional period and fetal death in Tianjin City. The results showed that SO_2 was associated with the occurrence of fetal death. For different SO_2 concentration exposure, pregnant women exposed to medium concentration groups were 1.36 times (OR, 1.36; 95% CI, 1.08–1.70) as high as that in the low exposure group in the first month of pregnancy, and pregnant women in the high concentration exposure group were 1.88 times (OR, 1.88; 95% CI, 1.44–2.44) of the low exposure group. Hwang et al. [45] also explored the relationship between a 1-ppb increase in SO_2 during different pregnancy and stillbirth. They got the conclusion that a 1-ppb increase in SO2 during the first trimester was significantly associated with stillbirth among all births (adjusted OR, 1.02; 95% CI, 1.00–1.04), but stratified estimates indicated that the association was present only among preterm births (adjusted OR, 1.04; 95% CI, 1.01–1.07). The results of these studies support the maternal SO_2 exposure can increase the risk of fetal death or stillbirth that occurred and that early pregnancy is the most susceptible stage. But Zhang's study [107] which was conducted in Taiyuan city found that SO_2 exposure was not associated with stillbirth. Although there is no consensus on the relationship between SO_2 exposure and stillbirth, it indicates that SO_2 may be a potential pollutant to maternal health.

8.2.3.4 The Relationships Between Nitrogen Oxide (NO_x) and Fetal Death and Stillbirth

Similarly, there was seldom special research on the relationship between NO_x and stillbirth, and the conclusions were not uniform. Hou's study [42] on the relationship between air pollution and fetal death in Tianjin showed that pregnant women with NO_2 exposure during the first and second month before pregnancy were significantly associated with fetal death; OR values were 152.05 (95% CI, 1.83–12,663.36) and 98.35 (95% CI, 2.96–3264.98), respectively. But in Hwang's study [45] which was conducted in Taiwan, we did not found that NO_2 exposure was associated with stillbirth. Some foreign researches showed that NO_2 exposure was closely related to stillbirths. Green's study [31] showed that a 10-ppb increase in NO_2 (OR, 1.08; 95% CI, 1.03, 1.13) during the entire pregnancy was associated with stillbirth, and NO_2 (RR, 1.03; 95% CI, 1.02–1.03) was also associated with stillbirth in the last third of pregnancy in Spain. Faiz et al. [24] conducted a study to explore the relationship between NO_2 exposure and stillbirth in New Jersey and found that a 10-ppb increase in NO_2 exposure in the first trimester could increase the risk of stillbirth. From the above, we can conclude that NO_2 exposure is associated with fetal death and stillbirth, but their causal relationship needs to be evaluated in the future.

8.2.3.5 The Relationships Between Ozone (O_3) and Fetal Death and Stillbirth

The study on relationship between O_3 exposure and stillbirth was rare and the relationship was not clear. In Hwang's population-based case-control study [45] which was conducted in Taiwan in 2011, they explored the association of O_3 exposure for each month of pregnancy, each trimester, and the whole pregnancy to elaborate the relevant gestational period for stillbirth. The results showed O_3 exposure was not related with stillbirth in their cohort study regardless of the timing of exposure or preterm versus term birth status. However, Green's study [31] which was conducted in California in 2014 found that a 10-ppb increase in O_3 exposure during the third trimester was significantly associated with stillbirth (OR, 1.03; 95% CI, 1.01–1.05), but neither the entire pregnancy exposure nor the exposures in the other two trimesters were still significant. Similarly, Arroyo's time series study [2] showed that O_3 had a statistically significant effect on stillbirth in the first trimester of pregnancy (RR, 1.16; 95% CI, 1.15–1.16). Although several foreign studies showed that O_3 exposure was associated with stillbirth, the relationship between O_3 exposure and stillbirth was not fully explored in China. A monographic study on the relationship between O_3 exposure and fetal death and stillbirth will be needed in the future.

From the above studies, air pollutants are closely related to fetal death and stillbirth; these conclusions are just based on a few case-control studies. In China, a monographic study on the relationship between air pollutants and fetal death and stillbirths are rare, especially for NO_x and O_3; thus, in order to better know about the relationship and pathogenesis between air pollutant exposure and fetal death and stillbirths and provide theoretical basis for the prevention and reduce the occurrence of fetal death and stillbirths, a longitudinal study on the relationship between air pollutants and fetal death should be carried out in the future.

8.2.4 *Air Pollutant and Risk of Birth Defects*

Common pollutants have been shown to be associated with the risk of preterm birth (PTB), low birth weight (LBW), and infant mortality [98, 105, 106]. However, few studies have explored the association between air pollution exposure during pregnancy and risk of congenital anomalies [93].

Congenital anomalies are considered as a main risk factor of stillbirth and neonatal and infant mortality. Worldwide, 3.2 million disabilities every year and an estimated 10% of deaths in children below 5 years old are caused by congenital anomalies [14]. In the United States, about 3% of births are associated with a birth defect [37]. In China, the incidence of congenital malformations is 5.6%, with about 900,000 new birth defect cases each year [104]. The precise etiology of most congenital anomalies is suggested to be multifactorial, with environmental exposures such as air pollution suspected to have a role [79]. The potential impact of environmental exposures to congenital anomalies has been recently reviewed [14, 93].

8.2.4.1 The Relationships Between Air Pollutants and Congenital Heart Defects (CHDs)

Congenital heart defects (CHDs) are the most common severe congenital anomalies and are the leading cause of infant mortality due to congenital anomalies [1]. A recent meta-analysis reported that NO_2 was significantly associated with coarctation of the aorta [14]. Another meta-analysis that combined results from four studies reported associations between tetralogy of Fallot (TF) and NO_2 exposure and between coarctation of the aorta and SO_2 exposure [93]. However, the summary risk estimates were based on a small number of four studies, and few specific types of CHDs were explored. Compared to those literatures published in the Europe and United States, a very limited study has been published on the association between CHDs and ambient air pollution in Asia.

Zhang [105] (Table 8.4) conducted a cohort study with a total of 105,988 live-born infants, stillbirths, and fetal deaths. They included all pregnant women who lived in the urban district of Wuhan during pregnancy over the 2-year period from 10 June 2011 to 9 June 2013. For each study participant, they assigned 1-week and 1-month averages of ambient air pollution exposure based on air pollution concentration obtained from the nearest exposure monitor to the living residence of mothers during their early pregnancy period. They observed an increased risk of CHDs, particularly ventricular septal defect (VSD), with increasing $PM_{2.5}$ exposure. Using 1-week averages, they found significant monotonically increasing associa-

Table 8.4 Prenatal air pollutant exposure and birth defects

Author/year	Location	Design	Sample size	Study period	Exposure assessment methods	Exposure period	Results					
							$PM_{2.5}$	PM_{10}	SO_2	NO_2	O_3	CO
Zhang et al. 2016	Wuhan	Cohort	105,988	2011–2013	1-month and 1-week averages of PM_{10} and $PM_{2.5}$ exposure from the nearest monitor	First month of pregnancy	CHD: OR, 1.01; 95% CI, 0.93–1.09; VSD: OR, 1.11; 95 CI, 0.98–1.25	CHD: OR, 0.94; 95% CI, 0.89–1.01; VSD: OR, 0.97; 95 % CI, 0.87–1.09	CHD: OR, 0.92; 95% CI, 0.81–1.04; VSD: OR, 0.90; 95% CI, 0.72–1.12; TF: OR, 0.91; 95 CI, 0.67–1.24	CHD: OR, 0.90; 95 CI, 0.79–1.02; VSD: OR, 0.83; 95%CI, 0.67–1.03; TF: OR, 0.72; 95% CI, 0.53–1.03	CHD: OR, 1.06; 95% CI, 1.00–1.13; VSD: OR, 1.07; 95 CI, 0.96–1.18; TF: OR, 1.15; 95 CI, 0.99–1.33	CHD: OR, 0.97; 95% CI, 0.90–1.06; VSD: OR, 0.99; 95 CI, 0.86–1.14; TF: OR, 1.01; 95 CI, 0.81–1.26
						Second month of pregnancy	CHD: OR, 1.10; 95% CI, 1.03–1.18; VSD: OR, 1.16; 95 CI, 1.03–1.30	CHD: OR, 0.99; 95% CI, 0.92–1.05; VSD: OR, 0.96; 95% CI, 0.86–1.07	CHD: OR, 0.87; 95% CI, 0.76–0.99; VSD: OR, 0.82; 95% CI, 0.65–1.03; TF: OR, 1.07; 95%CI, 0.78–1.46	CHD: OR, 0.94; 95%CI, 0.84–1.06; VSD: OR, 0.89; 95%CI, 0.73–1.09; TF: OR, 0.81; 95% CI, 0.60–1.11	CHD: OR, 1.10; 95% CI, 1.03–1.17; VSD: OR, 1.09; 95% CI, 0.97–1.21; TF: OR, 1.24; 95%CI, 1.07–1.44	CHD: OR, 0.92; 95% CI, 0.84–1.01; VSD: OR, 0.93; 95%CI, 0.81–1.07; TF: OR, 0.97; 95%CI, 0.77–1.21
						Third month of pregnancy	CHD: OR, 1.08; 95% CI, 1.01–1.16; VSD: OR, 1.21; 95%CI, 1.08–1.36	CHD: OR, 0.98; 95% CI, 0.93–1.05; VSD: OR, 0.99; 95% CI, 0.90–1.10	CHD: OR, 0.83; 95% CI, 0.73–0.95; VSD: OR, 0.81; 95% CI, 0.64–1.02; TF: OR, 0.97; 95% CI, 0.68–1.38	CHD: OR, 0.90; 95%CI, 0.81–1.01; VSD: OR, 0.91; 95%CI, 0.76–1.09; TF: OR, 0.80; 95% CI, 0.60–1.06	CHD: OR, 1.12; 95% CI, 1.05–1.19; VSD: OR, 1.17; 95%CI, 1.05–1.31; TF: OR, 1.31; 95%CI, 1.13–1.51	CHD: OR, 0.99; 95% CI, 0.91–1.08; VSD: OR, 1.12; 95%CI, 0.87–1.29; TF: OR, 1.05; 95%CI, 0.85–1.30

Jin et al. 2015 [50]	Lanzhou	Cross-sectional	8969	2010–2012	Daily average concentrations of ambient PM_{10}, SO_2, and NO_2 were obtained from four air monitoring stations	First trimester		PCHD: OR, 3.10; 95% CI, 1.28–7.51; CMGA: OR, 4.23; 95% CI, 1.53–11.65; ICPDA: OR, 3.96; 95% CI, 1.36–11.53				
						Second trimester		PCHD: OR, 4.23; 95% CI, 2.11–8.50; CMGA: OR, 4.00; 95% CI, 1.80–8.87; CMCS: OR, 5.51; 95% CI, 1.36–22.35; ICPDA: OR, 3.59; 95% CI, 1.57–8.22		PCHD: OR, 1.86; 95%CI, 1.18–2.93; CMGA: OR, 1.99; 95% CI, 1.18–3.36; ICPDA: OR, 1.92; 95% CI, 1.11–3.34		
						Entire pregnancy		PCHD: OR, 2.28; 95% CI, 1.45–3.60; CMGA: OR, 2.33; 95% CI, 1.37–3.97; CMCS: OR, 2.70; 95% CI, 1.11–6.57; ICPDA: OR, 2.09; 95% CI, 1.21–3.62	CMCS: OR, 5.16; 95%CI, 1.01–26.26	PCHD: OR, 2.26; 95%CI, 1.26–4.08; CMGA: OR, 2.16; 95% CI, 1.12–4.18; ICPDA: OR, 2.32; 95%CI, 1.14–4.71		

(continued)

Table 8.4 (continued)

Author/ year	Location	Design	Sample size	Study period	Exposure assessment methods	Exposure period	Results					
							$PM_{2.5}$	PM_{10}	SO_2	NO_2	O_3	CO
Liu et al. 2016 [61]	Fuzhou	Case-control	4634	2007–2013	10-day or 1-month averages of daily particulate matter using an air monitor-based inverse distance weighting method during early pregnancy.	First and second trimester		PDA: OR, 1.54, 1.63; 95% CIS, 1.17–2.23, 1.06–3.24; FCM: OR, 1.28; 95% CI, 1.03, 1.61, VSD: OR, 1.19; 95% CI, 1.00, 1.43; TF: OR, 1.44; 95% CI,1.01, 2.19				
Hwang et al. 2008 [44]	Taiwan	Case-control	7183	2001–2003	Complete monitoring data for the air pollutants SO_2, NO_x, O_3, CO, and PM_{10} and daily temperature and relative humidity are available for 72 Taiwan Environmental Protection Agency (EPA) monitoring stations	First month		CL: OR, 1.01; 95%CI, 0.96–1.06	CL: OR, 0.92; 95%CI, 0.63–1.35	CL: OR, 0.95; 95%CI, 0.81–1.12	CL: OR, 1.17; 95%CI, 1.01–1.36	CL: OR, 1.00; 95%CI, 0.96–1.04
						Second month		CL: OR, 1.00; 95%CI, 0.95–1.05	CL: OR, 0.84; 95%CI, 0.57–1.25	CL: OR, 0.96; 95%CI, 0.81–1.13	CL: OR, 1.22; 95%CI, 1.03–1.46	CL: OR, 1.00; 95%CI, 0.96–1.03
						Third month		CL: OR, 0.99; 95%CI, 0.95–1.05	CL: OR, 0.72; 95%CI, 0.47–1.08	CL: OR, 0.93; 95%CI, 0.79–1.09	CL: OR, 1.09; 95%CI, 0.93–1.26	CL: OR, 1.00; 95%CI, 0.96–1.03

tions between the risk of VSD and $PM_{2.5}$ exposure during weeks 7–10 of pregnancy, with adjusted OR ranging from 1.11 to 1.17 (95 % CI, 1.02–1.20, 1.03–1.22, 1.05–1.24, and 1.08–1.26 separately) per 10-μg/m^3 change of $PM_{2.5}$ concentration; significant associations had not been found between PM_{10} and CHD and VSD. Epidemiological evidence linking to CHDs and maternal $PM_{2.5}$ exposure is still inconsistent and limited. The California study explored the odds of CHDs during the first 2 months of pregnancy and reported positive associations between transposition of the great arteries and $PM_{2.5}$, but opposed relationships between pulmonary valve stenosis and $PM_{2.5}$ [72].

The Wuhan study [106] also observed increasing associations between CHDs overall, VSD, and TF individually and O_3 exposure. The results are consistent to other studies that have explored the associations between CHDs and ozone exposure. The Southern California study demonstrated a higher risk of valve defects and aortic artery in relation to increasing O_3 exposure during the second month of pregnancy [80]. Similarly, the Northeast England study reported an increased risk of congenital malformations of tricuspid valve and pulmonary with exposure after limiting to participants living within 16 kilometers of a monitor [17].

The Lanzhou cohort study [50] investigated 8969 singleton live births in China during 2010–2012. Using inverse distance weighting method, maternal exposures to PM_{10}, NO_2, and SO_2 were estimated as a combination of monitoring concentration for time spent in a work location and at home. They found significant positive associations between patent ductus arteriosus (PDA) and PM_{10} exposure during the first trimester, second trimester, and the entire pregnancy (first trimester OR, 3.96; 95% CI, 1.36–11.53; second trimester OR, 3.59; 95% CI, 1.57–8.22; entire pregnancy OR, 2.09; 95% CI, 1.21–3.62) and associations with NO_2 of the second trimester and the entire pregnancy (second trimester OR, 1.92; 95% CI, 1.11–3.34; entire pregnancy OR, 2.32; 95% Cl, 1.14–4.71). They also found positive associations between PM_{10} exposure and congenital malformations of cardiac septa in the second trimester and the entire pregnancy and SO_2 exposures with the entire pregnancy.

Liu [61] conducted a hospital-based case-control study in Fuzhou China during 2007–2013. Using a monitor-based inverse distance weighting method, the exposure was assigned 10-day or 1-month averages of daily PM_{10} during early pregnancy. They enrolled a total of 662 live birth and 3972 live birth controls. Their results showed that PM_{10} exposure was positively related with the risks of atrial septal defect (adjusted ORs ranging from 1.29 to 2.17), patent ductus arteriosus (aORs, 1.54, 1.63; 95% CIs, 1.17–2.23, 1.06–3.24), overall fetal cardiovascular malformations (aOR, 1.28; 95% CI, 1.03–1.61), VSD (aOR, 1.19; 95% CI, 1.00–1.43), and tetralogy of Fallot (aOR, 1.44; 95% CI, 1.01–2.19) in the various observed periods scaled by 10 days or 1 month in the first and second trimester. The strongest significant associations were found for PM_{10} exposure in the second quartile. No correlations of PM_{10} levels with CHDs in the other time periods of gestation were observed.

8.2.4.2 The Relationships Between Air Pollutants and Oral Cleft

Oral cleft is common but has received less attention than congenital heart defects in studies of air pollution research. Some previous epidemiologic studies have linked oral clefts with ambient air pollution, but the evidence is inconsistent.

The Hong Kong study [15] explored the possible correlation between the risk of oral clefts and NO, NO_2, SO_2, and O_3. The authors tabulated 48,804 births with 59 cleft examples occurring in Hong Kong from 2002 to 2009. And they found that the incidence of oral clefts is inversely correlated with NO_x at around 8 weeks postconception, particularly for cleft lip (CL). The Taiwan study [44] created a case-control study that examined 721,289 Taiwanese newborns from 2001 to 2003. They collected 653 cases of cleft lip (CL) and/or cleft palate (CP) and 6530 controls. This study explored the associations between exposure to SO_2, NO_x, O_3, CO, and PM_{10} and the risk of clefts during the first trimester of pregnancy. The authors found that exposure to O_3, but not the other pollutants, was associated with an increased cleft lip with cleft palate risk.

The Wuhan cohort study observed an increased risk of oral cleft with increasing $PM_{2.5}$ exposure in the most susceptible time periods in pregnancy. In contrast, higher CO exposure was associated with decreased odds of cleft palate of the third month of the pregnancy. A meta-analysis confirmed the association of ozone exposure and risk of oral cleft (OR,1.08;95%CI,1.01-1.16). NO_x was consistently associated with decreased risk of cleft palate and cleft lip with or without palate [78].

It is possible that inconsistent findings for risk of congenital heart defect and oral cleft and air pollution exposure across studies may be at least partially due to differences in the pollution concentration and range of pollution experienced in different countries. The measurement error in air pollution exposure assessment that was too great is another possibility of inconsistent findings. The previous studies generally use pure spatial modeling [14], pure temporal approaches [32], the nearest monitor approach, or spatiotemporal modeling [85], which all can cause measurement error for the distance between the monitor and the subject. Using the estimated date of conception may cause exposure misclassification of timing 14 days from the last menstrual period (LMP), because women may not recall LMP date accurately, and the LMP may be unreliable. Using residential information for the mother at the time of birth rather than during the first trimester may also raise exposure misclassification of timing. There would be exposure misclassification if local pollution sources existed, such as construction, traffic, or other spatially distributed risk factors. Measurement error would also be different for each air pollutants, as the patterns of their spatial distribution vary quite widely, and it may be caused by not considering the time spent in different microenvironments [8].

8.2.5 *Air Pollution and the Risk of Infertility*

Air pollution may have an impact on adverse reproductive outcomes in both humans and animals, resulting in infertility, abnormality, and spontaneous abortion. Evidence showed that exposure to air pollution has been linked to infertility [13, 27, 71].

Infertility is a complex disorder that is comprised of female factors (tubal, cervical, uterine, endometriosis, ovarian, hormonal), male factors (based on semen parameters), and other unexplained causes. The definition of infertility recommended by the World Health Organization (WHO) is that a non-contracepting couple who attempts conception for at least 1 year without success [73]. The incidence of infertility has been increasing in industrial countries from 7–8% in 1960 to 20–35% nowadays [90]. In China, the infertility rate is about 17.13 % [51]. Besides gene and congenital disease, environment and behavior could also lead to male and female infertility. Increasing infertility rate is a top priority for the WHO [87]. In addition, the impact of air pollution on infertility has attracted more and more attention worldwide.

As we know, there were two studies that analyzed the impact of air pollution on sterility or fertility rates in Spain [71] and USA [64], respectively. Three studies have reported a significant detrimental effect of high concentrations of air pollutants on clinical pregnancy rates and implantation rates [59, 76, 88]. However, topics above were not conducted in China, and there were studies only about the relationship between air pollution and spontaneous abortion and semen quality in Chinese population.

8.2.5.1 The Relationships Between Air Pollutants and Spontaneous Abortion

Liu carried out a survey among 200 early spontaneous abortions and 200 controls matched by age. Through investigating their state of living environment, they found that living nearly factories less than 500 m with air pollution was associated with risk of early spontaneous abortion, with an odds ratio of 4.58 and 95% CI of 1.69–12.39 [62]. Hou et al. studied the effects of air pollutant (SO_2, NO_2, PM_{10}, and total suspended particulates (TSP)) exposure on fetal loss. They collected 959 cases and 959 normal pregnancies within 14 weeks of pregnancy as controls in Nanjing, China. The data based on hospital records and national pollution monitor station records were collected. In their study, the mean monthly variation of the pollution level of SO_2, TSP, NO_2, and PM_{10} during the study period was higher in winter and spring (heating period) and lower in summer and autumn. The ratios of fetal loss were 2.04% during the heating period and 1.96% during the non-heating period. Besides, their results suggested that fetal loss within 14 weeks was associated with higher exposure to SO_2 (OR, 19.76; 95 % CI, 2.34–166.71), NO_2 (OR, 107.60; 95 % CI, 2.55–4544.61), and TSP (OR, 2.04; 95 % CI, 1.01–4.13) in their first month of pregnancy. However, no association has been found in any period of pregnancy for PM_{10} exposure [42].

Faiz et al. [25] and Mohorovic et al. [68] reported a significant increased risk of miscarriage in the general population with exposure to high levels of NO_2, SO_2, and products of coal combustion, respectively. However, Green et al. [30] did not found that there was statistically significant association between maximum annual average traffic within 50 m and increased risk of miscarriage in the general population.

8.2.5.2 The Relationships Between Air Pollutants and Semen Quality

Zhou et al. conducted a cross-sectional study in urban and rural areas of Chongqing, the largest of the four direct-controlled municipalities in China [111]. A group of 1346 healthy and eligible volunteers finally agreed to and completed all steps in the study. Sperm motion analysis was performed using a computer-aided semen analysis system (CASAS). CASAS outcomes included curvilinear velocity (VCL), straight-line velocity (VSL), average path velocity (VAP), beat cross frequency (BCF), and amplitude of lateral head displacement (ALH). Outdoor PM_{10}, SO_2, and NO_2 concentrations were collected from six ambient air quality monitoring stations in the studied districts or from counties that belonged to the national air pollutant detection network in 2007. Through their analysis, they found that the spatial distributions of PM_{10}, SO_2, and NO_2 were significantly higher in urban areas than that in rural areas. Further, they observed better male semen quality in the rural area of Chongqing than in the urban area, especially for sperm morphology and CASAS kinetic parameters. In addition, the outdoor exposure concentrations for PM_{10}, SO_2, and NO_2 were found to be significantly and negatively associated with a normal sperm morphology percentage (−0.212, −0.378, −0.381, respectively, $p < 0.001$). In contrast to the other parameters, the PM_{10} were positively associated with sperm concentration (0.075, $p = 0.031$). Nevertheless, they did not observe any statistically significant associations in the regression analysis between the air pollutants of SO_2 and NO_2 and routine semen quality parameters (sperm volume, sperm concentration, progressive motility, and total sperm motility). Vehicular exhaust is considered in great part responsible for the reproductive effects related to air pollutants. Zhang et al. [110] investigated the effects of motor tail gas on semen motility and the activity of succinate dehydrogenase (SDH) in 35 traffic policemen and 30 male teachers. They found that semen motility was decreased, and the activity of SDH was inhibited in traffic policemen with higher blood lead compared to male teachers. De Rosa et al. [18] evaluated semen quality in men exposure to traffic and observed that nitrogen oxide and Pb adversely affected semen quality.

The results above were in accordance with researches aboard. Selevan et al. [86] reported that periods of elevated air pollutants in Teolice, Czech Republic, were significantly associated with quality of semen, including proportionately fewer motile sperm, less sperm with normal morphology or normal head shape, and proportionately more sperm with abnormal chromatin. Calogero et al. [9] studied the effects of environment car exhaust pollution on human sperm. Their results showed that car exhaust exposure has a significantly genotoxic effect on human spermatozoa.

There is relatively little epidemiological data on this issue. In contrast, there are more experimental data available regarding the effect of pollutants on fertility in experimental mammals. These highlight the biological plausibility of the observed fertility decline in humans in relation to air pollution, which may operate through various mechanisms. Yan explored the effects of long-term exposure to $PM_{2.5}$ from automobile exhaust on the reproductive function of Sprague-Dawley rats. The rats in $PM_{2.5}$-exposed groups showed significantly disordered histological structure of the seminiferous tubules, reduced sperm count in the testicular lumen, some exfoliated secondary spermatocytes, downregulated Connexin 43 expression in the testis, and damaged blood-testis barrier [102]. Hong suggested that diesel exhaust particles could damage ovarian structure and function and make oocyte mitochondria vacuolization [38]. Veras et al. [92] pointed to endocrine disruption as the mechanism of action, by altering the normal function of the neuroendocrine-gonadal axis and producing hormonal imbalance. This study by Veras suggested that PM decreased the number of antral follicles, which can cause premature ovarian failure. Several studies have reported a deleterious effect of air pollution on sperm morphology [28], concentration [34], and motility [35]. Another possibility is damage, including immune-mediated injury, during critical stages of embryo development, where direct transfer of pollutants through the placenta leads to irreversible damage of dividing cells, potentially resulting in miscarriage [75]. In this regard, women exposed to CO in air pollution during pregnancy have been found to have increased levels of carboxyhemoglobin and circulating nucleated red blood cells, both of which are markers of fetal hypoxia [112]. Other investigators have also described a disruption of the normal pattern of segregation of the first two cell lines and trophectoderm [48]. Finally, changes in the vascular compartment or uterine environment in relation to air pollution before pregnancy have been reported in women who have miscarriage [92].

In summary, although there is a lack of studies about the impact of air pollution on infertility in China, air pollution indeed is an increased risk of human infertility. Further prospective researches are extremely essential, covering the effects of chronic exposures and short-term exposures of different air pollutants on infertility in Chinese population and using a better characterization of exposure models. Besides, there is a need for the implementation of control programs to reduce particulate and other pollutant emissions. Furthermore, it is necessary to strengthen education about pregnancy healthcare among childbearing-age women and to reduce exposure to air pollution.

In addition, there are rarely data addressing the relationship between air pollution and the risk of macrosomia (defining of birth weight more than 4000 g). A cohort study on whether high levels of air pollutant concentration are related to increased risk of fetus macrosomia was conducted in Wuhan, China, during 10 June 2011 to 9 June 2013. The exposure was estimated based on daily average concentrations of pollutants estimated using the data from the nine closest monitors in Wuhan, China. Logistic regressions were performed to explore the relationships between exposure to each of the pollutants during different pregnancy periods (first, second, and third trimesters) and macrosomia while controlling for key

Table 8.5 Adjusted ORs and 95% CIs for macrosomia attributable to maternal exposure to pollutants during the first, second, and third trimester of pregnancy in Wuhan, China, 2010–2013 (Data not published yet)

	First trimester			Second trimester			Third trimester		
Variates	aOR	95%CI		aOR	95%CI		aOR	95%CI	
PM_{10}	1.058	(1.028)	(1.089)	1.026	(0.996)	(1.057)	1.083	(1.046)	(1.121)
SO_2	0.919	(0.875)	(0.965)	0.965	(0.922)	(1.010)	1.007	(0.943)	(1.075)
NO	1.127	(1.072)	(1.184)	1.173	(1.118)	(1.231)	1.060	(1.004)	(1.118)
NO_2	0.960	(0.915)	(1.008)	0.895	(0.853)	(0.940)	1.000	(0.946)	(1.057)
CO	0.742	(0.551)	(1.001)	1.010	(0.784)	(1.300)	1.125	(0.861)	(1.470)
O_3	1.026	(1.007)	(1.044)	1.019	(0.999)	(1.039)	1.056	(1.034)	(1.079)
$PM_{2.5}$	0.982	(0.957)	(1.007)	0.987	(0.966)	(1.008)	0.966	(0.943)	(0.990)

aOR adjusted odds ratio, all logistic modes were adjusted by maternal age, history of diabetes, gestational age, gravidity, parity, weight gain during gestation, season of conception, smoking and passive smoking, infant sex, etc.

covariates (maternal age, gestational age, maternal baseline body mass index, weight gain during pregnancy, parity, maternal education, tobacco use, passive smoke, seasons of conception, infant sex). After controlling for some key covariates, we found 5.8% (OR, 1.058; 95% CI, 1.028, 1.089) and 8.3% (OR, 1.083; 95% CI, 1.046, 1.121) increase in risk of macrosomia with each 10-μg/m^3 increase in PM_{10} concentrations in the first trimester and the third trimester; 13% (OR, 1.127; 95% CI, 1.072, 1.184), 17% (OR, 1.173; 95% CI, 1.118, 1.231) and 6% (OR, 1.060; 95% CI, 1.004, 1.118) increase in risk of macrosomia with each 10-μg/m^3 increase in NO during the first, second, and third trimester; and 3% (OR, 1.026; 95% CI, 1.007, 1.044) and 6% (OR, 1.056; 95% CI, 1.034, 1.079) increase in risk of macrosomia with each 10-μg/m^3 increase in O_3 concentrations in the first and third trimester (Table 8.5).

The findings of the preliminary study suggest an association between air pollutants (e.g., PM_{10}, NO, O_3) and macrosomia in different exposure windows. The study could not elucidate the exact biological mechanism based on the present data, given that more and more studies demonstrated that gestational diabetes mellitus (GDM) is independently associated with macrosomia with an adjusted odds ratio (adjusted OR) of 1.71 (95% CI, 1.52–1.94) [36]. Several animal studies suggested that exposure to air pollution may induce profound metabolic effects through oxidative stress. In particular, $PM_{2.5}$ exposure was associated with glucose intolerance, decrease insulin sensitivity, and altered hepatic glucose and lipid metabolism [63, 101, 109]. Hence, the authors postulate that air pollutants might increase the risk of macrosomia by increasing insulin resistance or other biological process associated with adipose metabolism. However, more toxicological studies and prospective cohort studies with improved exposure assessments are needed to establish causality related to specific pollutants and increasing risk of macrosomia.

8.3 Mechanisms of Adverse Pregnancy Outcomes Caused by Air Pollution

Although the biologic mechanisms by which ambient air pollutants may affect fetal outcomes have not been fully explored, there are several mechanisms that have been proposed to explain associations between air pollutants and adverse pregnancy outcomes.

8.3.1 Mechanisms of Particular Matter (PM) on Adverse Pregnancy Outcomes

The aerodynamic diameter of particulate matter can decide its deposition in the parts of the respiratory tract: less than 10-um particles (PM_{10}) can enter the nasal cavity, less than or equal to 2.5 um ($PM_{2.5}$) can be in-depth and alveolar deposition, and then enter the blood circulation [33]. Several studies have shown that air pollutants can directly penetrate the placenta or produce an active product, which has a negative effect on the outcome of pregnancy [67, 91]. Some of the potentially toxic elements, such as plumbum, cadmium, nickel, manganese, vanadium, bromine, and zinc and polycyclic aromatic hydrocarbons (PAHs), are adsorbed on the $PM_{2.5}$. These small particles are easily deposited in the alveolar region and easily enter into the blood [69]. Toxicology experiments showed that plumbum exposure was one of the causes of congenital malformations and nervous system abnormalities of the fetus [3, 20]. Gerhard et al. [29] study showed that mercury might affect the luteal function and decrease the level of progesterone. They also found that plumbum, cadmium, and arsenic also had negative effects on ovarian function (especially luteal function) and further increased the risk of fetus death. Cai et al. [11] found that when the concentration of cobalt in the environment was more than 100 mu g/L, the survival rate of zebra fish embryos decreased and fetal malformation increased. Nakashima et al. [70] experiments found that high concentration of cadmium could increase mouse embryonic malformation rate and mortality in a dose-dependent manner. In addition, small dose of PAHs could activate NK cell in placenta and then NK cell attacked the placenta and might lead to fetal death [19]. $PM_{2.5}$ also has toxic effects on chromosome and DNA, including changing the number and structure of the chromosome, DNA mutation, and further causing fetal death or infertility. Animal experiments showed that particulate matter could cause the mice germ cell mutations and further pass the mutations to offspring [89]. Jurewicz et al. [52] found that $PM_{2.5}$ exposure was related to double Y chromosome and chromosome 21.

From the above, we can conclude that heavy metals and PAHs are the main toxic components of particulate matter. Particulate matter can increase adverse pregnancy outcome through the effect on maternal hormone metabolism and immune status; $PM_{2.5}$ and its components can lead to the oxidative stress in matrix and thus influence the fetal development; $PM_{2.5}$ has genetic toxicity and increases gene mutation. Besides, PM can combine placental growth factor receptor and cause inflammatory response of the placenta, further causing exchange barrier of oxygen nutrient of fetal placenta; particulate matter also alters coagulation function and hemodynamics. These mechanisms can be individually or in combination to affect the exchange of nutrients between placenta and fetus, resulting in adverse pregnancy outcomes [24].

8.3.2 Mechanisms of Gaseous Pollutants on Adverse Pregnancy Outcomes

CO can preponderantly compete with oxygen to combine with hemoglobin and also damage the white blood cell, platelet, and endothelial cell, further resulting in oxygen transport and electron transfer disorder. When pregnant women are exposed to CO, CO can penetrate into the placental barrier and interfere oxygenation of the fetus. In addition, CO can decrease maternal oxygen-carrying capacity, cause insufficient intrauterine oxygen supply and blood viscosity change and also disturb fetal and maternal nutrition communication [58, 112]. CO exposure in prenatal period can also affect fetal brain stem cholinergic and catecholaminergic pathways and also cause fetal brain pons and decreased medulla norepinephrine concentration [57]. All the above mechanisms can harass the fetal growth and development and lead to fetal malfunction. SO_2 as a water-soluble irritant gas can be transformed into sulfite or hydrosulfite in the blood and then quickly distribute to the whole body and could increase the risk of wheezing and shortness of breath, decrease resistance to infection, and affect the respiratory function or cause asthma. Embryological studies proved that SO_2 exposure in pregnancy (especially early pregnancy) had negative effect on the fetus. SO_2 could be transferred to the baby via the placenta, seriously disturbed the normal fetal growth and development, and even cause fetal malformation or death [22]. NO_x comes mainly from a variety of fuel combustion emissions, mainly in the form of NO, and easily converted to NO_2. NO_2 can enter into the blood in the form of nitrite and nitrate, which can transform low iron hemoglobin into methemoglobin and lead to tissue hypoxia, hindering nutrition and oxygen exchange between the mother and fetus and indirectly increase the risk of fetal adverse outcome [23]. On the mechanism of O_3 on adverse pregnancy outcomes, animal experiments confirmed that O_3 had immunotoxicity and embryonic toxicity [53]. O_3 can increase the concentration of lipid peroxidation and inflammatory factors in the circulation system of pregnant women and have a bad effect on the blood circulation of the placenta and ultimately affect the growth of the fetus and increase the risk of fetal loss or death [83].

8.4 Summary

From the above studies, it is clear that air pollution can lead to an increase risk of maternal and prenatal adverse outcomes. Pregnant women are especially vulnerable to the adverse impacts of air pollution, increasing the potential risk for adverse birth outcomes. Pregnant women had been exposed to air pollutants before the occurrence of adverse pregnancy outcomes, and therefore it was credible that results based on the records of previous exposure and causal timing were also reasonable. The results of these studies could explain the association between air pollutants and adverse pregnancy outcome in a certain extent. Therefore, we can conclude that exposure to air pollutants during pregnancy is a potential risk to the health of the pregnant women and fetus and increases the risk of adverse pregnancy outcome.

References

1. Agay-Shay K, Friger M, Linn S, Peled A, Amitai Y, Peretz C. Air pollution and congenital heart defects. Environ Res. 2013;124(3):28–34.
2. Arroyo V, Díaz J, Carmona R, Ortiz C, Linares C. Impact of air pollution and temperature on adverse birth outcomes: Madrid, 2001–2009. Environ Pollut. 2016;218:1154–61.
3. Basha R, Reddy G. Developmental exposure to lead and late life abnormalities of nervous system. Indian J Exp Biol. 2010;48(7):636–41.
4. Beck S, Wojdyla D, Say L, Betran AP, Merialdi M, Requejo JH, Rubens C, Menon R, Van Look PF. The worldwide incidence of preterm birth: a systematic review of maternal mortality and morbidity. B World Health Organ. 2010;88(1):31–8.
5. Beelen R, Raaschou-Nielsen O, Stafoggia M, Andersen ZJ, Weinmayr G, Hoffmann B, et al. Effects of long-term exposure to air pollution on natural-cause mortality: an analysis of 22 European cohorts within the multicentre ESCAPE project. Lancet. 2014;383(9919):785–95.
6. Bhutta ZA, Yakoob MY, Lawn JE, Rizvi A, Friberg IK, Weissman E, Buchmann E, Goldenberg RL. Stillbirths: what difference can we make and at what cost. Lancet. 2011;377(9776):1523–38.
7. Bobak M. Outdoor air pollution, low birth weight, and prematurity. Environ Health Pers. 2000;108:173–6.
8. Briggs D. The role of GIS: coping with space (and time) in air pollution exposure assessment. J Toxicol Environ Health. 2016; Part A 68:1243–61.
9. Calogero AE, La Vignera S, Condorelli RA, Perdichizzi A, Valenti D, Asero P, Carbone U, Boggia B, De Rosa N, Lombardi G, D'Agata R, Vicari LO, Vicari E, De Rosa M. Environmental car exhaust pollution damages human sperm chromatin and DNA. J Endocrinol Investig. 2011;34(6):e139–43.
10. Capobussi M, Tettamanti R, Marcolin L, Piovesan L, Bronzin S, Gattoni ME, Polloni I, Sabatino G, Tersalvi CA, Auxilia F, Castaldi S. Air pollution impact on pregnancy outcomes in Como, Italy. J Occup Environ Med. 2016;58(1):47–52.
11. Cai G, Zhu J, Shen C, Cui Y, Du J, Chen X. The effects of cobalt on the development, oxidative stress, and apoptosis in zebrafish embryos. Biol TraceElem Res. 2012;150(1–3):200–7.
12. Chang H, Reich B, Miranda M. A spatial time-to-event approach for estimating associations between air pollution and preterm birth. J R Stat Soc C Appl Stat. 2013;62(2):1–17.
13. Checa VM, Gonzalez-Comadran M, Jacquemin B. Outdoor air pollution and human infertility: a systematic review. Fertil Steril. 2016;106(4):897–904.

14. Chen EK, Zmirounavier D, Padilla C, Deguen S. Effects of air pollution on the risk of congenital anomalies: a systematic review and meta-analysis. Int J Environ Res Public Health. 2014;11:7642–68.
15. Chung MK, Lao TT, Ting YH, Leung TY, Lau TK, Wong TW. Environmental factors in the first trimester and risk of oral-facial clefts in the offspring. Reprod Sci. 2013;20:797–803.
16. Coker E, Ghosh J, Jerrett M, Gomez-Rubio V, Beckerman B, Cockburn M, Liverani S, Su J, Li A, Kile ML, Ritz B, Molitor J. Modeling spatial effects of PM (2.5) on term low birth weight in Los Angeles County. Environ Res. 2015;142:354–64.
17. Dadvand P, Rankin J, Rushton S, Pless-Mulloli T. Association between maternal exposure to ambient air pollution and congenital heart disease: a register-based spatiotemporal analysis. Am J Epidemiol. 2011;173:171–82.
18. De Rosa M, Zarrilli S, Paesano L, Carbone U, Boggia B, Petretta M, Maisto A, Cimmino F, Puca G, Colao A, Lombardi G. Traffic pollutants affect fertility in men. Hum Reprod. 2003;18(5):1055–61.
19. Detmar J, Jurisicova A. Embryonic resorption and polycyclicaromatic hydrocarbons: putative immune-mediated mechanisms. Syst Biol Reprod Med. 2010;56(1):3–17.
20. Dias M, Partington M. Congenital brain and spinal cord malformations and their associated cutaneous markers. Pediatrics. 2015;135(4):e1105–19.
21. Dibben C, Clemens T. Place of work and residential exposure to ambient air pollution and birth outcomes in Scotland, using geographically fine pollution climate mapping estimates. Environ Res. 2015;140:535–41.
22. Dugandzic R, Dodds L, Stieb D, Smith-Doiron M. The association between low level exposures to ambient air pollution and term low birth weight: a retrospective cohort study. Environ Health. 2006;5:3.
23. Estarlich M, Ballester F, Aguilera I, Fernández-Somoano A, Lertxundi A, Llop S, Freire C, Tardón A, Basterrechea M, Sunyer J, Iñiguez C. Residential exposure to outdoor air pollution during pregnancy and anthropometric measures at birth in a multicenter cohort in Spain. Environ Health Perspect. 2011;119(9):1333–8.
24. Faiz A, Rhoads G, Demissie K, Lin Y, Kruse L, Rich DQ. Does ambient air pollution trigger stillbirth? Epidemiology. 2013;24(4):538–44.
25. Faiz A, Rhoads G, Demissie K, Kruse L, Lin Y, Rich DQ. Ambient air pollution and the risk of stillbirth. Am J Epidemiol. 2012;176(4):308–16.
26. Fan L. Study on the associations between air pollutants exposure and adverse pregnancy outcomes in Hainan Province. Hainan: Huazhong University of Science & Technology; 2013.
27. Frutos V, Gonzalez-Comadran M, Sola I, Jacquemin B, Carreras R, Checa V. Impact of air pollution on fertility: a systematic review. Gynecol Endocrinol. 2015;31(1):7–13.
28. Gaspari L, Chang SS, Santella RM, Garte S, Pedotti P, Taioli E. Polycyclic aromatic hydrocarbon-DNA adducts in human sperm as a marker of DNA damage and infertility. Mutat Res. 2003;535(2):155–60.
29. Gerhard I, Waibel S, Daniel V, Runnebaum B. Impact of heavy metals on hormonal and immunological factors in women with repeated miscarriages. Hum Reprod Update. 1998;4(3):301–9.
30. Green RS, Malig B, Windham GC, Fenster L, Ostro B, Swan S. Residential exposure to traffic and spontaneous abortion. Environ Health Perspect. 2009;117(12):1939–44.
31. Green R, Sarovar V, Malig B, Basu R. Association of stillbirth with ambient air pollution in a California cohort study. Am J Epidemiol. 2015;181(11):874–82.
32. Groot AC, Bartelings MM, Deruiter MC, Poelmann RE. Basics of cardiac development for the understanding of congenital heart malformations. Pediatr Res. 2005;57:169–76.
33. Guo X. Environmental health science [M]. Beijing: Higher Education Press; 2011. p. 62–93.
34. Guven A, Kayikci A, Cam K, Arbak P, Balbay O, Cam M. Alterations in semen parameters of toll collectors working at motorways: does diesel exposure induce detrimental effects on semen. Andrologia. 2008;40(6):346–51.

35. Hammoud A, Carrell DT, Gibson M, Sanderson M, Parker-Jones K, Peterson CM. Decreased sperm motility is associated with air pollution in Salt Lake City. Fertil Steril. 2010;93(6):1875–9.
36. He XJ, Qin FY, Hu CL, Zhu M, Tian CQ, Li L. Is gestational diabetes mellitus an independent risk factor for macrosomia: a meta-analysis? Arch Gynecol Obstet. 2015;291(4):729–35.
37. Home C. Update on overall prevalence of major birth defects – Atlanta, Georgia, 1978–2005. Morb Mortal Wkly Rep. 2008;57:1–5.
38. Hong XR, Fang F, Song YF, Chen XQ, Zheng L, Yu HY. Instillation of diesel exhaust particles on the posterior wall of pharynx on reproductive function in female mice. Chin J Prevent Med. 2011;45(11):1026–30.
39. Hou H, Chen Y, Wang D, Zou XP. Epidemiology research on relationship between mainly air pollutants and embryo diapause in Yianjin. J Reproduct Med. 2009;18(2):370–5.
40. Hou H, Chen Y, Wang D, Zou XP. Association between ambient air particulate matter and missed abortion in early pregnancy in Tianjin. J Int Obstet Gynecol. 2012;39(6):609–12.
41. Hou H, Yang Z, Zhu SP, Wei H. The relationship between maternal blood BaP-DNA adducts and embryonic diapauses. J Reproduct Med. 2009;18(2):81–5.
42. Hou H, Wang D, Zou XP, Yang ZH, Li TC, Chen YQ. Does ambient air pollutants increase the risk of fetal loss? A case–control study. Arch Gynecol Obstet. 2014;289(2):285–91.
43. Huang C, Nichols C, Liu Y, Zhang Y, Liu X, Gao S, Li Z, Ren A. Ambient air pollution and adverse birth outcomes: a natural experiment study. Popul Health Metrics. 2015;13(1):1–7.
44. Hwang BF, Jaakkola JJ. Ozone and other air pollutants and the risk of oral clefts. Environ Health Perspect. 2008;116:1411–5.
45. Hwang B, Lee Y, Jaakkola J. Air pollution and stillbirth: a population-based case–control study in Taiwan. Environ Health Perspect. 2011;119(9):1345–9.
46. Jacobs M, Zhang G, Chen S, Mullins B, Bell M, Jin L, Guo Y, Huxley R, Pereira G. The association between ambient air pollution and selected adverse pregnancy outcomes in china: a systematic review. Sci Total Environ. 2016;579:1179.
47. James P. Preterm birth: causes, consequences and prevention. J Obstet Gynaecol. 2008;28(5):559.
48. Januario DA, Perin PM, Maluf M, Lichtenfels AJ, Nascimento SP. Biological effects and dose-response assessment of diesel exhaust particles on in vitro early embryo development in mice. Toxicol Sci. 2010;117(1):200–8.
49. Jiang LL, Zhang YH, Song GX, Chen GH, Chen BH, Zhao NQ. Kan HD.A time series analysis of outdoor air pollution and preterm birth in Shanghai, China. Biomed Environ Sci. 2007;20(5):426–31.
50. Jin L, Qiu J, Zhang Y, Qiu W, He X, Wang Y, Sun Q, Li M, Zhao N, Cui H, Liu S, Tang Z, Chen Y, Yue L, Da Z, Xu X, Huang H, Liu Q, Bell M, Zhang Y. Ambient air pollution and congenital heart defects in Lanzhou, China. Environ Res Lett. 2015;10
51. Jun G, Ersheng G. Infertility rate of Chinese childbearing women and its influencing factors. Chin J Health Stat. 2005;01:27–9.
52. Jurewicz J, Radwan M, Sobala W, Polańska K, Radwan P, Jakubowski L, Ulańska A, Hanke W. The relationship betweenexposure to air pollution and sperm disomy. Environ Mol Mutagen. 2015;56(1):50–9.
53. Kavlock R, Daston G, Grabowski CT. Studies on the developmental toxicity of ozone. I Prenatal effects. Toxicol Appl Pharm. 1979;48(1):19–28.
54. Kramer M. The epidemiology of adverse pregnancy outcomes: an overview. J Nutr. 2003;133(Suppl 2):1592S–6S.
55. Lai H, Tsang H, Wong C. Meta-analysis of adverse health effects due to air pollution in Chinese populations. BMC Public Health. 2013;13:360.
56. Lamichhane DK, Leem JH, Lee JY, Kim HC. A meta-analysis of exposure to particulate matter and adverse birth outcomes. Environ Health Toxicol. 2015;3(30):e2015011.

57. Lavezzi A, Ottaviani G, Mingrone R, Matturri L. Analysis of the human locus coeruleus in perinatal and infant sudden unexplained deaths. Possible role of the cigarette smoking in the development of this nucleus. Brain Res Dev Brain Res. 2005;154(1):71–80.
58. Leem J, Kaplan B, Shim Y, Pohl HR, Gotway CA, Bullard SM, Rogers JF, Smith MM, Tylenda CA. Exposures to air pollutants during pregnancy and preterm delivery. Environ Health Perspect. 2006;114(6):905–10.
59. Legro RS, Sauer MV, Mottla GL, Richter KS, Li X, Dodson WC, Liao D. Effect of air quality on assisted human reproduction. Hum Reprod. 2010;25(5):1317–24.
60. Li X, Huang S, Jiao A, Yang X, Yun J, Wang Y, et al. Association between ambient fine particulate matter and preterm birth or term low birth weight: An updated systematic review and meta-analysis. Environ Pollut. 2017; pii: S0269–7491(16)32550–7.
61. Liu CB, Hong XR, Shi M, Chen XQ, Huang HJ, Chen JH, Yang K, Chen SQ, Chen HQ, Kan HD, Sun QH. Effects of prenatal PM_{10} exposure on fetal cardiovascular malformations in Fuzhou, China: a retrospective case-control study. Environ Health Perspect. 2016;057001:1–8.
62. Liu X. Etiologic study of early spontaneous abortion. Beijing: Peking Union Medical College; 2007.
63. Liu C, Xu X, Bai Y, Wang TY, Rao X, Wang A, Sun L, Ying Z, Gushchina L, Maiseyeu A, Morishita M, Sun Q, Harkema JR, Rajagopalan S. Air pollution–mediated susceptibility to inflammation and insulin resistance: influence of CCR2 pathways in mice. Environ Health Perspect. 2013;122(1):17–26.
64. Mahalingaiah S, Hart JE, Laden F, Farland LV, Hewlett MM, Chavarro J, Aschengrau A, Missmer SA. Adult air pollution exposure and risk of infertility in the Nurses' Health Study II. Hum Reprod. 2016;31(3):638–47.
65. Maroziene L, Grazuleviciene R. Maternal exposure to low-level air pollution and pregnancy outcomes: a population-based study. Environ Health. 2002;1(1):6.
66. Mattingly P. Evaluation of fetal death, 2013. Available online: http://emedicine.medscape.com/article/259165-overview. Accessed on 13 Mar 2016.
67. Michikawa T, Morokuma S, Yamazaki S, Fukushima K, Kato K, Nitta H. Exposure to air pollutants during the early weeks of pregnancy, and placenta praevia and placenta accreta in the western part of Japan. Environ Int. 2016;92-93:464–70.
68. Mohorovic L, Petrovic O, Haller H, Micovic V. Pregnancy loss and maternal methemoglobin levels: an indirect explanation of the association of environmental toxics and their adverse effects on the mother and the fetus. Int J Environ Res Public Health. 2010;7(12):4203–12.
69. Montiel-Davalos A, Alfaro-Moreno E, Lopez-Marure R. $PM_{2.5}$ and PM_{10} induce the expression of adhesion molecules and the adhesion of monocytic cells to human umbilical vein endothelial cells. Inhal Toxicol. 2007;19(Suppl1):91–8.
70. Nakashima K, Wakisaka T, Fujiki Y. Dose-response relationship of cadmium embryotoxicity in cultured mouse embryos. ReprodToxicol. 1987–1988; 1(4): 293–298.
71. Nieuwenhuijsen MJ, Basagana X, Dadvand P, Martinez D, Cirach M, Beelen R, Jacquemin B. Air pollution and human fertility rates. Environ Int. 2014;70:9–14.
72. Padula AM, Tager IB, Carmichael SL, Hammond SK, Lurmann F, Shaw GM. The Association of Ambient air Pollution and Traffic Exposures with Selected Congenital Anomalies in the San Joaquin Valley of California. Am J Epidemiol. 2013;177:1074–85.
73. Patrick J, Rowe FH. WHO manual for the standardized investigated investigation and diagnosis of the infertile couple. In.; 1993.
74. Pedersen M, Stayner L, Slama R, Sorensen M, Figueras F, MJ N, Raaschou-Nielsen O, Dadvand P. Ambient air pollution and pregnancy-induced hypertensive disorders: a systematic review and meta-analysis. Hypertension. 2014;64(3):494–500.
75. Perera FP, Jedrychowski W, Rauh V, Whyatt RM. Molecular epidemiologic research on the effects of environmental pollutants on the fetus. Environ Health Perspect. 1999;107(Suppl 3):451–60.

76. Perin PM, Maluf M, Czeresnia CE, Januario DA, Saldiva PH. Impact of short-term preconceptional exposure to particulate air pollution on treatment outcome in couples undergoing in vitro fertilization and embryo transfer (IVF/ET). J Assist Reprod Genet. 2010;27(7):371–82.
77. Qian N, Yu H, Cai R, Han M, Qian Y, Wang C. Association between maternal air pollution exposure during first trimester and birth weight in shanghai. J Environ Occup Med. 2016;33(9):827–32.
78. Rao A, Ahmed MK, Taub PJ, Mamoun JS. The correlation between maternal exposure to air pollution and the risk of orofacial clefts in infants: a systematic review and meta-analysis. J Oral Max Res. 2016;7(1):e2.
79. Ritz B. Air pollution and congenital anomalies. Occup Environ Med. 2010;67:221–2.
80. Ritz B, Yu F, Fruin S, Chapa G, Shaw GM, Harris JA. Ambient air pollution and risk of birth defects in Southern California. Am J Epidemiol. 2002;155:17–25.
81. Ruan H, Zhao Q, Chen L. A case-control study on air pollution and preterm birth in Panyu, Guangzhou. Invest Res. 2010;17(21):140–1.
82. Ruan H, Zhao Q, Chen L. A case-control study on the relationship between air pollution and premature delivery in Panyu District, Guangzhou. Chin Modern Med. 2007;17:140–1.
83. Salam MT, Millstein J, Li YF, Lurmann FW, Margolis HG, Gilliland FD. Birth outcomes and prenatal exposure to ozone, carbon monoxide, and particulate matter: results from the Children's Health Study. Environ Health Perspect. 2005;113(11):1638–44.
84. Savitz DA, Harmon Q, Siega-Riz AM, Herring AH, Dole N, Thorp JM. Behavioral influences on preterm birth: integrated analysis of the pregnancy, infection, and nutrition study. Matern Child Health J. 2012;16(6):1151–63.
85. Schembari A, Nieuwenhuijsen MJ, Salvador J, de Nazelle A, Cirach M, Dadvand P, Beelen R, Hoek G, Basagaña X, Vrijheid M. Traffic-related air pollution and congenital anomalies in Barcelona. Environ Health Perspect. 2014;122:317–23.
86. Selevan SG, Borkovec L, Slott VL, Zudova Z, Rubes J, Evenson DP, Perreault SD. Semen quality and reproductive health of young Czech men exposed to seasonal air pollution. Environ Health Perspect. 2000;108(9):887–94.
87. Sharma R, Biedenharn KR, Fedor JM, Agarwal A. Lifestyle factors and reproductive health: taking control of your fertility. Reprod Biol Endocrinol. 2013;11:66.
88. Slama R, Bottagisi S, Solansky I, Lepeule J, Giorgis-Allemand L, Sram R. Short-term impact of atmospheric pollution on fecundability. Epidemiology. 2013;24(6):871–9.
89. Somers C, McCarry B, Malek F, Quinn JS. Reduction of particulate air pollution lowers the risk of heritable mutations in mice. Science. 2004;304(5673):1008–10.
90. Thompson ST. Preventable causes of male infertility. World J Urol. 1993;11(2):111–9.
91. Triche E, Hossain N. Environmental factors implicated in the causation of adverse pregnancy outcome. Semin Perinatol. 2007;31(4):240–2.
92. Veras MM, Damaceno-Rodrigues NR, Caldini EG, Maciel RA, Mayhew TM, Saldiva PH, Dolhnikoff M. Particulate urban air pollution affects the functional morphology of mouse placenta. Biol Reprod. 2008;79(3):578–84.
93. Vrijheid M, Martinez D, Manzanares S, Dadvand P, Schembari A, Rankin J, Nieuwenhuijsen M. Ambient air pollution and risk of congenital anomalies: a systematic review and meta-analysis. Environ Health Perspect. 2011;119:598–606.
94. Wang D. Exploration of relationship between maternal exposure to ambient sulfur dioxide and missed abortion. Shijiazhuang: Hebei Medical University; 2008.
95. Wang X, Ding H, Ryan L, Xu X. Association between air pollution and low birth weight: a community-based study. Environ Health Perspect. 1997;105(5):514–20.
96. WHO. Preterm birth. http://www.who.int/mediacentre/factsheets/fs363/en/. Accessed Nov 2016.
97. Woodruff T, Grillo J, Schoendorf K. The relationship between selected causes of postneonatal infant mortality and particulate air pollution in the United States. Environ Health Perspect. 1997;105(6):608–12.

98. Wu J, Wilhelm M, Chung J, Ritz B. Comparing exposure assessment methods for traffic-related air pollution in an adverse pregnancy outcome study. Environ Res. 2011;111:685–92.
99. Xu P, Chen Y, Ye X. Haze, air pollution, and health in China. Lancet. 2013;382(9910):2067.
100. Xu X, Ding H, Wang X. Acute effects of total suspended particles and sulfur dioxides on preterm delivery: a community-based cohort study. Arch Environ Health. 1995;50:407–15.
101. Xu X, Liu C, Xu Z, Tzan K, Zhong M, Wang A, Lippmann M, Chen LC, Rajagopalan S, Sun Q. Long-term exposure to ambient fine particulate pollution induces insulin resistance and mitochondrial alteration in adipose tissue. Toxicol Sci. 2011;124(1):88–98.
102. Yan C, Cao X, Shen L, Dong D, Peng J, Chen J, Zhou Y, Long C, Lin T, He D. Long-term exposure to $PM_{2.5}$ from automobile exhaust results in reproductive dysfunction in male rats. National J Androl. 2016;22(2):104–9.
103. Yang Z. Study of Bbp level and missed abortion in early pregnancy's correlative factors. Shijiazhuang: Hebei Medical University; 2008.
104. Yu M, Ping Z, Zhang S, He Y, Dong R, Guo X. The survey of birth defects rate based on birth registration system. Chin Med J (Engl). 2015;128(1):7–14.
105. Zhang B, Liang S, Zhao J, Qian Z, Bassig BA, Yang R, Zhang Y, Hu K, Xu S, Zheng T, Yang S. Maternal exposure to air pollutant PM2.5 and PM10 during pregnancy and risk of congenital heart defects. J Expo Sci Environ Epidemiol. 2016;26:422–7.
106. Zhang B, Zhao J, Yang R, Qian Z, Liang S, Bassig BA, Zhang Y, Hu K, Xu S, Dong G, Zheng T, Yang S. Ozone and other air pollutants and the risk of congenital heart defects. Sci Rep. 2016;6:34852.
107. Zhang Y, Liu X, Ren Z. Effects of air pollution on pregnancy outcome in Taiyuan, China. J Environ Health. 2007;24(3):128–31.
108. Zhang YP, Zhang ZQ, Wu YC, Xu LZ. Association of air Pollution with preterm birth in Taiyuan. J Environ Health. 2008;25(3):194–7.
109. Zheng Z, Xu X, Zhang X, Wang A, Zhang C, Hüttemann M, Grossman LI, Chen LC, Rajagopalan S, Sun Q, Zhang K. Exposure to ambient particulate matter induces a NASH-like phenotype and impairs hepatic glucose metabolism in an animal model. J Hepatol. 2012;58(1):148–54.
110. Zhang ZM, Pei Q, Lu Z. Effect of tail gae from motor petrol on sperm motility in human. Chin Environ Sci. 1995;15(5):391–2.
111. Zhou N, Cui Z, Yang S, Han X, Chen G, Zhou Z, Zhai C, Ma M, Li L, Cai M, Li Y, Ao L, Shu W, Liu J, Cao J. Air pollution and decreased semen quality: a comparative study of Chongqing urban and rural areas. Environ Pollut. 2014;187:145–52.
112. Ziaei S, Nouri K, Kazemnejad A. Effects of carbon monoxide air pollution in pregnancy on neonatal nucleated red blood cells. Paediatr Perinat Epidemiol. 2005;19(1):27–30.
113. Zou X. Relationship between missed abortion and exposure to carbon monoxide in the air during peri-conception period. Shijiazhuang: Hebei Medical University; 2008.

Chapter 9
Comparison of Health Impact of Air Pollution Between China and Other Countries

Linwei Tian and Shengzhi Sun

Abstract Air pollution is the world's largest single environmental risk according to the World Health Organization (WHO), which caused around seven million deaths in 2012. Extensive epidemiological studies have been carried out worldwide to examine the health impacts of ambient air pollution, consistently demonstrating significant health impacts of ambient air pollution. Air pollution problem in China is especially serious; it has become the fourth biggest threat to the health of the Chinese people. In this review, we summarized existing literature, compared health impact of air pollution between China and other countries, and found substantial heterogeneity in the risk estimates of air pollution. The effect heterogeneities may be due to the differences in the characteristics of populations (e.g., the proportion of the elder population and people with preexisting diseases), exposure profile (e.g., air pollution concentrations and composition), and regional climate. Although the magnitude of relative risk estimates of air pollution is generally similar with that in other parts of the world, air pollution is one of China's most serious environmental health problems given the huge number of people exposed to high concentration levels of air pollution in China.

Keywords Particulate matter • Concentration-response functions • Effect heterogeneity

9.1 Introduction

Extensive epidemiological studies have been conducted worldwide to examine the health impacts of air pollution, and the evidence for the adverse effects of ambient air pollution on morbidity and mortality is convincing in both China [11, 40] and

---

L. Tian (✉) • S. Sun
School of Public Health, Li Ka Shing Faculty of Medicine, The University of Hong Kong, Hong Kong, China
e-mail: linweit@hku.hk

G.-H. Dong (ed.), *Ambient Air Pollution and Health Impact in China*, Advances in Experimental Medicine and Biology 1017, DOI 10.1007/978-981-10-5657-4_9

other countries ([21]; Isabelle [30, 36]) (details discussed in Chaps. 4 and 5). However, the concentration-response functions (CRFs), the percent change of a given health outcome per μg/m^3 increment in concentration of air pollution, may differ according to geographical regions. These effect heterogeneities may be due to the difference in the characteristics of populations, exposure profile (e.g., air pollution concentration, particulate matter size distribution and composition), and regional climate. Insights into these modifying factors might help to understand the health effects of air pollution and could lead to different air quality guidelines for different parts of the world.

China is a developing country and is in the process of fast industrialization and rapid economic development. With fast development over the past three decades, China had become the world's second largest economy in terms of gross domestic product (GDP) in 2010 and was the world's biggest energy consumer in 2009. Coal is the major source of energy; the amount of coal consumption in China constitutes about 50% of the world's total coal consumption in 2012. At the same time, the number of on-road civilian vehicles increased dramatically from 16.09 million in 2000 to 93.56 million in 2011 [12]. China's extensive industrial development, substantial coal-dependent energy consumption, and increasing number of vehicles have led to a rise in emissions of air pollutants. Consequently, air pollution is one of China's most serious environmental problems. Ambient air pollution has become the fourth biggest threat to the health of the Chinese people according to the Global Burden of Disease Study 2010 [46].

China has its unique characteristics when compared to the developed countries (e.g., the United States and the European countries). Being the largest producer and consumer of coal in the world, together with outdated technologies and poor maintenance of coal burning facilities, China is one of the few countries with the highest air pollution levels (both outdoor and indoor) in the world [41, 42]. In addition, sociodemographic characteristics of residents (e.g., age structure, disease pattern, and socioeconomic status) in China also differ from those of developed countries. Thus, comparing the health effects of air pollution between China and other countries may provide some insights into the potential modifiers of the adverse health effects of air pollution.

In this chapter, we focus on the short-term health effects of particulate matter (PM) and ozone (O_3), comparing the findings from China and other countries, and discussing the potential effect modifiers for the adverse health effect of air pollution.

9.2 Comparison of CRFs Between China and Other Countries

Case-crossover and time-series analysis are the two main methods to analyze the short-term health effects of air pollution. The case-crossover study design, however, has its own disadvantages: it does not account for over-dispersion and may underestimate the variance of health estimates in each city, and it can induce larger estimates for heterogeneity than does time-series analysis [21]. On the other hand, there is a broad spectrum of model specification in the time-series analysis, including the methods used for smoothing (e.g., natural splines or penalized spline) and the number of degrees of freedom for smoothing, but effect estimates remained fairly stable [21]. Thus, we restricted our comparative analysis to those estimates from time-series studies in order to reduce the complexities introduced by differing analytic strategies.

Although numerous effect estimates have been generated from single-city studies, these results are hard to synthesize. Multi-city studies combining estimates from various locations using a common protocol may reduce potential bias (e.g., analytic bias, lag selection bias, and publication bias) and increase precision, thus can be used to evaluate the heterogeneity of the air pollution effect across cities. In addition, there are relatively few studies in China which have examined effects of air pollution on hospital admissions, and few studies have examined long-term effects of air pollution. Currently, the China Kadoorie Biobank (CKB) cohort study is under way and has not produced findings on air pollution and health yet. This large, well-established study should generate new findings for China. Thus, in this chapter, we mainly compare results from multi-city studies in the world, supplemented by meta-analysis results in the literature.

9.2.1 Particulate Matter (PM)

Airborne particulate matter (PM) is a complex mixture of solid and liquid particles of primary and secondary origin, which contain a wide range of inorganic and organic components. It is created by both natural and man-made sources and is found both outdoors and indoors. From the health perspectives, PM is typically defined by size, with the smaller particles having larger health impacts. PM with aerodynamic diameter less than 10 μm is called $PM_{10,}$ and PM with aerodynamic diameter less than 2.5 μm is called $PM_{2.5}$.

Numerous large-scale multi-city epidemiological studies have been carried out to estimate the health effects of air pollution worldwide. Air pollution and health: A European and North American Approach (APHENA) was a Health Effects Institute (HEI)-sponsored project consisting of three components: (1) the Air Pollution and Health: A European Approach Phase 2 (APHEA-2) study involving 32 European cities; (2) the National Morbidity, Mortality, and Air Pollution Study (NMMAPS), conducted in 90 largest US cities; and (3) multi-city research on the health effects of air pollution in 12 Canadian cities [21]. Estudio de Salud y Contaminación del Aire en Latinoamérica (ESCALA) was another HEI-sponsored project that aimed to estimate the health effects of air pollution in 9 Latin American cities [30]. Simpson et al. conducted a multi-city study to estimate the short-term mortality effects of air pollution in 4 Australian cities (Melbourne, Sydney, Brisbane, and Perth), which account for about 50% of the Australian population [36]. The China Air Pollution and Health Effect Study (CAPES) was the largest multi-city study conducted in China which included 16 Chinese cities coving about 69 million population [11]. These multi-city epidemiological studies generally applied two stages of analysis to estimate city-specific and national average associations of PM_{10} with daily mortality. In the first stage of the analysis, city-specific estimate was estimated followed by a common protocol for fitting the Poisson regression models to the air pollution and mortality time-series data in each city. In the second stage, hierarchical model or meta-analysis was used to pool city-specific estimates and obtain the national average estimates of the association of PM_{10} with mortality. Potential effect modification by sociodemographic characteristics and regional climate was also assessed across study regions.

The associations between PM_{10} and all-cause mortality were positive and statistically significant for the all-age group in Canada, Europe, the United States, Latin America, and China, but not in Australia. When pooling estimates from cities, the percentage excess risks per 10 μg/m^3 increase of PM_{10} concentration for all-cause mortality were 0.84% (95% confidence interval: 0.30%, 1.40%), 0.26% (0.15%, 0.36%), 0.26% (0.15%, 0.37%), 0.77% (0.60%, 1.00%), 0.20% (−0.80%, 1.20%), and 0.35% (0.18%, 0.52%) for Canada, Europe, the United States, Latin American, Australia, and China, respectively. The effect estimates for Canada (0.84%) and Latin America (0.77%) were two or three times higher than those for Europe (0.26%), United States (0.26%), Australia (0.20%) and China (0.35%).

APHENA studied the health effects of PM_{10} on cardiovascular mortality for people aged less than 75 and people aged 75 years and older separately, while other multi-city studies examined the effects of PM_{10} on cardiovascular mortality for all ages. The risk estimates of PM_{10} on cardiovascular mortality were positive and statistically significant in Canada (people ≥75 years), Europe (people ≥75 years), the United States (people ≥75 years), Latin America (all-age group), and China (all-age group).

Effects of PM_{10} on respiratory mortality were less consistent across regions. A 10 μg/m^3 increase in PM_{10} was associated with respiratory mortality of −1.4% (−3.2%, 0.5%), 0.37% (0.03%, 0.73%), 0.15% (−0.28%, 0.59%), 1.39% (0.98%, 1.81%), and 0.56% (0.31%, 0.81%) in Canada, Europe, the United States, Latin America, and China, respectively. Those associations were statistically significant in Europe, Latin America, and China only.

In summary, there is indeed heterogeneity for mortality effects of PM_{10}. The relative risks estimated in Canadian cities were much larger than in other regions. The effect sizes in China were comparable with those in Europe and the United States (Fig. 9.1 and Table 9.1).

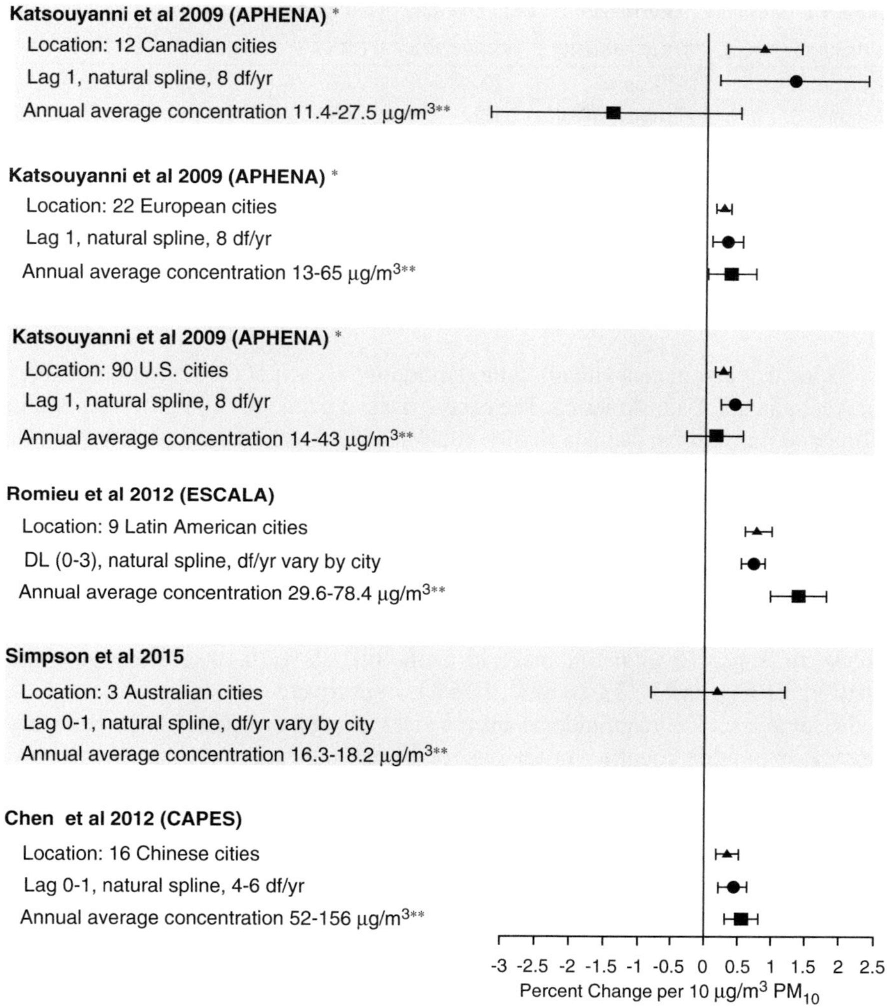

Fig. 9.1 Percentage excess risk (ER %) of all-natural, cardiovascular, and respiratory mortality per 10 μg/m^3 increase in PM_{10} in multi-city studies. Estimates with ▲ indicate all-natural mortality, estimates with • represent cardiovascular mortality, and estimates with ▪ denote respiratory mortality. *Estimates are for all ages, except estimates of cardiovascular mortality are for people ≥75 years. **Concentration ranges across cities (Source: Adapted from Romieu et al. [30]; p. 100)

9.2.2 Gaseous Pollutants

Multi-city studies of APHENA, ESCALA, and Simpson et al. [36] also examined the mortality effects of ozone (O_3). A four-city time-series study investigated the short-term effects of O_3 in the Pearl River Delta (PRD) of southern China [40].

Table 9.1 Mortality risk estimates for PM_{10} exposure based on meta-analyses results

Country or region	Daily mortality	Reference	ER (%) per 10 μg/m^3	95% CI
Europe	All-cause	[2]	0.6%	0.4–0.8%
Europe	Cardiovascular	[2]	0.9%	0.5–1.3%
Europe	Respiratory	[2]	1.3%	0.5–2.09%
Asia	All-cause	[14]	0.49%	0.23–0.76%
China	All-cause	[34]	0.32%	0.28–0.35%
China	Cardiovascular	[34]	0.43%	0.37–0.49%
China	Respiratory	[34]	0.32%	0.23–0.40%

ER(%) percentage excess risk, *CI* confidence interval

Ozone was associated with all-natural mortality in each of the six regions, except for Australia and Latin America. The excess risks of ozone per 10 μg/m^3 increase in all-natural mortality in Canada [0.52% (0.24%, 0.80%)] and China [0.81% (0.63%, 1.00%)] were greater compared to effects in Europe [0.13% (0.05%, 0.21%)] and the United States [0.18% (0.01%, 0.35%)]. For cardiovascular mortality, the risk estimates varied substantially. The effect estimates are positive and statistically significant for Canada (people ≥75 years), Latin America (all-age group), and China (all-age group). The association between ozone and respiratory mortality was generally nonsignificant, except Latin American, Australia, and China with percentage excess risks per 10 μg/m^3 increase in ozone of 0.21% (0.10%, 0.31%), 1.10% (0.10%, 2.10%), and 1.33% (0.89%, 1.76%), respectively.

In summary, the magnitude of excess risks of ozone in China was generally larger than in other countries or regions for all-natural, cardiovascular, and respiratory mortality among the multi-city studies (Fig. 9.2 and Table 9.2). The underlying reasons should be explored in future studies.

9.3 Modifying Factors for CRFs

Sources of heterogeneity in the risk estimates for PM and O_3 may include different exposure profiles, regional climate, and population characteristics.

9.3.1 Concentration Ranges of Air Pollutants

Cities of the developing countries currently have the highest air pollution concentration (Figs. 9.3 and 9.4), particularly those cities in Asia with extensive coal burning [29]. For example, the ranges of annual mean concentrations of PM_{10} and O_3 in 74 Chinese cities were from 42 to 233 μg/m^3 and 69–200 μg/m^3, compared with the mean concentrations of 105 μg/m^3 and 145 μg/m^3 in those 74 Chinese cities in 2014, respectively [13], which is far exceeding the limits of both national air quality

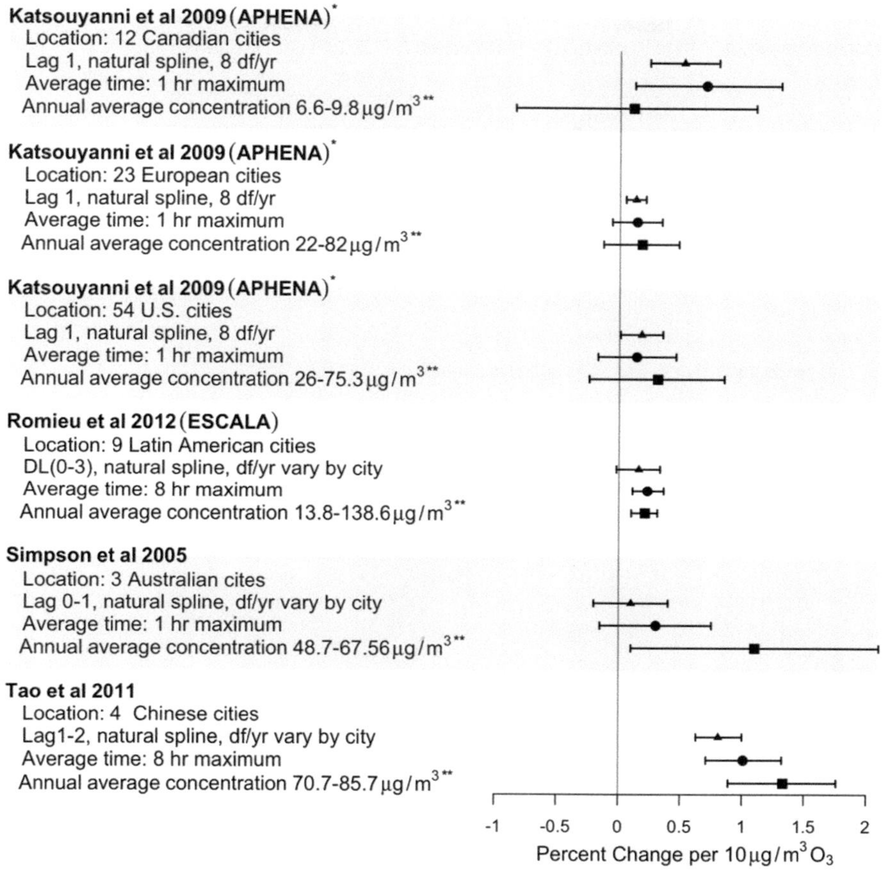

Fig. 9.2 Percentage excess risk (ER %) of all-natural, cardiovascular, and respiratory mortality per 10 μg/m^3 increase in O_3 in multi-city studies. Estimates with ▲ indicate all-natural mortality, estimates with ● represent cardiovascular mortality, and estimates with ■ denote respiratory mortality. *Estimates are for all ages, except estimates of cardiovascular mortality are for people ≥75 years. **Concentration ranges across cities (Source: Adapted from Romieu et al. [30]; p. 101)

standards and the air quality guidelines recommended by the World Health Organization (WHO). Cities of the developed countries seem to have better air quality in terms of PM_{10} and O_3 concentration compared to developing countries. For instance, the median concentration of PM_{10} ranges from 14.0 μg/m^3 to 43.7 μg/m^3 and 13.1 μg/m^3 to 65.0 μg/m^3 in the United States and European cities, respectively. The median concentration of O_3 ranges from 26.0 μg/m^3 to 75.3 μg/m^3 and 37.5 μg/m^3 to 82.2 μg/m^3 in the United States and European cities, respectively [21].

The CRF for PM_{10} with mortality was found to be linear in *NMMAPS* [17] and *APHEA-2* project [31, 32] at low air pollution concentrations levels with no threshold identified. The CRF outside the range of concentration in North America and

Table 9.2 Mortality risk estimates for ozone exposure based on meta-analyses results

Country or region	Daily mortality	Reference	ER (%) per 10 μg/m³	95 % CI
Europe	All-cause	[2]	0.3%	0.1–0.4%
Europe	Cardiovascular	[2]	0.4%	0.3–0.5%
Europe	Respiratory	[2]	0%	−0.4–0.5%
The United States	All-cause	[4]	0.84%	0.48–1.20%
The United States	Cardiovascular	[4]	0.85%	−0.66–2.39%
The United States	Respiratory	[4]	0.65%	−1.84–3.21%
China	All-cause	[34]	0.48%	0.38–0.58%
China	Cardiovascular	[34]	0.45%	0.29–0.60%
China	Respiratory	[34]	0.73%	0.49–0.97%

ER (%) percentage excess risk, *CI* confidence interval

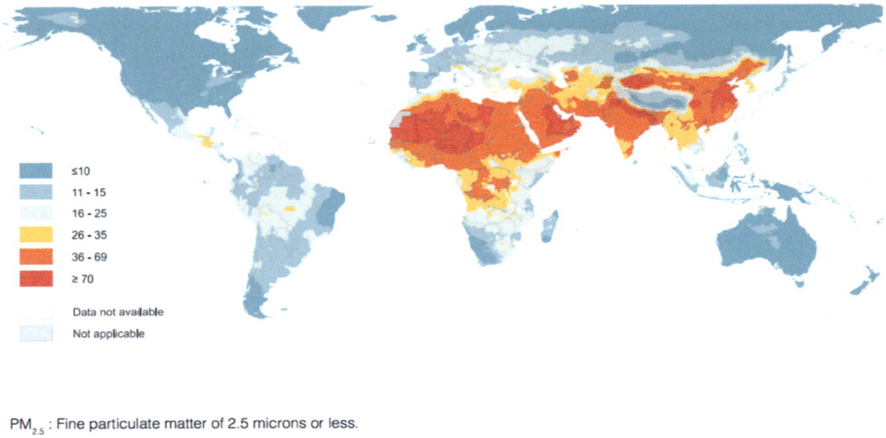

Fig. 9.3 Global map of modeled annual median concentration of $PM_{2.5}$, in μg/m³ (Source: World Health Organization, website: http://who.int/phe/publications/air-pollution-global-assessment/en/)

Europe is less clear. It is likely that the CRF will begin to flatten at very high concentrations as studies suggested [3, 45] (Fig. 9.5); therefore, variation in air pollution levels could contribute to heterogeneity in CRF, particularly if the underlying curves are nonlinear.

9.3.2 The Concentration of Copollutants

The concentration of copollutants (e.g., nitrogen dioxide) is the highest in China (Fig. 9.4). It might be an effect modifier for the mortality effects of PM. For example, higher nitrogen dioxide (NO_2) concentrations were found to be associated with larger PM mortality effect estimates in single-city studies [16, 18]. A multi-city study [21]

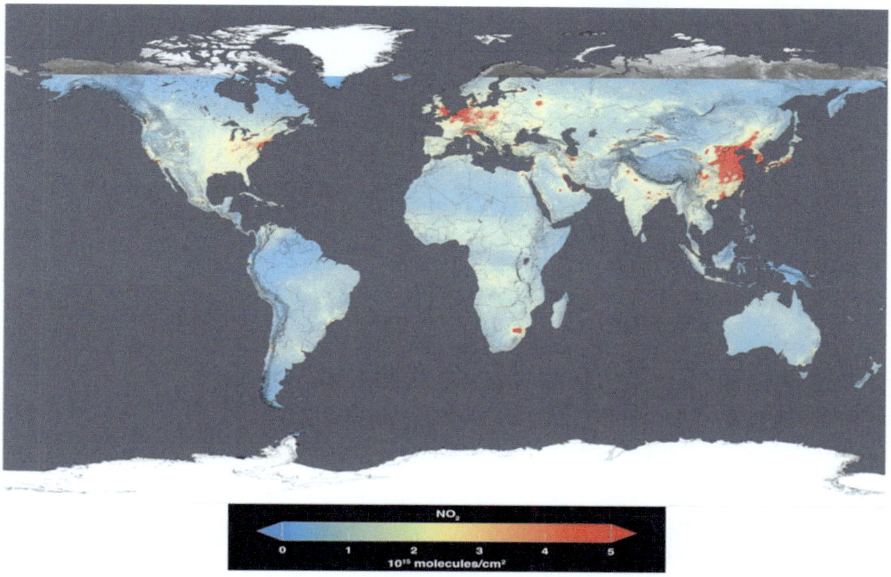

Fig. 9.4 Global map of the 2014 annual average concentration of nitrogen dioxide (Source: The National Aeronautics and Space Administration (NASA), website: https://www.nasa.gov/press-release/new-nasa-satellite-maps-show-human-fingerprint-on-global-air-quality)

also found that in cities with higher mean NO_2 levels and mean NO_2/PM_{10} ratios, the risk estimates of mortality were higher [0.44% (cities with higher mean NO_2 levels) versus 0.17% (cities with lower mean NO_2 levels) and 0.42% (cities with higher mean NO_2/PM_{10} ratios) versus 0.17% (cities with lower mean NO_2/PM_{10} ratios) increase per 10 μg/m^3 change in PM_{10} for all-cause mortality in Europe, respectively, and 0.28% versus 0.01% and 0.27% versus 0.16% in the United States, respectively]. This may suggest an interaction between NO_2 and PM. The reason for this interaction may be that exposure to long-term NO_2 could deteriorate lung function and induce oxidative stress, which may increase the vulnerability to the short-term effects of PM [18]. NO_2 is an indicator of pollution originating from traffic. It may also indicate that PM originating from traffic is more toxic than those from other sources.

9.3.3 Chemical Composition and Emission Sources

As discussed in Chap. 2, PM is a complex mixture of different components including elemental carbon (EC), organic carbon (OC), sulfate (SO_4), nitrate (NO_3), and trace elements. Variation in estimated health effects could be driven by regional variation in the chemical composition of PM.

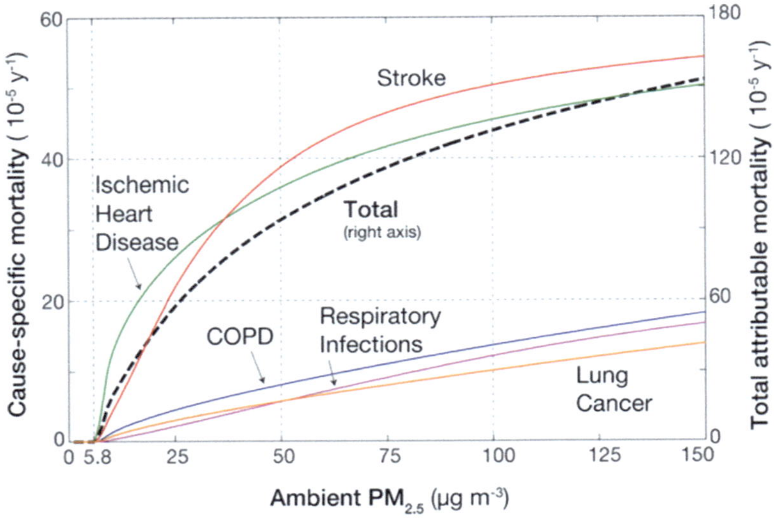

Fig. 9.5 Global concentration-mortality relationships for ambient $PM_{2.5}$ for five individual endpoints (solid lines, left axis) and total of five causes (dashed line, right axis) based on integrated exposure response curves developed for the Global Burden of Disease (GBD) studies [9]. Vertical axes indicate per-capita mortality rates attributable to $PM_{2.5}$ for a hypothetical global population uniformly exposed to a given level of $PM_{2.5}$. Plotted data illustrate the relative contribution of individual disease endpoints to total mortality for a typical population exposed at a given concentration by incorporating concentration-response curves and global disease incidence data (Source: Apte et al. [3])

Numerous studies have reported adverse health effects of each PM chemical constituents. Levy et al. conducted a multi-city time-series analysis to estimate the association between $PM_{2.5}$ constituents and hospital admissions in a population of 12 million US Medicare enrollees [23]. They found that EC was associated with an elevated risk of cardiovascular hospital admission. EC and NO_3 were also linked with emergency type 2 diabetes hospitalizations in Hong Kong elderly population [38]. NO_3, sodium ion, chloride ion, magnesium, and nickel were associated with cardiovascular hospitalizations, while sodium ion, aluminum, and magnesium were found to be associated with respiratory hospitalizations [27]. Chemical components of PM were also associated with mortality [25, 51] and low birthweight [5].

Several recent multi-city studies have examined whether variability in proportions of PM constituents can explain the effect heterogeneity. Bell et al. [6] examined associations between elderly hospital admissions and 52 chemical components of $PM_{2.5}$ for 187 counties in the United States and found that variation in EC, nickel, and vanadium could explain the variations in $PM_{2.5}$ mass effect estimates for both cardiovascular and respiratory hospitalizations [6]. The association between PM_{10} and mortality was also modified when the PM was high in nickel and vanadium in the *NMMAPS* project [24].

Different chemical constituents of PM are generated by various pollutant sources. A few epidemiological studies have used source apportionment analysis (e.g., posi-

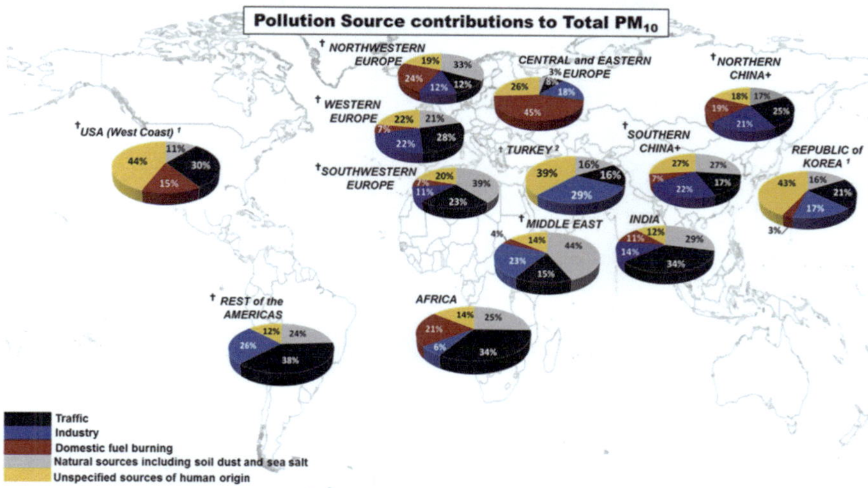

Fig. 9.6 Population-weighted averages for relative source contributions to total PM_{10} in urban cities. †regions in which domestic fuel burning have not been assessed. [1]Based only on one study. [2]Based only on one study including traffic (Source: Karagulian et al. [19])

tive matrix factorization) to explore the health effects of PM sources and found some PM sources were more harmful than others. For example, Stanek et al. [37] conducted a literature review summarizing findings from 29 studies in the United States [37]. The authors found that PM from motor vehicle emission was associated with cardiovascular mortality, while PM from coal combustion was associated with total mortality. However, limited number of studies was conducted in China to estimate source-specific health risks of PM. Pun et al. [28] estimated PM_{10} sources contributed from 19 chemical components by positive matrix factorization in Hong Kong and found that respiratory hospitalizations were associated with vehicle exhaust, regional combustion, and secondary nitrate [28].

Figure 9.6 shows population-weighted averages for relative source contributions to total PM_{10}in urban sites utilizing the global source apportionment database at the website of the World Health Organization [44]. This database included a total of 529 source apportionment records in cities of 51 countries for a total of 560 million people. This global review [19] found that the contributions by source category varied substantially across regions. For example, traffic was the main contributor to urban ambient PM_{10} in Africa and India (34%), the United States (30%), Western Europe (28%), and Northern China (28%); industry was an important contributor in Turkey (29%), Western Europe (22%), and China (21–22%); domestic fuel burning was the main contributor in Central and Eastern Europe (45%), Northwestern Europe (24%), Africa (21%), and the Northern China (19%). Contributions by source category in China are comparable with other regions.

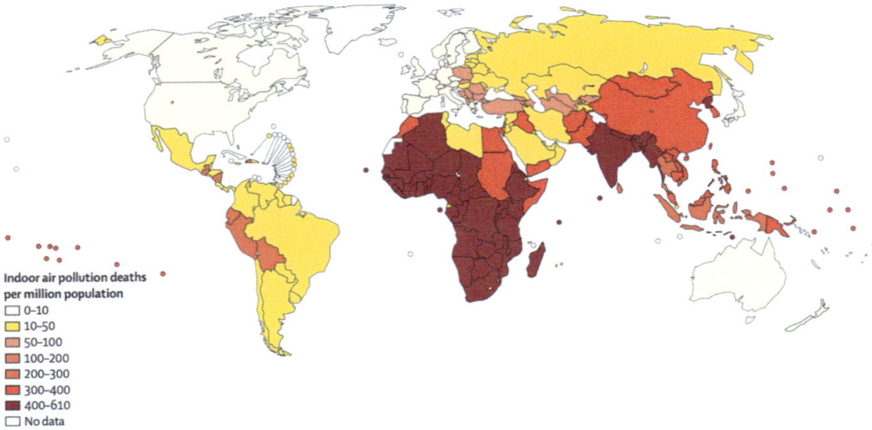

Fig. 9.7 WHO map of household air pollution and mortality (Source: World Health Organization (WHO), website: http://www.who.int/heli/risks/indoorair/en/iapmap.pdf)

9.3.4 Indoor Air Pollution

High ambient air pollution concentration does not necessarily result in high personal exposure. Human exposure is determined by the concentration of air pollution in the environments and by the amount of time people spend there. The majority of epidemiological studies utilized outdoor concentrations of air pollutants measured in fixed monitoring stations as a surrogate for personal exposure. Although outdoor air pollutants may penetrate indoors, the concentration of indoor and outdoor air pollutants may differ substantially due to outdoor-to-indoor migration and source of indoor pollutants. Humans spend much more time indoors than outdoors. Indoor air pollution would contribute substantially to personal exposure, especially in countries in which cooking and heating depend on solid fuels (e.g., coal and biomass) with open or poorly ventilated stoves, such as in Africa, India, and China (Fig. 9.7). During cooking or heating, on average, the levels of PM are about 10–20 times higher than the levels in international health guidelines [43]. Thus, simply utilizing outdoor air pollution concentration as a surrogate for personal exposure may result in bias of personal exposure assessment [35]. This bias may vary among countries, which could be one of the reasons to explain effects heterogeneity among countries.

9.3.5 Responsiveness of the Population

Individuals respond differently when exposed to air pollution. Health effects might be exacerbated in susceptible subgroups when exposed to the same levels of pollutants as the average population. This susceptibility to air pollution has been linked to personal characteristics, such as inherent personal characteristics (e.g., age, socioeconomic status, and chronic diseases) and aspects of lifestyle that may enhance the adverse health effects of air pollutants (e.g., cigarette smoking and exercise).

Population with lower **socioeconomic status** tends to have higher risks of both mortality and morbidity when exposed to the same levels of air pollution. For example, higher susceptibility is found in persons with the least educated, lower income, lower occupational status, and unemployment [8, 50]. The multi-city study of the *APHENA* project [21] found cities with higher percentage of unemployment were associated with a greater effect of PM_{10} on all-cause mortality (0.57% versus 0.27% and 0.23% versus 0.11% increase per 10 μg/m^3 change of PM_{10} in Europe and the United States, respectively). Higher mortality risk estimates were also linked to low educational status in the multi-city study of the *CAPES* study in China [11].

Higher risk estimates were also found in individuals already affected by **preexisting cardiopulmonary disease** [26, 39, 49] as well as **diabetes** [48]. For example, the risk estimates of PM_{10} for people with respiratory diseases have almost doubled for those with cardiovascular diseases [47]. People with diabetes have 0.7% higher risks (2.0% versus 1.3%) per 10 μg/m^3 increase in PM_{10} for cardiovascular hospital admissions than people who do not have diabetes [48].

Age is another effect modifier for the health effects of PM. Stratified analyses have consistently reported the increased risk of hospital admissions [15] and mortality [10, 33] among **older adults** compared with younger population when exposed to air pollutants. Multi-city study of the *APHENA* project [21] also found cities with higher percentage of older people (≥ 75 years) were associated with a greater effect of PM_{10} on all-cause mortality (0.32% versus 0.25% and 0.24% versus 0.03% increase in mortality per 10 μg/m^3 change in PM_{10} for all-cause mortality in Europe and the United States, respectively). Both the meta-analysis of 144 effect estimates from 39 time-series studies and *NMMAPS* of 95 US cities showed that elders had much higher ozone-associated mortality risk than younger groups [4]. A literature review summarized 23 related studies and concluded that age is the most consistent effect modifier of the association between short-term exposure to PM and mortality and hospitalization with risk estimates to be 0.30% (0.11%, 0.49%) higher for older than for younger population [7]. The multi-city study of the *CAPES* study also reported that older people (≥ 65 years) tended to be more vulnerable than younger population (5–64 years) in China [11].

Figure 9.8 shows the age distribution by countries. The proportion of the older population (≥ 65 years) is higher in developed countries or regions (e.g., European areas, Australia, and the United States). China has the lowest proportion of older among the four countries or regions (China, European areas, Australia, and the United States) in 2011. However, China's population is growing old at a faster rate than almost all other countries. It is expected that more than a quarter of China's population will be over 65 in 2050, which is almost 500 million people [1] (Fig. 9.9). With population aging in China, the effect sizes of air pollution are expected to increase.

Fig. 9.8 Age distribution by country and area in 2011 (Source: World Bank, World Development Indicators, website: https://www.indexmundi.com/facts/visualizations/age-distribution/#country=au:cn:xc:us)

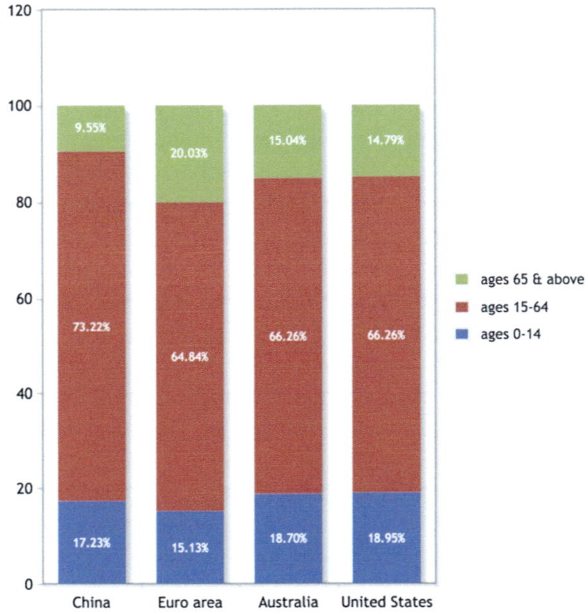

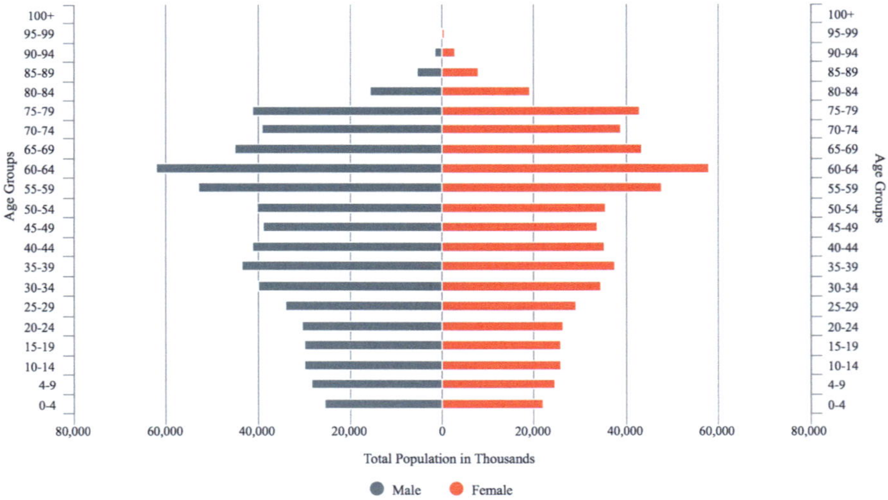

Fig. 9.9 Population projections for China in 2050 (Source: United Nations World Population Prospects, website: http://chinapower.csis.org/aging-problem/)

9.3.6 *Regional Climate*

Many multi-city studies have consistently shown that the effect estimates of air pollution were higher in warmer cities. Katsouyanni et al. [20] reported that the short-term mortality effect of PM_{10} per 10 μg/m^3 increase was 0.29% in cities with relatively cold climate, whereas it was 0.82% in warm climate in the *APHEA-2* projects with a total population of more than 43 million [20]. Kioumourtzoglou et al. [22] assessed whether community-level variables modify the association between long-term $PM_{2.5}$ exposures and mortality in a population of more than 35 million Medicare enrollees from 207 US cities. They found that temperature was the most consistent effect modifications [22]. Larger estimates of the effects of particles on mortality were found in warmer cities (e.g., 0.8% versus 0.3% increase in mortality per 10 μg/m^3 change in PM_{10}). One possible explanation is that people tend to spend more time outdoors in warmer cities and keep their windows open so that the concentrations of air pollution measured at the outdoor fixed monitoring stations may better represent the average population exposure.

9.4 Summary

Health risk estimates of air pollution are heterogeneities across region. Differences in responsiveness of the population, exposure profiles such as air pollution concentration and composition, and regional climate may be contributable to this effect heterogeneity. Although the relative risk estimates of air pollution are generally similar between China and elsewhere, air pollution is one of China's most serious environmental health problems given the large number of people exposed to high concentration levels of air pollution in China.

Acknowledgements This chapter was supported by grants from the National Natural Science Foundation of China (No. 81571130090, 91543112) and the National Key Research and Development Program of China (2016YFC0206506, 2017YFC0702701).

References

1. Center for Strategic & International Studies (CSIS). Does china have an aging problem? Available: http://chinapower.csis.org/aging-problem/ (n.d.). Accessed 12 June 2017.
2. Anderson HR, Atkinson RW, Peacock JL, Marston L, Konstantinou K, World Health Organization. Meta-analysis of time-series studies and panel studies of particulate matter (PM) and ozone (O3): report of a WHO task group. WHO Regional Office for Europe: Copenhagen; 2004.
3. Apte JS, Marshall JD, Cohen AJ, Brauer M. Addressing global mortality from ambient $PM_{2.5}$. Environ Sci Technol. 2015;49(13):8057–66.

4. Bell ML, Dominici F, Samet JM. A meta-analysis of time-series studies of ozone and mortality with comparison to the national morbidity, mortality, and air pollution study. Epidemiology. 2005;16(4):436–45.
5. Bell ML, Belanger K, Ebisu K, Gent JF, Lee HJ, Koutrakis P, et al. Prenatal exposure to fine particulate matter and birth weight: variations by particulate constituents and sources. Epidemiology. 2010;21(6):884–91.
6. Bell ML. Assessment of the health impacts of particulate matter characteristics. Health Eff Ins. 2012;161:5–38.
7. Bell ML, Zanobetti A, Dominici F. Evidence on vulnerability and susceptibility to health risks associated with short-term exposure to particulate matter: a systematic review and meta-analysis. Am J Epidemiol. 2013;178(6):865–76.
8. Bell ML, Zanobetti A, Dominici F. Who is more affected by ozone pollution? A systematic review and meta-analysis. Am J Epidemiol. 2014;180(1):15–28.
9. Burnett R, Pope C III, Ezzati M, Olives C, Lim S, Mehta S, et al. An integrated risk function for estimating the global burden of disease attributable to ambient fine particulate matter exposure. Environ Health Perspect. 2014;122:397–403.
10. Cakmak S, Dales RE, Vidal CB. Air pollution and mortality in Chile: susceptibility among the elderly. Environ Health Perspect. 2007:524–7.
11. Chen R, Kan H, Chen B, Huang W, Bai Z, Song G, et al. Association of particulate air pollution with daily mortality: the china air pollution and health effects study. Am J Epidemiol. 2012;175(11):1173–81.
12. Chen Z, Wang JN, Ma GX, Zhang YS. China tackles the health effects of air pollution. Lancet. 2013;382(9909):1959–60.
13. Fu L, Wan W, Zhang WH. China air 2015: air pollution prevention and control progress in Chinese cities. Manila: Clean Air Asia; 2015.
14. HEI International Oversight Committee. Health effects of outdoor air pollution in developing countries of Asia: a literature review. Boston: Health Eff Inst; 2004.
15. D'Ippoliti D, Forastiere F, Ancona C, Agabiti N, Fusco D, Michelozzi P, et al. Air pollution and myocardial infarction in Rome: a case-crossover analysis. Epidemiology. 2003;14(5):528–35.
16. Deguen S, Petit C, Delbarre A, Kihal W, Padilla C, Benmarhnia T, et al. Neighbourhood characteristics and long-term air pollution levels modify the association between the short-term nitrogen dioxide concentrations and all-cause mortality in paris. PLoS One. 2015;10(7):e0131463.
17. Dominici F, Daniels M, Zeger SL, Samet JM. Air pollution and mortality: estimating regional and national dose-response relationships. J Am Stat Assoc. 2002;97(457):100–11.
18. Faustini A, Stafoggia M, Renzi M, Cesaroni G, Alessandrini E, Davoli M, et al. Does chronic exposure to high levels of nitrogen dioxide exacerbate the short-term effects of airborne particles? Occup Environ Med. 2016;73(11):772–8.
19. Karagulian F, Belis CA, Dora CFC, Prüss-Ustün AM, Bonjour S, Adair-Rohani H, et al. Contributions to cities' ambient particulate matter (PM): a systematic review of local source contributions at global level. Atmos Environ. 2015;120:475–83.
20. Katsouyanni K, Touloumi G, Samoli E, Gryparis A, Le Tertre A, Monopolis Y, et al. Confounding and effect modification in the short-term effects of ambient particles on total mortality: results from 29 European cities within the APHEA2 project. Epidemiology. 2001;12(5):521–31.
21. Katsouyanni K, Samet JM, Anderson H, Atkinson R, Le Tertre A, Medina S, et al. Air pollution and health: a European and North American approach (APHENA). Health Eff Inst. 2009;142:5–90.
22. Kioumourtzoglou MA, Schwartz J, James P, Dominici F, Zanobetti A. $PM_{2.5}$ and mortality in 207 U.S. cities: modification by temperature and city characteristics. Epidemiology. 2016;27(2):221–7.
23. Levy JI, Diez D, Dou Y, Barr CD, Dominici F. A meta-analysis and multisite time-series analysis of the differential toxicity of major fine particulate matter constituents. Am J Epidemiol. 2012;175(11):1091–9.

24. Lippmann M, Ito K, Hwang JS, Maciejczyk P, Chen LC. Cardiovascular effects of nickel in ambient air. Environ Health Perspect. 2006;114(11):1662–9.
25. Ostro B, Feng WY, Broadwin R, Malig B, Green R, Lipsett M. The impact of components of fine particulate matter on cardiovascular mortality in susceptible subpopulations. Occup Environ Med. 2008;65(11):750–6.
26. Park SK, O'Neill MS, Vokonas PS, Sparrow D, Schwartz J. Effects of air pollution on heart rate variability: the VA normative aging study. Environ Health Perspect. 2005;113(3):304–9.
27. Pun VC, Yu ITS, Qiu H, Ho KF, Sun Z, Louie PK, et al. Short-term associations of cause-specific emergency hospitalizations and particulate matter chemical components in Hong Kong. Am J Epidemiol. 2014;179(9):1086–95.
28. Pun VC, Tian L, Yu IT, Kioumourtzoglou MA, Qiu H. Differential distributed lag patterns of source-specific particulate matter on respiratory emergency hospitalizations. Environ Sci Technol. 2015;49(6):3830–8.
29. Romieu I, Samet JM, Smith KR, Bruce N. Outdoor air pollution and acute respiratory infections among children in developing countries. J Occup Environ Med. 2002;44(7):640–9.
30. Romieu I, Gouveia N, Cifuentes LA, de Leon AP, Junger W, Vera J, et al. Multicity study of air pollution and mortality in Latin America (the ESCALA study). Health Eff Inst. 2012;171:5–86.
31. Samoli E, Touloumi G, Zanobetti A, Le Tertre A, Schindler C, Atkinson R, et al. Investigating the dose-response relation between air pollution and total mortality in the APHEA-2 multicity project. J Occup Environ Med. 2003;60(12):977–82.
32. Samoli E, Analitis A, Touloumi G, Schwartz J, Anderson HR, Sunyer J, et al. Estimating the exposure-response relationships between particulate matter and mortality within the APHEA multicity project. Environ Health Perspect. 2005;113(1):88–95.
33. Sandström T, Frew A, Svartengren M, Viegi G. The need for a focus on air pollution research in the elderly. Eur Respir J. 2003;21(40 suppl):92s–5s.
34. Shang Y, Sun Z, Cao J, Wang X, Zhong L, Bi X, et al. Systematic review of Chinese studies of short-term exposure to air pollution and daily mortality. Environ Int. 2013;54(2013):100–11.
35. Shi S, Chen C, Zhao B. Modifications of exposure to ambient particulate matter: tackling bias in using ambient concentration as surrogate with particle infiltration factor and ambient exposure factor. Environ Pollut. 2017;220(2017):337–47.
36. Simpson R, Williams G, Petroeschevsky A, Best T, Morgan G, Denison L, et al. The short-term effects of air pollution on daily mortality in four Australian cities. Aust N Z J Public Health. 2005;29(3):205–12.
37. Stanek LW, Sacks JD, Dutton SJ, Dubois JJB. Attributing health effects to apportioned components and sources of particulate matter: an evaluation of collective results. Atmos Environ. 2011;45(32):5655–63.
38. Sun S, Qiu H, Ho KF, Tian L. Chemical components of respirable particulate matter associated with emergency hospital admissions for type 2 diabetes mellitus in Hong Kong. Environ Int. 2016;97(2016):93–9.
39. Sunyer J, Schwartz J, Tobías A, Macfarlane D, Garcia J, Antó JM. Patients with chronic obstructive pulmonary disease are at increased risk of death associated with urban particle air pollution: a case-crossover analysis. Am J Epidemiol. 2000;151(1):50–6.
40. Tao Y, Huang W, Huang X, Zhong L, Lu SE, Li Y, et al. Estimated acute effects of ambient ozone and nitrogen dioxide on mortality in the Pearl River Delta of southern China. Environ Health Perspect. 2012;120(3):393–8.
41. van Donkelaar A, Martin RV, Brauer M, Kahn R, Levy R, Verduzco C, et al. Global estimates of ambient fine particulate matter concentrations from satellite-based aerosol optical depth: development and application. Environ Health Perspect. 2010;118(6):847–55.
42. WHO. Guidelines for air quality. Geneva: World Health Organization; 2000.
43. WHO. Air quality guidelines: global update 2005: particulate matter, ozone, nitrogen dioxide, and sulfur dioxide. Geneva: World Health Organization; 2006.
44. WHO. Database on source apportionment studies for particulate matter in the air (PM_{10} and $PM_{2.5}$). Available: http://www.who.int/quantifying_ehimpacts/global/source_apport/. n.d. Accessed 6 June 2017.

45. Xie W, Li G, Zhao D, Xie X, Wei Z, Wang W, et al. Relationship between fine particulate air pollution and ischaemic heart disease morbidity and mortality. Heart. 2015;101(4):257–63.
46. Yang G, Wang Y, Zeng Y, Gao GF, Liang X, Zhou M, et al. Rapid health transition in China, 1990–2010: findings from the Global Burden of Disease Study 2010. Lancet. 2013;381(9882):1987–2015.
47. Zanobetti A, Schwartz J, Gold D. Are there sensitive subgroups for the effects of airborne particles? Environ Health Perspect. 2000;108(9):841–5.
48. Zanobetti A, Schwartz J. Cardiovascular damage by airborne particles: are diabetics more susceptible? Epidemiology. 2002;13(5):588–92.
49. Zanobetti A, Schwartz J. The effect of particulate air pollution on emergency admissions for myocardial infarction: a multicity case-crossover analysis. Environ Health Perspect. 2005;113(8):978–82.
50. Zeka A, Zanobetti A, Schwartz J. Individual-level modifiers of the effects of particulate matter on daily mortality. Am J Epidemiol. 2006;163(9):849–59.
51. Zhou J, Ito K, Lall R, Lippmann M, Thurston G. Time-series analysis of mortality effects of fine particulate matter components in Detroit and Seattle. Environ Health Perspect. 2011;119(4):461–6.

Chapter 10
Air Pollution, Disease Burden, and Health Economic Loss in China

Yue Niu, Renjie Chen, and Haidong Kan

Abstract As the largest developing country in the world, China is now facing one of the severest air pollution problems. The objective of this section is to evaluate the disease burden and corresponding economic loss attributable to ambient air pollution in China. We reviewed a series of studies by Chinese or foreign investigators focusing on the disease burden and economic loss in China. These studies showed both the general air pollution and haze episodes have resulted in substantial disease burden in terms of excess number of premature deaths, disability-adjusted life-year loss, and years of life lost. The corresponding economic loss has accounted for an appreciable proportion of China's national economy. Overall, the disease burden and health economic loss due to ambient air pollution in China is greater than in the remaining parts of the world, for one of the highest levels of air pollution and the largest size of exposed population. Consideration of both health and economic impacts of air pollution can facilitate the Chinese government to develop environmental policies to reduce the emissions of various air pollutants and protect the public health.

Keywords Air pollution • Disease burden • Economic loss • China

10.1 Introduction

As the largest developing country in the world, China is experiencing rapid industrialization and urbanization and subsequently is also facing deteriorating air quality. In 2016, 81% of population of China was exposed to $PM_{2.5}$ concentrations higher than the Interim Target-1 limit (35 μg/m^3) proposed by the World Health

Y. Niu • R. Chen • H. Kan (✉)
Department of Environmental Health, School of Public Health, Fudan University, Shanghai 200032, China
e-mail: kanh@fudan.edu.cn

G.-H. Dong (ed.), *Ambient Air Pollution and Health Impact in China*, Advances in Experimental Medicine and Biology 1017, DOI 10.1007/978-981-10-5657-4_10

Organization, and 50% of the population was exposed to annual average $PM_{2.5}$ concentrations above 48.2 μg/m^3 [19].

Numerous studies have demonstrated that ambient air pollution was significantly associated with increased risks of adverse health outcomes. Evaluating the disease burden and corresponding economic loss attributable to ambient air pollution in China is of great importance to providing continued rationales for Chinese government to mitigate air pollution.

10.2 Air Pollution and Disease Burden

10.2.1 Estimation Methods

Disease burden can be deemed as the gap between current health status and an ideal situation in which everyone lives into life expectancy without disease and disability. Causes of the gap are premature mortality, disability, and illness.

Disease burden can be quantified through the following indexes:

10.2.1.1 Attributable Fraction (AF) and Population-Attributable Fraction (PAF)

The AF measures the contribution of a risk factor to disease or mortality. Since the entire population is always exposed to the ambient air pollution, AF can be interpreted as the PAF. The equations are shown as follows:

$$AF = \frac{RR - 1}{RR}$$

where AF represents the fraction of mortality for a specific disease attributable to air pollution and RR is relative risk for a specific disease comparing a level of air pollution to a reference level.

$$PAF = \frac{Pe * (RR - 1)}{Pe * (RR - 1) + 1}$$

where PAF measures the population fraction for a specific disease attributable to a level of air pollution and Pe is the proportion of population exposed to this level of air pollution, the value of which is always assumed to be 100%.

10.2.1.2 Attributable Mortality (AM)

AM was often used to estimate the number of premature deaths due to air pollution.

$$\mathrm{AM} = y_0 * \mathrm{AF} * \mathrm{Pop}$$

where AM is also called excess death, y_0 is the baseline mortality rate of a specific disease in a given population, and Pop is the size of total population exposed to a level of air pollution.

10.2.1.3 DALYs, YLLs, and YLDs

One disability-adjusted life year (DALY) can be regarded as one lost year of "healthy" life. DALYs for a disease or health condition are calculated as the sum of the years of life lost (YLL) due to premature mortality in the population and the years lost due to disability (YLD) for incident cases of the health condition:

$$\mathrm{DALYs} = \mathrm{YLLs} + \mathrm{YDLs}$$

where YLLs and YLDs are estimated following the below formula, respectively:

$$\mathrm{YLL} = \mathrm{N} * \mathrm{L}$$

where N refers to the number of deaths and L is standard life expectancy at age of death.

$$\mathrm{YDL} = \mathrm{I} * \mathrm{DW} * \mathrm{L}$$

where I represents the number of incident cases, DW is disability weight (ranging from 0 to 1), and L refers to the average duration of the case until remission or death (years).

10.2.2 Mortality Attributable to Ambient air Pollution

Mortality is most widely used to evaluate the disease burden of air pollution. According to the report of Global Burden of Disease (GBD) project, ambient $PM_{2.5}$ was identified as a leading risk factor for global disease burden with estimates of 3.3, 2.9, and 4.2 million attributable deaths worldwide in the year 2010, 2013, and 2015, respectively [3, 4]. The fraction of excess deaths attributable to $PM_{2.5}$ increased from 4.7% in 1990 to 5.3% in 2013 worldwide. The rising trend in total deaths accelerated remarkably after 2000, driven largely by China [24]. In 2015, ambient particulate matter pollution was the fifth leading risk factor in China and contributed to 1.1 million deaths estimated by GBD 2015, which made up 26% of the global deaths (4.2 million) attributable to particulate matter pollution [3].

It is clear that, with the deterioration of air quality and rapid population expansion, the disease burden from ambient air pollution is growing gradually in China. In 2006, the annual average concentrations of PM_{10} in all of 113 major Chinese cities were above the Chinese Ambient Air Quality Standard of 40 μg/m^3 except Guilin, causing 299,700 premature deaths [2]. Based on the estimated results of the Global Burden of Diseases Study 2010, Liu et al. used PAF as the index to evaluate the disease burden attributable to ambient particulate matter pollution in 1990 and 2010 in China [16]. They found that ambient air pollution resulted in 1.235 million deaths in 2010, increasing by 33.4% compared with that in 1990 (0.926 million). Among people over 25 years old, 35.0% of stroke, 29.9% of ischemic heart disease, 27.2% of lung cancer, and 21.0% of chronic obstructive pulmonary disease were due to ambient particulate matter pollution. Utilizing the disease-specific mortality in China in 2015, Song et al. estimated that $PM_{2.5}$ in 2015 contributed as much as 15.5% to all-cause deaths. There were about 1.5 million total deaths attributable to $PM_{2.5}$ exposure, which is higher than estimates in 2010 (1.235 million). 40.3%, 33.1%, 26.8%, 23.9%, and 18.7% of total stroke deaths, acute lower respiratory infection deaths, ischemic heart disease deaths, lung cancer deaths, and chronic obstructive pulmonary disease deaths can be attributed to $PM_{2.5}$ exposure, respectively [19].

Strong spatial variations of health burden were observed in China, with high attributable deaths concentrated in northern, eastern, and southern parts of China, including the Beijing-Tianjin-Hebei Metropolitan Region, Yangtze River Delta, and Pearl River Delta, which have severe air pollution problems, high population density, or both [14, 15].

Beijing, the capital of China located in northern China, has suffered the heavy air pollution especially in winter over the past decades. The annual average concentration of $PM_{2.5}$ reached 87 μg/m^3 in 2013, greatly exceeding China's class II standards of 35 μg/m^3 [22]. Wang et al. reported that 20,043 premature deaths as well as about one million other related medical cases could be attributed to $PM_{2.5}$ pollution in Beijing in 2013 [21]. Similarly, from 2001 to 2012, the estimated average total mortality due to $PM_{2.5}$ is about 22,000–30,000 individuals per year in Beijing, a quarter of which concentrated in the center of Beijing for its higher concentration of $PM_{2.5}$ [30].

The Pearl River Delta region, a major engine of economic growth in China, is one of the most urbanized regions in the world. However, rapid development has brought increasing emissions of ambient pollutants, causing considerable air pollution. Lu et al. used satellite-derived PM_{10} data and $PM_{2.5}$ data to quantify the potential health damages attributable to the long-term exposure to air pollution in the Pearl River Delta region. The results showed that the estimated total mortality (cerebrovascular disease + chronic obstructive pulmonary disease + ischemic heart disease + lung cancer) attributable to $PM_{2.5}$ was 45,000 in 2012 [17]. Spatial mapping revealed that the total mortality attributable to $PM_{2.5}$ was high in the centers of Guangzhou, Foshan, and Hong Kong while low in rural areas with lower particulate matter concentrations, such as Huizhou and Zhaoqing. Guangzhou had the most mortality cases in this region over the 10-year period. The attributable mortality in Guangzhou reached 10,000 in 2013, which contributed around 25% of all-cause mortality in the entire Pearl River Delta region [17].

10.2.3 Mortality Attributable to Haze Episode

Haze has been one of the most disastrous air pollution events in China in recent years, especially in the Mideastern regions. A dust-haze day was defined as having daily visibility within 10 km, no rainfall, and relative humidity less than 80% [26].

Some studies have been conducted to estimate the acute health effects of haze. In 2013, three serious haze episodes occurred in Beijing-Tianjin-Hebei region. From January 10 to January 31, the daily mean concentrations of $PM_{2.5}$ in Beijing, Tianjin, Shijiazhuang, Tangshan, Handan, Xingtai, Baoding, Langfang, and Hengshui exceeded 400 $\mu g/m^3$, with the highest concentration even reaching 718 $\mu g/m^3$. During this period, Zhang et al. found that the short-term $PM_{2.5}$ exposure caused 2725 excess deaths (846 excess deaths due to respiratory disease and 1878 excess deaths due to circulatory disease) in Beijing-Tianjin-Hebei region [28]. In Beijing, an estimated 690 premature deaths, 45,350 acute bronchitis, and 23,720 asthma cases occurred due to this event [5].

10.2.4 DALY and YLL Attributable to Ambient Air Pollution

DALY is a more comprehensive index to quantify the health burden resulting from ambient air pollution, because it takes into account both fatal and nonfatal health loss. In China, Liu et al. found that ambient air pollution caused 25.23 million DALYs loss in 2010, increasing by 4.0% compared with that in 1990 (24.26 million person years) [16]. For PM_{10} air pollution, around 5.26 million person years of DALYs loss was calculated in urban areas in 2006 [1].

Another rounded and accurate index is YLL. The calculation of it is under the assumption that deaths occurring at different ages result in unequal potential YLL, as the life expectancy of young people is longer than that of elderly people. Several time-series analyses have evaluated the short-term impact of air pollution on YLL. He et al. found that an increase of 10 $\mu g/m^3$ in PM_{10} and $PM_{2.5}$ was associated with 4.27 and 2.97 person years increase for YLL, respectively [8]. Similar findings were also reported in Beijing that an interquartile range increase in either $PM_{2.5}$ or PM_{10} was related to YLL increases of 15.8 person years [7].

10.2.5 The Impact of Pollution Sources on Disease Burden

Understanding the information on air pollution sources of higher health risks is significant for policy makers to implement actions more effectively to reduce air pollution. The Global Burden of Disease from major air pollution sources has been established to estimate the disease burden attributable to ambient air pollution from major $PM_{2.5}$ sources in China, India, and Eastern Europe. A recently completed study has evaluated deaths attributable to ambient $PM_{2.5}$ pollution in China, by source of emissions [6]. The results suggested that coal burning was the first contributor to ambient $PM_{2.5}$, responsible for an estimated 366,000 premature deaths in 2013 in China, accounting for approximately 40% of the total deaths due to $PM_{2.5}$ exposure estimated by GBD 2013 (916,000 premature deaths). In addition, industrial sources and household solid fuel combustion also had great contribution to disease burden attributable to ambient $PM_{2.5}$ in China, causing 250,000 and 177,000 premature deaths, respectively. Lelieveld et al. also found that residential and commercial energy use is the largest source category in China, explaining 32% of premature mortality linked to ambient air pollution in 2010 [12].

10.3 Health Economic Loss

In 1967, the economic loss caused by air pollution in the USA was firstly calculated by the American economist Ridker through the human capital approach. In 1981, the concept, theory, and method of environmental pollution economic loss assessment were proposed and discussed in the congress of the National Symposium on Environmental Economics. Thereafter, the quantification of health-based economic loss caused by air pollution is becoming a critical component in evaluating the total economic loss and formulating policies relevant.

10.3.1 Estimation Methods

The process of evaluating the health economic loss from environmental pollution can be divided into two major stages: the establishment of dose-response relationship between health effects and air pollution and the monetization of the value of health effects [18]. The four main monetization approaches are listed as follows:

10.3.1.1 Human Capital Method (HCM)

HCM is designed to estimate the value of human capital under the hypothesis that the economic loss due to morbidity and mortality is equal to the value of that individual's future contribution to production if he or she had continued to work in full health. However, this method overestimates the value of production because of its controversial assumption that a worker will never lose his job in the future [10].

10.3.1.2 Friction Cost Method (FCM)

Since HCM cannot deal well with the condition when another person from the unemployment pool replaces the present value of a worker's future earnings until the sick or impaired worker returns or is eventually replaced, FCM is put forward as an alternative to HCM to estimate the value of human capital during this short-term period defined as "friction period." FCM assumes that impairment or premature death will not affect the total productivity following the friction period. Hence it is highly controversial and even paradoxical to jump into the conclusion that illness, injuries, and premature deaths would reduce the total unemployment [10].

10.3.1.3 Cost of Illness (COI)

COI method is often used to measure the direct cost of illness imposed on the whole society, including the loss in earning due to illness; medical costs such as hospital care, home health care, medicine, and services of the doctors and nurses; and other related out-of-pocket expenditures [18]. COI always underestimate the illness value because it is difficult to calculate overall cost due to disease. Besides, the method ignores individual's willingness to pay [13], which varies in different subgroups.

10.3.1.4 Willingness to Pay (WTP)

WTP is an indirect evaluation method, which is currently being widely recognized and accepted. This method calculates how much money that an individual is eager to pay in order to reduce the risk of illness or mortality [10]. From the perspective of economy, the WTP method is the most reasonable because it can reflect the true value of all goods and utilities completely [13].

The WTP method is always used to calculate the value of a statistical life (VOSL), which is the monetized benefit of reduced mortality risk. We can have a better understanding of VOSL through the following example [20]: imagine a population of 1000 individuals where everyone faces the same annual mortality risk of 1/1000. There would be one death in a year defined as one statistic life. If each individual is willing to pay $1000 to avoid mortality risk, the total willingness to pay for the population is $1 million, which is the VOSL. For China, a meta-analysis found that residents in the urban areas and rural areas were willing to pay 1.59, 0.32 million yuan for saving one statistic life, respectively [25].

10.3.2 Health Economic Loss Due to Ambient Air Pollution

In the past decades, accompanying with the rapid economic growth in China, the health economic loss due to air pollution was increasing dramatically. Between 1990 and 2013, the global cost of health effects due to ambient $PM_{2.5}$ exposure has raised by 63%, from US$ 2176 billion in1990 to US$ 3552 billion. In China, the total welfare loss from ambient $PM_{2.5}$ grew from US$ 126.6 billion to US$ 1589.8 billion, and the total forgone labor output increased from US$ 12.5 billion to US$ 44.6 billion during this period [24]. According to the report of World Bank published in 2007, the annual total health cost related to air pollution amounted to 157 billion yuan in urban areas of China calculated through human capital approach. If the WTP method was used, the estimated economic loss would be even larger (520 billion yuan) [23]. Zhang et al. reported that in 2004, the total economic cost due to particulate matter pollution in 111 cities of China was US$ 29.18 billion, which was lower than estimates from World Bank. The top three cities contributing the greatest economic costs were Beijing (US$ 2.77 billion), Shanghai (US$ 2.52 billion), and Tianjin (US$ 1.20 billion), constituting more than one fifth of the total costs of the 111 cities [27]. Another study on the economic loss in multiple cities concluded that around 310 billion yuan loss could be attributed to the exposure to air pollution in 74 cities in 2015, which constituted about 1.63 % of the 74 cities' gross domestic product (GDP). The top three cities were Chongqing, Beijing, and Baoding, while Tianjin and Shanghai ranked the fourth and fifth, respectively [13].

For some major cities, the similar change trend was also observed. In Beijing, Hou et al. found that the total health economic loss of PM_{10} increased from US$ 8.52 billion in 2008 to US$ 10.41 billion in 2012, while the GDP grew from US$ 171 billion in 2008 to US$ 275 billion in 2012. However, the proportion of the average economic cost that occupied Beijing's average GDP dropped by 1.19% from 2008 to 2012 [9]. In Shanghai, it was estimated that the total economic cost caused by particulate air pollution in urban areas was approximately US$ 625.4 million in 2001, making up 1.03% of GDP of the city [11]. In 2009, the air pollution-related economic loss increased to 7248 million yuan (around US$ 1000 million), while the proportion of it accounting for GDP dropped to 0.49% [29].

There was obvious spatial distribution in health economic loss due to air pollution. Larger population, higher resident incomes, and more serious air pollution were three potential factors that could result in higher health economic costs due to air pollution. It was found that compared with inland cities, the economic costs were higher in coastal cities, such as Dalian, Shenzhen, and Qingdao, because of its rapid economic growth and higher resident incomes. Since population is another important factor, cities with large population may have more costs. Generally, the northern cities may have higher economic costs than the southern cities, because the air pollution issue was more serious in northern cities [27].

10.4 Summary

With rapid growth of national economy driven by energy consumption (especially coal consumption), China is now facing unprecedented air pollution problems. The disease burden and health economic loss due to ambient air pollution in China are greater than in the remaining parts of the world, for one of the highest levels of air pollution and the largest size of exposed population. Consideration of both health and economic impacts of air pollution can facilitate the Chinese government to develop environmental policies to reduce the emissions of various air pollutants and protect the public health.

Acknowledgment This chapter was supported by Grants from Public Welfare Research Program of National Health and Family Planning Commission of China (201502003), National Natural Science Foundation of China (91643205), China Medical Board Collaborating Program (13-152), and Cyrus Tang Foundation (CTF-FD2014001).

References

1. Chen R, Chen B, Kan H. Application of disability adjusted life years in evaluating health effects due to particulate air pollution in Chinese cities (in Chinese). Chin J Prev Med. 2010a;44(2):140–3.
2. Chen R, Chen B, Kan H. A health-based economic assessment of particulate air pollution in 113 Chinese cities (in Chinese). China Environ Sci. 2010b;30(3):410–5.
3. GBD 2015 Risk Factors Collaborators. Global, regional, and national comparative risk assessment of 79 behavioural, environmental and occupational, and metabolic risks or clusters of risks, 1990–2015: a systematic analysis for the Global Burden of Disease Study 2015. Lancet. 2016;388(10053):1659–724.
4. GBD 2013 Risk Factors Collaborators. Global, regional, and national comparative risk assessment of 79 behavioural, environmental and occupational, and metabolic risks or clusters of risks in 188 countries, 1990–2013: a systematic analysis for the Global Burden of Disease Study 2013. Lancet. 2015;386(10010):2287–323.
5. Gao M, Guttikunda SK, Carmichael GR, Wang Y, Liu Z, Stanier CO, Saide PE, Yu M. Health impacts and economic losses assessment of the 2013 severe haze event in Beijing area. Sci Total Environ. 2015;511:553–61.
6. GBD MAPS Working Group. Burden of disease attributable to coal-burning and other major sources of air pollution in China, Special Report 20. Boston: Health Effects Institute; 2016.
7. Guo Y, Li S, Tian Z, Pan X, Zhang J, Williams G. The burden of air pollution on years of life lost in Beijing, China, 2004–08: retrospective regression analysis of daily deaths. BMJ. 2013;347:f7139.
8. He T, Yang Z, Liu T, Shen Y, Fu X, Qian X, Zhang Y, Wang Y, Xu Z, Zhu S, Mao C, Xu G, Tang J. Ambient air pollution and years of life lost in Ningbo, China. Sci Rep. 2016;6:22485.

9. Hou Q, An X, Tao Y, Sun Z. Assessment of resident's exposure level and health economic costs of PM_{10} in Beijing from 2008 to 2012. Sci Total Environ. 2016;563–564:557–65.
10. Jo C. Cost-of-illness studies: concepts, scopes, and methods. Clin Mol Hepatol. 2014;20(4):327–37.
11. Kan H, Chen B. Particulate air pollution in urban areas of Shanghai, China: health-based economic assessment. Sci Total Environ. 2004;322(1–3):71–9.
12. Lelieveld J, Evans JS, Fnais M, Giannadaki D, Pozzer A. The contribution of outdoor air pollution sources to premature mortality on a global scale. Nature. 2015;525(7569):367–71.
13. Li L, Lei Y, Pan D, Yu C, Si C. Economic evaluation of the air pollution effect on public health in China's 74 cities. Springerplus. 2016;5:402.
14. Liu J, Han Y, Tang X, Zhu J, Zhu T. Estimating adult mortality attributable to $PM_{2.5}$ exposure in China with assimilated $PM_{2.5}$ concentrations based on a ground monitoring network. Sci Total Environ. 2016;568:1253–62.
15. Liu M, Huang Y, Ma Z, Jin Z, Liu X, Wang H, Liu Y, Wang J, Jantunen M, Bi J, Kinney PL. Spatial and temporal trends in the mortality burden of air pollution in China: 2004–2012. Environ Int. 2017;98:75–81.
16. Liu S, Zhou M, Wang L, Li Y, Liu Y, Liu J, You J, Yin P. Burden of disease attributable to ambient particulate matter pollution in 1990 and 2010 in China (in Chinese). Chin J Prev Med 2015;(4):327–33.
17. Lu X, Lin C, Li Y, Yao T, Fung JC, Lau AK. Assessment of health burden caused by particulate matter in southern China using high-resolution satellite observation. Environ Int. 2017;98:160–70.
18. Quah E, Boon TL. The economic cost of particulate air pollution on health in Singapore. J Asian Econ. 2003;14(1):73–90.
19. Song C, He J, Wu L, Jin T, Chen X, Li R, Ren P, Zhang L, Mao H. Health burden attributable to ambient $PM_{2.5}$ in China. Environ Pollut. 2017;223:575–86.
20. Svensson M. The value of a statistical life in Sweden: estimates from two studies using the "certainty approach" calibration. Accid Anal Prev. 2009;41(3):430–7.
21. Wang G, Gu S, Chen J, Wu X, Yu J. Assessment of health and economic effects by $PM_{2.5}$ pollution in Beijing: a combined exposure-response and computable general equilibrium analysis. Environ Technol. 2016;37(24):3131–8.
22. Wang Y, Ying Q, Hu J, Zhang H. Spatial and temporal variations of six criteria air pollutants in 31 provincial capital cities in China during 2013-2014. Environ Int. 2014;73:413–22.
23. World Bank and State Environmental Protection Administration, P. R. China. Cost of pollution in China: economic estimates of physical damages. Washington, DC. 2007.
24. World Bank. The cost of air pollution: strengthening the economic case for action. Washington, DC. 2016.
25. Xu X, Chen R, Kan H, Ying X. Meta-analysis of contingent valuation studies on air pollution-related value of statistical life in China (in Chinese). Chin Health Resour. 2013;16(1):64–7.
26. Yang J, Zhou M, Yin P, Li M, Ou CQ, Gu S, Liu Q. Mortality as a function of dust-haze in China: a multi-city time-series study. Lancet. 2016;388(Suppl 1):S19.
27. Zhang M, Song Y, Cai X, Zhou J. Economic assessment of the health effects related to particulate matter pollution in 111 Chinese cities by using economic burden of disease analysis. J Environ Manag. 2008;88(4):947–54.
28. Zhang Y, Ma G, Yu W, Cao D. Health damage assessment due to $PM_{2.5}$ exposure during haze pollution events in Beijing-Tianjin-Hebei region in January 2013 (in Chinese). Nat Med J China. 2013;93(34):2707–10.
29. Zhao W. Health risks and economic loss on urban citizens caused by air pollution. Shanghai: Shanghai Jiao Tong University; 2012.
30. Zheng S, Pozzer A, Cao C, Lelieveld J. Long-term (2001–2012) concentrations of fine particulate matter ($PM_{2.5}$) and the impact on human health in Beijing, China. Atmos Chem Phys. 2015;15:5715–25.

Chapter 11
Air Pollution Prevention and Control Policy in China

Cunrui Huang, Qiong Wang, Suhan Wang, Meng Ren, Rui Ma, and Yiling He

Abstract With rapid urbanization and development of transport infrastructure, air pollution caused by multiple-pollutant emissions and vehicle exhaust has been aggravated year by year in China. In order to improve air quality, the Chinese authorities have taken a series of actions to control air pollution emission load within a permissible range. However, although China has made positive progress on tackling air pollution, these actions have not kept up with its economy growth and fossil-fuel use. The traditional single-pollutant approach is far from enough in China now, and in the near future, air pollution control strategies should move in the direction of the multiple-pollutant approach. In addition, undesirable air quality is usually linked with the combination of high emissions and adverse weather conditions. However, few studies have been done on the influence of climate change on atmospheric chemistry in the global perspective. Available evidence suggested that climate change is likely to exacerbate certain kinds of air pollutants including ozone and smoke from wildfires. This has become a major public health problem because the interactions of global climate change, urban heat islands, and air pollution have adverse effects on human health. In this chapter, we first review the past and current circumstances of China's responses to air pollution. Then we discuss the control challenges and future options for a better air quality in China. Finally, we begin to unravel links between air pollution and climate change, providing new opportunities for integrated research and actions in China.

Keywords Air quality • Climate change • Control policy • Human health • Co-benefits

---

C. Huang (✉) • Q. Wang • S. Wang • M. Ren • R. Ma • Y. He
Department of Health Policy and Management, School of Public Health, Sun Yat-sen University, 74 Zhongshan 2nd Road, Yuexiu District, Guangzhou 510080, China
e-mail: huangcr@mail.sysu.edu.cn

G.-H. Dong (ed.), *Ambient Air Pollution and Health Impact in China*, Advances in Experimental Medicine and Biology 1017, DOI 10.1007/978-981-10-5657-4_11

11.1 Introduction

China's economy has developed rapidly during the past three decades and now ranks as the second largest economy in the world. The country has also become the second largest energy consumer on a global scale, which fossil-fuel consumption occupied the major proportion of total energy use [5]. In many cities, energy consumption is the leading contributor of anthropogenic air pollutants [30]. With rapid urbanization and development of transport infrastructure, air pollution caused by multiple-pollutant emissions and vehicle exhaust has been aggravated year by year [29].

Today, beyond 75% of urban population are under exposure of multiple air pollutants that fail to meet the national ambient air quality standards. In particular, the severe pollution haze that affected northern China in early 2013 has drawn public's intense concern. The Chinese government has taken a variety of actions to control air pollution and improve air quality [27, 31]. In this chapter, we will first review the past and current circumstances of China's responses to air pollution. Then we discuss the control challenges and future options for a better air quality in China. Finally, we begin to unravel links between air pollution and climate change, providing new opportunities for integrated research and actions in China.

11.2 China's Responses to Air Pollution

11.2.1 Legislation for Air Pollution Prevention and Control

China has gradually developed a holistic legislation framework for air pollution control since the People's Republic of China was established. In order to reduce the harmful effects of silica dust, the State Council promulgated the *Decision on Preventing Silica Dust in Mining Companies* in 1956. It was the first legal document issued by the government aiming at air pollution control. Afterward, the Cultural Revolution had completely affected China's society, leading to the temporary stagnancy of legislation. When China restored its legitimate seat at the United Nations in 1971, the government was aware of the importance of environmental protection and recognized serious problems in nationwide air quality [7]. Then the government began to take strict actions in prevention and control of air pollution. The main structure of China's environmental control system is based on a "two committees, one bureau" model, and the structure is similar in every level of China's administrative system (Fig. 11.1).

In 2000, China revised the *Air Pollution Prevention and Control Law*, and this revision strengthened control on SO_2 emissions particularly caused by coal burning. It also reinforced control of emissions from vehicles, including the prohibition of production and trading of vehicles that did not meet the discharge standards [7]. With the automobile ownership elevated steadily, China's air pollution is currently

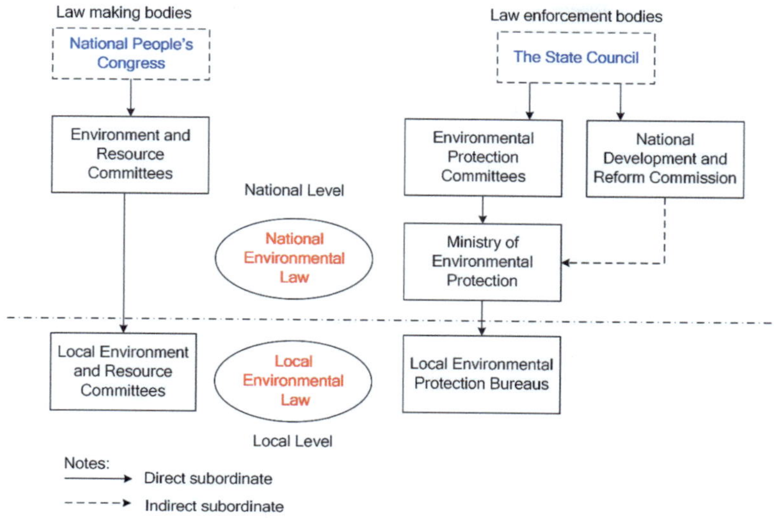

Fig. 11.1 Environmental governance structure in China (Adapted from Ref. [3])

caused by a combination of emissions from coal combustion and voiture exhaust. Since 2013, frequently occurring haze events have made the government strengthen the control of dust haze, particulate matters, and nitrogen oxides. The severe haze pollution was also one major factor promoting the 2014 revision of the *Environmental Protection Law*. The 2014 revision added measures such as information disclosure, public participation, and daily penalties for air pollution control. Other supporting legal documents include the *Rules on the Standard for Compulsory Retirement of Motor Vehicles*, 2012 *Ambient Air Quality Standards*, and Ministry of Environmental Protection's Guidelines for *Joint Prevention and Control of Air Pollution in Key Regions*.

11.2.2 Ambient Air Quality Standards

In 1982, *Ambient Air Quality Standards* in China was originally formulated, which was China's first official document aiming to protect human health and ecological environment. The Ministry of Environmental Protection issued the revised *Ambient Air Quality Standards* in 1996, 2000, and 2012, respectively (Fig. 11.2). The first amendment in 1996 included three grade standards (Grades I, II, and III) and recommended the guideline values of SO_2, NO_x, NO_2, CO, O_3, PM_{10}, and TSP. The second amendment in 2000 deleted the standard of NO_x pollutant and relaxed the guideline values of NO_2 and O_3 [31]. The third amendment was made in 2012, which set new control standards of $PM_{2.5}$ and O_3 and frapped the limit values of NO_2 and PM_{10}.

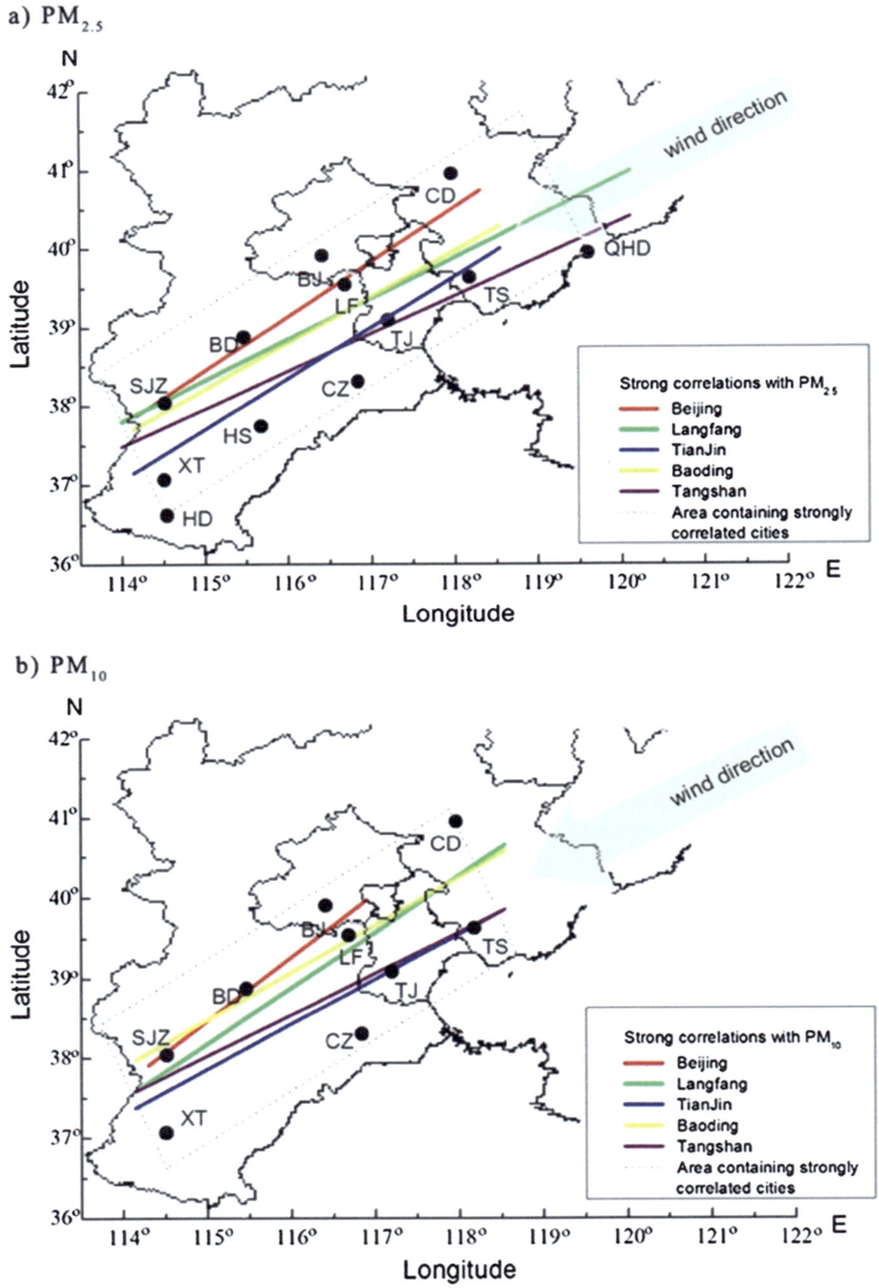

Fig. 11.2 The distribution of cities with strongly correlated (**a**) $PM_{2.5}$ and (**b**) PM_{10} values. The wind direction arrow represents the dominant wind direction during the autumn study period (Source: Ref. [26])

Table 11.1 National ambient air quality standards in China (GB3095-2012, third amendment)

Pollutants	Time interval	NAAQS guideline value (GB3095-2012)		Unit
		Grade-I	Grade-II	
SO_2	1-h	150	500	μg/m³
	Daily	50	150	
	Annual	20	60	
NO_2	1-h	200	200	μg/m³
	Daily	80	80	
	Annual	40	40	
CO	1-h	10	10	mg/m³
	Daily	4	4	
O_3	1-h	160	200	μg/m³
	Max 8-h	100	160	
$PM_{2.5}$	Daily	35	75	μg/m³
	Annual	15	35	
PM_{10}	Daily	50	150	μg/m³
	Annual	40	70	

Table 11.2 $PM_{2.5}$ and PM_{10} guideline values of National Ambient Air Quality Standards in China, WHO, the USA, and the EU

Pollutants	Time interval	China (2012)		WHO (2005)			USA (2006)	EU (2008)
		Grade-I	Grade-II	Target-I	Target-II	Target-III		
$PM_{2.5}$	Daily	35	75	75	50	37.5	35	25
	Annual	15	35	35	25	15	15	20
PM_{10}	Daily	50	150	150	100	75	150	50
	Annual	40	70	70	50	30		40

In the 2012 Standards, the guideline values of Max 8-h O_3, daily $PM_{2.5}$, and daily PM_{10} are 100, 35, and 50 μg m^{-3} for Grade-I and 160, 75, and 150 μg m^{-3} for Grade-II, respectively (Table 11.1). Currently, the $PM_{2.5}$ and PM_{10} Grade-II standards are in consistent with the WHO's recommended interim target-I standard, and $PM_{2.5}$ Grade-I standard is also in accordance with US recommended guideline values of $PM_{2.5}$ (Table 11.2) [31]. Because fine particles $PM_{2.5}$ have a severe effect on human health, China would have far-reaching benefits to its air quality if the 2012 Standards can be properly implemented. However, achieving this goal will be challenging, as $PM_{2.5}$ concentrations in a great number of cities are well above the recommended standard at present [33].

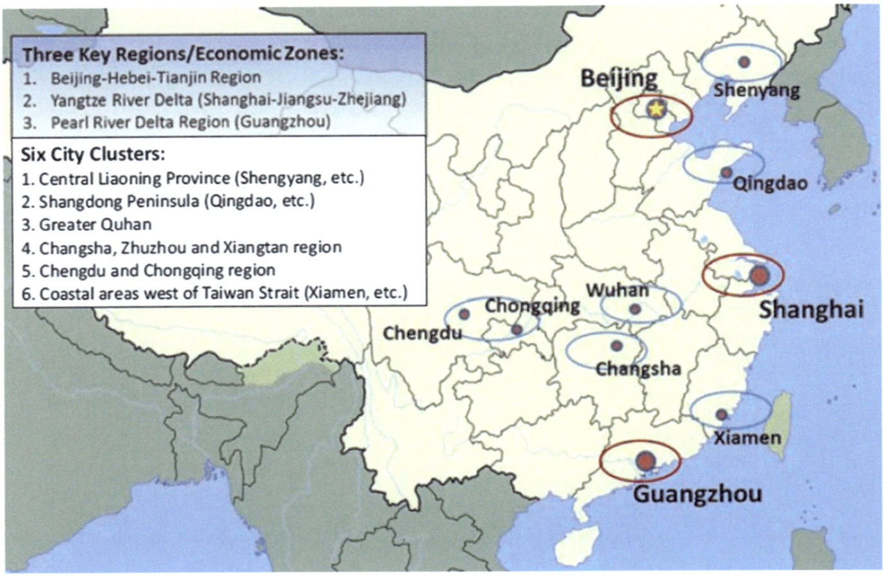

Fig. 11.3 Three key regions and six city clusters identified for regional air pollution control (Source: Ref. [27])

11.2.3 Policies on Air Pollution Prevention And Control

Air pollution and its reduction are often regional (i.e., transboundary) rather than local (Fig. 11.2) [26]. In May 2010, the State Council issued the *Joint Prevention and Control of Air Pollution*, aiming to establish a cooperative prevention and control system for regional air pollution. As shown in Fig. 11.3, this regulation designated three key regions and six city clusters [26].

In addition, the State Council issued the first national *Action Plan on Air Pollution Prevention and Control (2013–2017)* in September 2013. The Action Plan requires that by 2017, PM_{10} levels to be decreased by ≥10% compared with 2012 levels in urban areas and the blue sky days to be increased year by year [4]. Another target of the plan is annual $PM_{2.5}$ concentrations to be reduced by ≥25%, 20%, and 15% in the Beijing-Tianjin-Hebei, Yangtze River Delta, and Pearl River Delta region, respectively [31].

In order to accomplish the proposed targets, the Action Plan indeed sets some important mandates and initiatives. It manages to establish the mechanism of regional coordination and general layout for regional environmental management. It also tries to clarify how the government, enterprise, and society should take responsibilities and encourage the public to involve in environmental protection [31]. In particular, the Action Plan proposes less share of coal and more share of non-fossil fuels in total energy consumption. It also sets specific goals for $PM_{2.5}$ in the three key regions, and PM_{10} shall be considered as a compulsory index in the development of our society and economy [7].

11.2.4 Air Pollution Control Challenges

In recent years, China has made positive progress on tackling air pollution. However, the prevention and control of air pollution are still in a race with the economy [6]. The country has maintained an annual economic growth rate of more than 8% for many years, and its per-capita GDP energy consumption was 1.4 times higher than average level of the world. There has been a steep rise in emissions of air pollutants in the past decades, from extensive industrial development, coal-dependent energy consumption, and increasing number of vehicles [4]. Although the government has made remarkable efforts to control air pollution, these actions have not kept up with its economy growth and fossil-fuel use [32].

In particular, the low priority given to environmental protection and the lack of cooperation among various government agencies hindered progress on air pollution control. The Ministry of Environmental Protection manages pollutant discharge, but it is a weak player within the government system. Currently, local governments still prefer to promote heavy industry to stimulate their economies, and in some areas, even factories owned by governments might be the largest polluters. Local environmental protection agencies are also in difficult situations because they are affiliated to local governments [33].

It is generally believed that critical factors affecting the effectiveness of air pollution control measures are issues concerning implementation of laws and policies. Nonetheless, the underlying problems of legislation and policy should not be overlooked [7]. The *Air Pollution Prevention and Control Law* requires local governments to be accountable for local air quality in the administrative areas under their jurisdiction. However, it is not clear how the local governments take responsibility for air quality, how such responsibilities should be assessed, and who should take responsibility if air quality failed to meet air quality standards. Therefore, local government officials who favor economic development may disregard their responsibilities for environmental protection [7].

At present, there is still a lack of cost-effective evaluation of air pollution control policies in China. The design of targets and measures for air pollution control should be accompanied by corresponding assessment methods, in order to track and evaluate the implementation and effectiveness of the measures. At this stage, China has not established a comprehensive or complete system for air quality management and also has not applied the cost-effective evaluation mechanism in environmental planning [6]. International experience is a valuable guideline for China to innovate management methods and gradually enrich and develop air pollution prevention and control strategy in the country.

Furthermore, there is a defective integration among laws and policies. To some extent, they are not mutually supportive. For instance, the proposed goals of Action Plan have been profoundly difficult to achieve, as some concrete measures (e.g., increasing clean energy supply, accelerating technological transformation, linking $PM_{2.5}$ reduction targets to performance evaluations of local governments) will be less effective if they remain at the policy level. In addition, there is a consistency problem among legislation and policies. For example, in current energy-use policies, total air pollutant load goals have no specific regional emission restrictions, unlike those in the *Air Pollution Prevention and Control Law*. Therefore, in order to make them mutually supportive, certain items of policies should be established through legislation [7].

Over the next decade, China will complete the capital-intensive industrialization and experience population boom. Pushed by the public, the willingness of Chinese government for addressing the problem of air pollution has never been so strong. A range of laws and policies were established and have already played important roles in protecting air quality [6]. Hence, there is a golden opportunity to make a change—to free China from the dilemma between economic development and environmental protection [33].

11.3 Future Options for a Better Air Quality in China

11.3.1 Moving Toward a Multiple-Pollutant Approach to Air Quality

In recent years, air pollution in China has become growing complicated, and the problem of secondary pollutants (i.e., NO_x, PM_{10}, $PM_{2.5}$, O_3, and CO_2) is becoming much more urgent. Actually, many of the pollutants share the same emission source, for example, emissions of SO_2, NO_x, and CO_2 are all strongly associated with fossil-fuel consumption. Huang et al. [12] investigated the chemical components during severe pollution periods in four typical cities in China and observed that secondary aerosol contributed 50–75% of $PM_{2.5}$ in eastern cities (Beijing, Shanghai, and Guangzhou) and 30% of $PM_{2.5}$ in a western city (Xi'an).

At the very beginning, China's air pollution control policies gave priority to dealing with the air pollutants one by one, covering a few major pollutants. Every Five-Year Plan starting with the twentieth century targeted to reduce the national NO_x, PM, SO_2, and CO_2 emission, and each set a different demand separately. On the basis of the US Environmental Protection Agency Air Quality Index (AQI), the *Ambient AQI Technical Provisions (Trial)* (HJ 633-2012) was released in 2012, and it was implemented in the *Ambient Air Quality Standard* (GB 3095-2012). The AQI level is assessed by the concentrations of six criteria pollutants including SO_2, NO_2, CO, O_3, $PM_{2.5}$, and PM_{10}. Basically, the AQI acts as a forewarning index for the government to inform the public to take proper health protection actions.

Currently, the traditional single-pollutant approach is far from enough in China. Figure 11.4 shows the number of days of severe and extremely serious pollution ($AQI > 200$) and the multiple pollutants in 31 provincial capital cities [11]. Previous epidemiological studies reported the negative effects of air pollutants on human health, usually on the account of a single air pollutant while adjusting for other air pollutants exposure [15, 17, 28]. The AQI system has its defectiveness in practice,

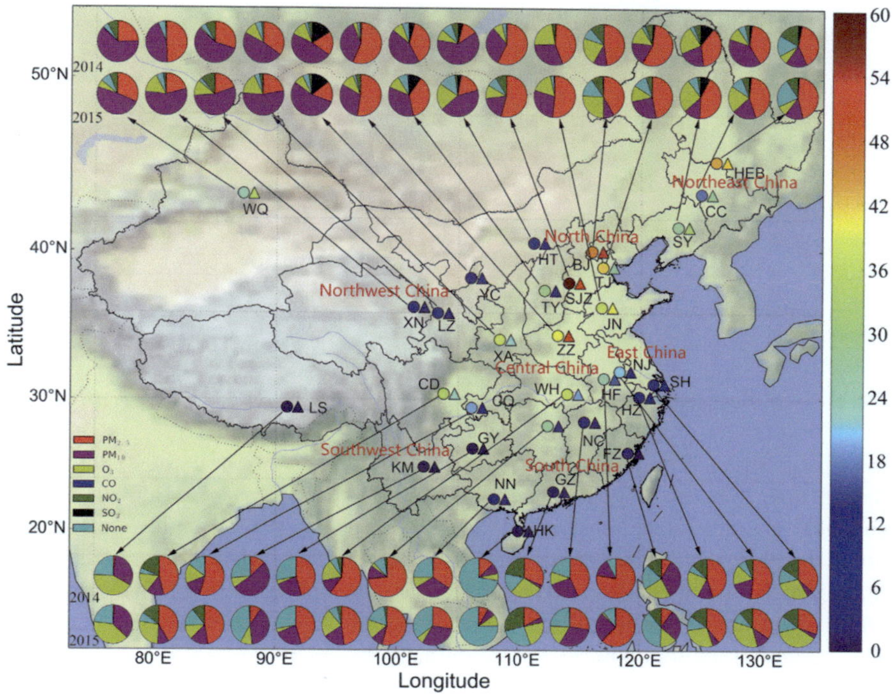

Fig. 11.4 The number of days of severe and very serious pollution and the rate of major pollutant in 31 provincial capital cities in China (2014–2015). The *circles* and *triangles* represent the total days of severe and very serious pollution in 2014 and 2015, respectively. The category "None" means the rate of AQI less than 50 in total days (Source: Ref. [11])

for it simply sums each unit ignoring the combined health effects of exposure to multiple air contaminants [14]. Thus, future epidemiological studies are urgently required to estimate the health impacts associated with multi-pollutant exposure and revise more scientific air quality index standards.

In the near future, China's air pollution control strategies should move in the direction of the multiple-pollutant approach. "Co-control" is now commonly used in the Chinese government documents, which refers to "coordinated control" or "synergetic control." There is no doubt that SO_2, NO_x, and PM emission loads, as well as VOC and CO_2 discharge, should be reduced simultaneously to mitigate the harmful health and environmental effects of air pollution in China over the next few years (Fig. 11.5). Emissions controls of NO_x, SO_2, NH_3, and NMVOC synergistically are highly needed in the future because multiple pollutants impact on $PM_{2.5}$ and O_3 concentrations nonlinearly, as well as atmospheric oxidation capacity. The key problems to be addressed are the acid rain, haze, photochemical smog, etc.

As meteorological conditions are the primary factors causing the day-to-day variations in pollutant concentrations [10], an effective global response to climate change also should be taken into consideration. This can be considered as a kind of

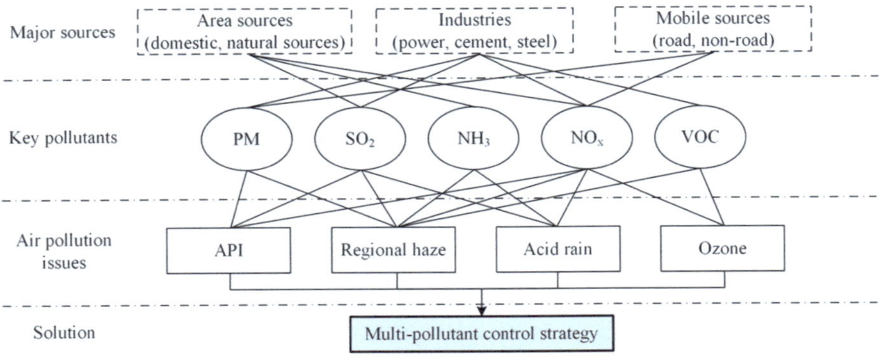

Fig. 11.5 Schematic diagram of the multi-pollutant approach (Adapted from Ref. [27])

multiple-pollutant control strategy. The purpose of co-control strategy is to tackle the adverse impacts of general air pollutants in the company of global climate change mitigation. Generally, policymakers in China pay close attention to multiple-pollutant approach due to it can reduce costs and increase effectiveness at the same time. Moreover, they believe that multiple-pollutant approach should focus on synergies between air pollution and energy use for more cost-effective co-control. Thus, contribution to the mitigation of climate change by the control on fossil-fuel use and expand the use of clean fuels should be taken into account.

11.3.2 Promoting Climate-Friendly Air Pollution Control Strategies

Climate change is a challenging problem in the new century, one to which the whole world has attached great importance. Actions were taken around the world to enhance resilience to the global climate change, such as the *United Nations Framework Convention on Climate Change* (UNFCC), the *Kyoto Protocol*, the *Copenhagen Declaration*, and the *Paris Agreement*. Climate-friendly air pollutant controlling strategy calls for the combination among techniques, policies, and regulations. This joint controlling measure aims at promoting reductions of toxic pollutants and greenhouse gas (GHG) emission loads at the same time. This strategy puts forward based on the theory that climate change and air pollution are closely interrelated problems.

Recent studies demonstrated that significant cost-effective and synergistic benefits could be achieved by joining the reduction local air pollution and GHGs together (Fig. 11.6). Climate-friendly measures, such as improvements of energy efficiency, combined of heat and electricity generation, fuel replacement, and integrated coal gasification combined cycle (IGCC) plants, can reduce SO_2, NO_x, and

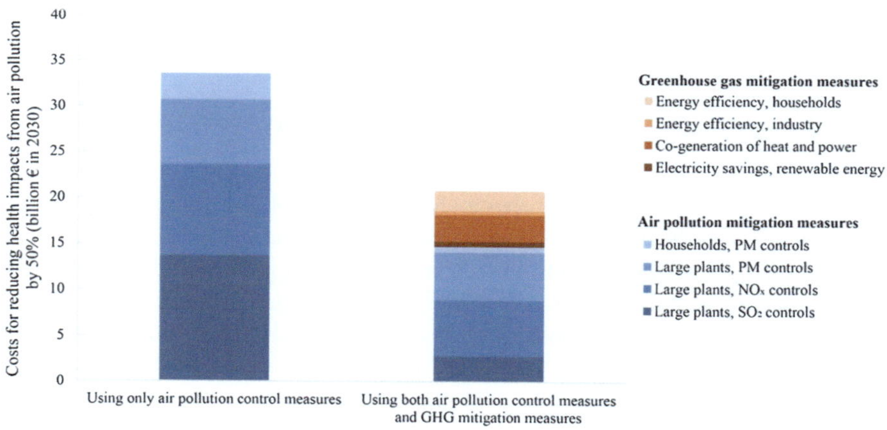

Fig. 11.6 Cost reduction from controlling air pollution and greenhouse gases simultaneously (Source: Ref. [16])

$PM_{2.5}$ emissions without any extra spending. An estimated 1 % of adverse impact by PM would significantly decrease when reducing the CO_2 emission by 1% [16].

Many countries or cities have put proposals to promote the climate-friendly air pollution control strategy. In the USA, the joint administration of GHG discharge and air pollutants has been put forward over a decade ago when the first guideline for state and local air quality officials (ALAPCO 1999) was released by the national association of air pollution regulators (then called STAPPA/ALAPCO). What's more, *New York's Air Quality Management Plan* adopted a multi-pollutant approach, involving GHG emission, to underline the importance of energy conservation and emission reduction by the state's improvement of energy efficiency and implementation of renewable energy plan.

Introduced by the Ministry of Environmental Protection, the regional air quality management rule was approved by China's state Council on May 11, 2010. Along with the *Joint Prevention and Control of Air Pollution to Promote Regional Air Quality*, *Guidelines for 12th Five-Year Plans for Joint Prevention and Control of Air Pollution in Key Regions* was approved by the Ministry of Environmental Protection successively to demonstrate the government's determinations to implement the climate-friendly air quality controlling. The coordination and cooperation of environmental and energy policy formulation is the foundation to realize the climate-friendly air pollution management. It's a cost- and time-saving way to reduce pollutant discharge. Therefore, an ingenious integration of measures to lessen air pollutants and GHG emissions in parallel will help tackle air pollution and climate change less expensively than dealing with either issue separately.

11.4 Climate Change and Air Pollution: Measures with Co-benefits

11.4.1 Links Between Climate Change and Air Pollution

Climate change and air pollution are closely linked (Fig. 11.7), for instance, a large extent of CO_2 and the main air pollutants stem from the same sources. Nevertheless, they were studied separately in most literatures, due to their complex and highly variable relationships. In the past few years, climate change researches have been primarily concerned with the roles of CO_2, CH_4, N_2O, HFCs, PFCs, and SF_6, and some of them only make up less than 1% of the atmosphere but contributing much to climate change. Few researches investigated the effects of major air pollutants due to their temporal effects on climate change, e.g., ozone and particulate matter. On the other hand, pollutants at local and regional scales draw more and more attentions in the field of air pollution research. On the whole, there have been relatively few studies on the contribution of background concentrations and the influence of climate change on atmospheric chemistry in the global perspective [23]. However, undesirable air quality is usually linked with the combination of high emissions and severe weather conditions. As to the rapid climate change now we are facing, further research is needed to better understand the inextricable overlap between climate change and air quality.

The two major air pollutants influencing human health are ground-level ozone and particulate matter in terms of climate change. Ozone is a double-edged sword for human health. The ozone layer located in the stratosphere protects human from harmful ultraviolet radiation; however, ground-level ozone has innumerable adverse health effects. As a powerful oxidant, it can impair lung function and irritate the respiratory system. Summer is at high risk of ozone pollution because strong ultraviolet light and high temperature are vital to the generation of ozone. Several observations have confirmed the strong correlation of elevated ozone with temperature in

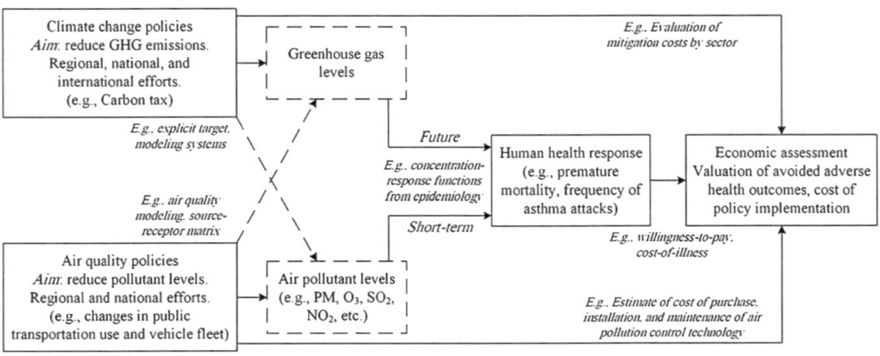

Fig. 11.7 Policies framework between climate change and air quality (Adapted from Ref. [1])

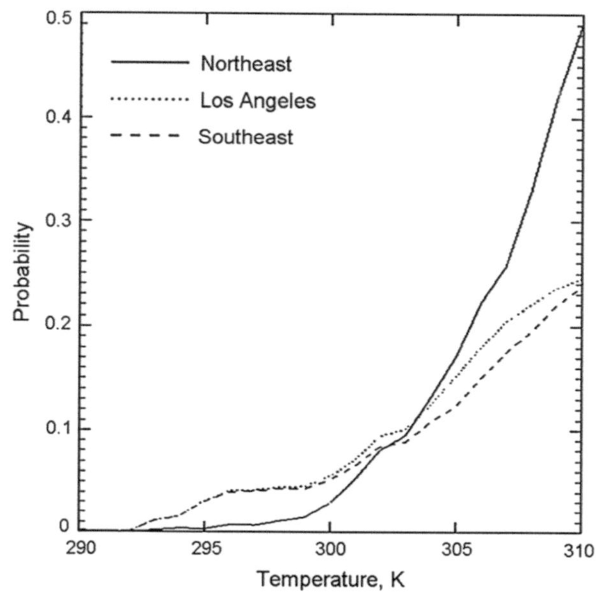

Fig. 11.8 The probability of ozone exceeding the 80 ppb as a function of daily maximum temperature for three US regions (Source: Ref. [13])

polluted regions (Fig. 11.8). The correlation is mostly showed under certain conditions, i.e., when ozone is above 60 ppb, and significant correlation with lower background ozone concentrations has not been observed [13].

The principal components of particulate matter are usually common. For example, sulfate, nitrate, organic carbon, and elemental carbon are mostly present as fine particulate matter ($PM_{2.5}$), and they have drawn much concern about human health. The size of the particulate matter is the main determinant of its health effects. Current studies have recognized the effects of inhaling particles on respiratory diseases, cardiovascular disease, etc. Sulfate, nitrate, and organic carbon are the oxidation products of SO_2, NO_x, and NMVOCs. Process of combustion can also emit carbon particles directly. Nitrate and organic carbon have both solid and gas phases, depending on environmental temperature. Precipitation is the main and efficiently way of atmospheric sinking of PM; thus the lifetime of PM varies with cumulative rainfall (Fig. 11.9), and pattern changes of precipitation could also influence PM distribution. Particles have either a cooling effect on the atmosphere through scattering of shortwave radiation or a warming effect through absorption of shortwave radiation according to their nature and thus have a bidirectional effect on temperature. This shows the complexity of PM compositions and the diverse effects of its components [13].

Although further research is needed, available evidence suggested that climate change was likely to exacerbate certain kinds of air pollutants including ozone and smoke from wildfires (Fig. 11.10). Meanwhile, O_3 and PM can interact with solar and terrestrial radiation, and the importance of these as climate forcing agents has been recognized. A study based on the panel of the United Nations Framework Convention on Climate Change (UNFCCC) classified countries shows that both the

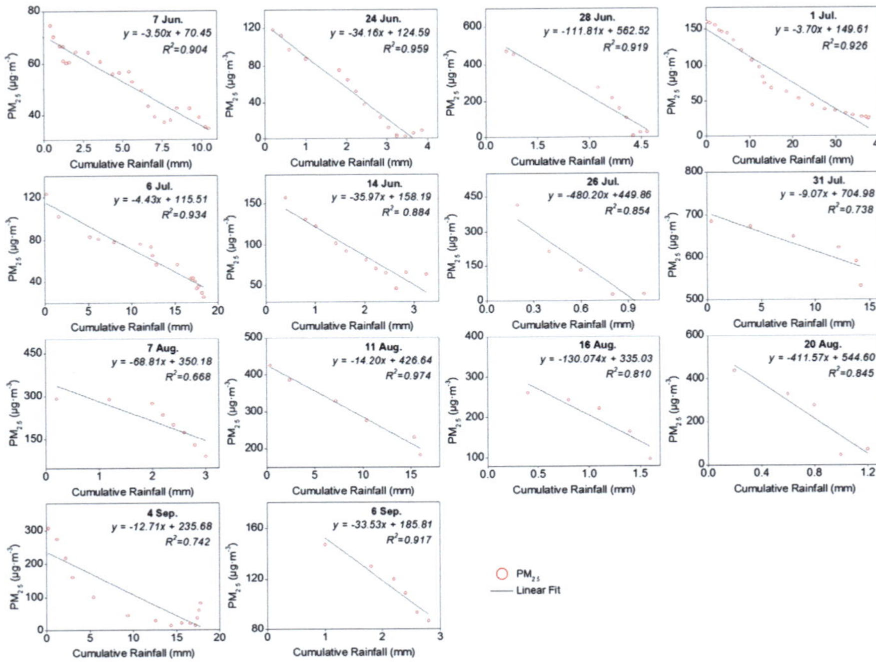

Fig. 11.9 Correlations between $PM_{2.5}$ concentrations and accumulated rainfall (Source: Ref. [18])

energy consumption and air quality indicators have a positively significant relationship with the climate change [19]. Because of the dual effect, the relationship between climate change and air pollution should be discussed in a broader context of chemistry-climate interactions. The variation of atmospheric chemistry puts an effect on both air quality (O_3 and PM) and climate (O_3, PM, and methane). Again, climate change can also influence air quality via affecting natural emissions (biosphere, dust, fires, lightning) [22]. Chemistry-climate interactions include several potential pathways, and further studies need to validate the current results [6].

11.4.2 Health Co-benefits of Improved Air Quality as a Result of Climate Change Mitigation

Excess morbidity and mortality related to extremely hot weather and poor air quality have been widely investigated in cities worldwide. This has become a major public health problem because the interactions of global climate change, urban heat islands, and air pollution have adverse effects on human health [9]. A large extent of air pollutants (e.g., PM, NO_2, and SO_2) and greenhouse gases (e.g., CO_2) stem from the same fossil-fuel combustion processes. Thus, reducing emissions of air

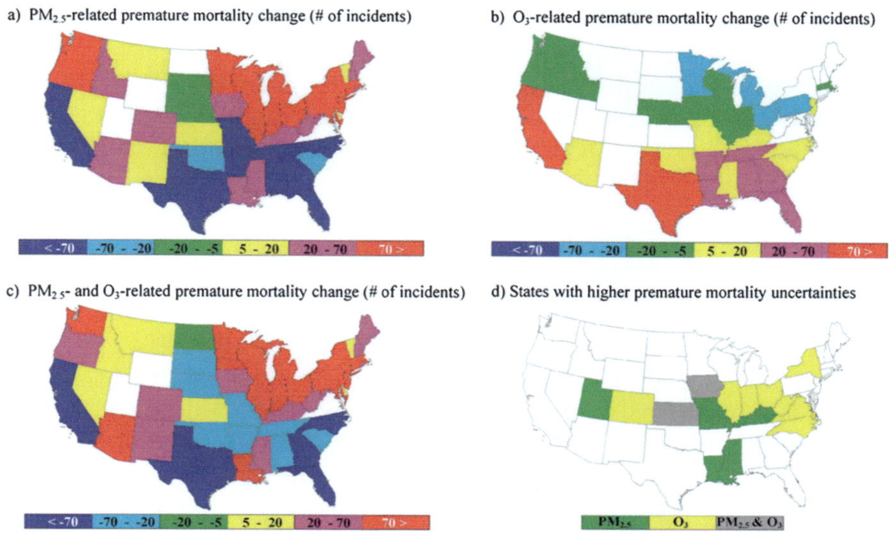

Fig. 11.10 Potential impact of climate change on air pollution-related human health effects in the USA. Estimated changes of (**a**) $PM_{2.5}$-related, (**b**) O_3-related, and (**c**) both pollutants-related premature mortality in 2050 compared to 2001. (**d**) States with higher premature mortality uncertainties due to $PM_{2.5}$- and O_3-related effects from uncertainties in meteorology forecasting (Source: Ref. [24])

pollutants to slow the pace of climate change will benefit public health. The health benefits and air pollution reduction benefited from GHG mitigation have been widely investigated. According to current studies, energy scenarios, emission control, and climate change and air quality modeling systems are closely related to the short-term gains in public health, and environment could benefit from GHG mitigation policies [1].

Several studies suggested reducing GHG has significant health benefits. By taking the control policy scenario over the 20-year period, as compared to current pollution practices, a study identified almost 153,000 adult deaths and more than 3700 infant deaths are avoidable in three Latin-American cities [2]. A synthesis of research on China's Clean Development Mechanism potential showed domestic environmental benefits of saving 3000–40,000 lives annually through co-benefits of abated air pollution [1]. A study focusing on Beijing estimated population exposure to air pollutants under various energy scenarios and calculated chronic and short-term excess deaths [20].

It is a challenging task to estimate the health benefits of improved air quality resulted from GHG mitigation policies. Due to existing of unquantified effects, current studies may underestimate the benefits, because clear exposure-response associations only show in a portion of the health impacts. The US EPA noted numerous unquantified effects in 1999; since then, some of the effects have been identified [21]. There are already some available approaches to estimate health co-benefits of improved air quality as a result of GHG mitigation. Although there is no well-recognized measurement of their health effects, consistent results from previous studies suggested that the health co-benefits from improved air quality are substantial. These published studies will provide useful information for the scope, design, and timing of climate change policies [1], and further researches need to investigate the added benefit of improving health outcomes related to air pollution.

11.4.3 Implementing the Paris Agreement to Tackle Climate Change and Air Pollution

The 2015 UNFCC conference was held in Paris, France. It was the 21st annual session of the Conference of the Parties (COP21) from the 1992 United Nations Framework Convention on Climate Change (UNFCCC). Stakeholders in the public health and government arenas hammered out strategies to reduce emissions of influential climate pollutants. On 22 April 2016 (Earth Day), 174 countries signed the Paris Agreement in New York and began adopting it within their own legal systems. The key result of the Paris Agreement was to limit global temperature rise to 2 °C comparing to preindustrial levels and to pursue efforts to limit the increase to 1.5°C. To achieve this, countries submitted Intended Nationally Determined Contributions (INDCs) outlining actions they intended to take and the level they intended to reach individually. The Paris Agreement entered into force on 4 November 2016—and the INDCs became the Nationally Determined Contributions (NDCs). To raise the level of ambition over time, a 5-year submission cycle for NDC was introduced, and new NDC must be more ambitious than previous one.

The Paris Agreement is the first climate agreement that covered all countries and won unanimous consent. The agreement is expected to serve as a solid basis for further cooperation. According to the agreement, developed and developing countries must assume their respective obligations and make their respective contributions in a unified institutional framework and a differentiated manner. China is part of the landmark Paris Agreement. The transition to cleaner energy will contribute to tackling air pollution challenges and put the country's future growth on a low-carbon pathway [25].

Climate change mitigation strategies, such as reducing the dependency on fossil fuels and increasing the portion of clean and renewable energies usage, have synergy with environmental protection and public health. China has nationally determined its actions by 2030, such as to achieve the peaking of carbon dioxide emissions around 2030 and to lower carbon dioxide emissions per unit of GDP by 60–65% from the 2005 level, according to submitted INDC on 30 June 2015, *Enhanced actions on climate change: China's intended nationally determined contributions*. Therefore, the implementation of the Paris Agreement will further promote China's actions in addressing global climate change and also improvement of air quality [8].

11.4.4 Summary

China has maintained an annual economic growth rate of more than 8% for many years, and its per-capita GDP energy consumption was 1.4 times higher than average level of the world. There has been a steep rise in emissions of air pollutants in the past decades, from extensive industrial development, coal-dependent energy consumption, and increasing number of vehicles. Today, most of the urban population are under exposure of multiple air pollutants whose concentration exceeded discharge standard.

In order to improve air quality, the Chinese authorities have taken a series of actions to control air pollution emission load within a permissible range. The *Air Pollution Prevention and Control Law* and *Ambient Air Quality Standards* were formulated and revised for multiple times aiming to protect human health and ecological environment. Although the government has made remarkable efforts to control air pollution, these actions have not kept up with its economy growth and fossil-fuel use. In particular, the low priority given to environmental protection and the lack of cooperation among various government agencies hindered progress on air pollution control. The design of targets and measures for air pollution control should be accompanied by corresponding assessment methods, in order to track and evaluate the implementation and effectiveness of the measures. Furthermore, there is defective integration among laws and policies.

At the very beginning, China's air pollution controlling policies gave priority to dealing with the air pollutants one by one, covering a few major pollutants. The traditional single-pollutant approach or local air pollutant control is far from enough in China. In the near future, China's air pollution control strategies should move in the direction of the multiple-pollutant approach. "Co-control" is now commonly used in the Chinese government documents, which refers to "coordinated control" or "synergetic control." Policymakers in China seem to be particularly interested in multiple-pollutant approach due to its potential to reduce costs and increase efficiency, and they also believe that multiple-pollutant approach should focus on synergies between air pollution and energy use for more cost-effective co-control.

Undesirable air quality is usually linked with the combination of high emissions and adverse weather conditions. However, there have been relatively few studies on the influence of climate change on atmospheric chemistry in the global perspective. Although further research is needed, available evidence suggested that climate change was likely to exacerbate certain kinds of air pollutants including ozone and smoke from wildfires. This has become a major public health problem because the interactions of global climate change, urban heat islands, and air pollution have adverse effects on human health. Reducing emissions of air pollutants to slow the pace of climate change will benefit public health. This can be considered as a kind of multiple-pollutant control strategy, for the purpose of it is to tackle the adverse impacts of general air pollutants in the company of global climate change mitigation.

Acknowledgment This work was supported by the Asia-Pacific Network for Global Change Research (CRRP2016-10MY-Huang) and the Nature Science Foundation of Guangdong Province (2016A030313216), National Natural Science Foundation of China (81602819), and 2017 Doctoral Fund of Guangdong Science and Technology Program (Wang S).

References

1. Bell M, Davis D, Cifuentes L, Krupnick A, Morgenstern R, Thurston G. Ancillary human health benefits of improved air quality resulting from climate change mitigation. Environ Health. 2008;7(1):41.
2. Bell ML, Davis DL, Gouveia N, Borja-Aburto VH, Cifuentes LA. The avoidable health effects of air pollution in three latin american cities: Santiago, sao paulo, and mexico city. Environ Res. 2006;100(3):431–40.
3. Chang YC, Wang NN. Environmental regulations and emissions trading in China. Energy Policy. 2010;38(7):3356–64.
4. Chen Z, Wang JN, Ma GX, Zhang YS. China tackles the health effects of air pollution. Lancet. 2013;382(9909):1959–60.
5. Department of Energy Statistics of National Bureau of Statistics of China. China energy statistical yearbook of 2012. Beijing: China Statistical Press; 2013.
6. Fang M, Chan CK, Yao XH. Managing air quality in a rapidly developing nation: China. Atmos Environ. 2009;43(1):79–86.
7. Feng L, Liao W. Legislation, plans, and policies for prevention and control of air pollution in china: achievements, challenges, and improvements. J Clean Prod. 2016;112(2):1549–58.
8. Gao Y. China's response to climate change issues after paris climate change conference. Adv Clim Chang Res. 2016;7(4):235–40.
9. Harlan SL, Ruddell DM. Climate change and health in cities: impacts of heat and air pollution and potential co-benefits from mitigation and adaptation. Curr Opin Environ Sustain. 2011;3:126–34.
10. He JJ, Yu Y, Xie YC, Mao HJ, Wu L, Liu N, Zhao SP. Numerical model-based artificial neural network model and its application for quantifying impact factors of urban air quality. Water Air Soil Pollut. 2016;227:235.
11. He JJ, Gong SL, Yu Y, Yu LJ, Wu L, Mao HJ, Song CB, Zhao SP, Liu HL, Li XY, Li RP. Air pollution characteristics and their relation to meteorological conditions during 2014–2015 in major Chinese cities. Environ Pollut. 2017;223:484–96.
12. Huang RJ, Zhang Y, Bozzetti C, Ho KF, Cao JJ, Han Y, Daellenbach KR, Slowik JG, Platt SM, Canonaco F, Zotter P, Wolf R, Pieber SM, Bruns EA, Crippa M, Ciarelli G, Piazzalunga A, Schwikowski M, Abbaszade G, Schnelle-Kreis J, Zimmermann R, An Z, Szidat S, Baltensperger U, El Haddad I, Prevot AS. High secondary aerosol contribution to particulate pollution during haze events in china. Nature. 2014;514(7521):218–22.
13. Jacob DJ, Winner DA. Effect of climate change on air quality. Atmos Environ. 2009;43(1):51–63.
14. Kyrkilis G, Chaloulakou A, Kassomenos PA. Development of an aggregate air quality index for an urban mediterranean agglomeration: relation to potential health effects. Environ Int. 2007;33(5):670–6.
15. Lobdell DT, Isakov V, Baxter L, Touma JS, Smuts MB, Ozkaynak H. Feasibility of assessing public health impacts of air pollution reduction programs on a local scale: new haven case study. Environ Health Perspect. 2011;119(4):487–93.
16. Markus A, Jiang KJ, Hao JM, Wang SX, Zhuang X, Wei W, Deng YX, Liu H, Xing J and Zhang CY, Imrich B, Jens B, Janusz C, Chris H, Lena H, Zbigniew K, Pallav P, Peter R, Wolfgang S, Geza T, Fabian W, Wilfried W. Gains asia: Scenarios for cost-effective control of air pollution and greenhouse gases in China. 2008.

17. Meng X, Wang CC, Cao DC, Wong CM, Kan HD. Short-term effect of ambient air pollution on copd mortality in four Chinese cities. Atmos Environ. 2013;77:149–54.
18. Ouyang W, Guo B, Cai G, Li Q, Han S, Liu B, Liu X. The washing effect of precipitation on particulate matter and the pollution dynamics of rainwater in downtown Beijing. Sci Total Environ. 2015;505:306–14.
19. Ozturk I. Measuring the impact of energy consumption and air quality indicators on climate change: evidence from the panel of unfcc classified countries. Environ Sci Pollut Res Int. 2015;22(20):15459–68.
20. Pan X, Yue W, He K, Tong S. Health benefit evaluation of the energy use scenarios in Beijing. China Sci Total Environ. 2007;374:242–51.
21. USEPA's Office of Air and Radiation. The benefits and costs of the clean air act from 1990 to 2020. 2011.
22. Romps DM, Seeley JT, Vollaro D, Molinari J. Projected increase in lightning strikes in the united states due to global warming. Science. 2014;346(6211):851–4.
23. Swart R, Amann M, Raes F, Tuinstra W. A good climate for clean air: linkages between climate change and air pollution. An editorial essay. Clim Chang. 2004;66(3):263–9.
24. Tagaris E, Liao KJ, Delucia AJ, Deck L, Amar P, Russell AG. Potential impact of climate change on air pollution-related human health effects. Environ Sci Technol. 2009;43(13):4979–88.
25. Tambo E, Wang DQ, Zhou XN. Tackling air pollution and extreme climate changes in china: implementing the paris climate change agreement. Environ Int. 2016;95:152–6.
26. Wang H, Zhao L, Xie Y, Hu Q. "Apec blue" – the effects and implications of joint pollution prevention and control program. Sci Total Environ. 2016;553:429–38.
27. Wang SX, Hao JM. Air quality management in china: issues, challenges, and options. J Environ Sci. 2012;24(1):2–13.
28. Wong CM, Vichit-Vadakan N, Kan H, Qian Z. Public health and air pollution in Asia (PAPA): a multicity study of short-term effects of air pollution on mortality. Environ Health Perspect. 2008;116(9):1195–202.
29. Wu X, Wu Y, Zhang S, Liu H, Fu L, Hao J. Assessment of vehicle emission programs in china during 1998–2013: achievement, challenges and implications. Environ Pollut. 2016;214:556–67.
30. Xu CQ, Hong JL, Ren YX, Wang QS, Yuan XL. Approaches for controlling air pollutants and their environmental impacts generated from coal-based electricity generation in China. Environ Sci Pollut Res. 2015;22(16):12399–410.
31. Zhang H, Wang S, Hao J, Wang X, Wang S, Chai F, Li M. Air pollution and control action in Beijing. J Clean Prod. 2016;112:1519–27.
32. Zhang J, Mauzerall DL, Zhu T, Liang S, Ezzati M, Remais JV. Environmental health in china: progress towards clean air and safe water. Lancet. 2010;375(9720):1110–9.
33. Zhang Q, He K, Huo H. Policy: cleaning China's air. Nature. 2012;484(7393):161–2.

Chapter 12
Perspective for Future Research Direction About Health Impact of Ambient Air Pollution in China

Guang-Hui Dong

Abstract Air pollution has become one of the major risks to human health because of the progressive increase in the use of vehicles powered by fossil fuels. Although lots of works on the health impact of ambient air pollution have been done in China, the following recommendations for future research were identified in this chapter: (1) the synergistic effect of indoor air pollution with climate change; (2) develop new technologies to improve accurate assessment of air pollution exposure; (3) well-designed cohort study of sensitive populations including children, older people, and people with chronic health problems; (4) multi-omics technologies in the underlying mechanisms study; and (5) benefits evaluation of improvement of air quality. In conclusion, China is becoming a suitable study site, providing an ideal opportunity to evaluate the effects of environmental pollution, including air pollution, on human health, which might serve as an example for developing countries where health impacts of air pollution are as serious as in China.

Keywords Air pollution • Health impact • Recommendations • Future research • China

China is now suffering from serious environmental pollution caused by economic development, which has led to serious adverse health effects across the whole life course—from conception to old age—impairing overall fetal growth, especially lung growth which persists across childhood, increasing the risks of developing new asthma which might not occur in its absence, and affecting the heart and lungs

G.-H. Dong (✉)
Guangzhou Key Laboratory of Environmental Pollution and Health Risk Assessment, Department of Preventive Medicine, School of Public Health, Sun Yat-sen University, 74 Zhongshan 2nd Road, Yuexiu District, Guangzhou 510080, Guangdong, China
e-mail: donggh5@mail.sysu.edu.cn; donggh512@hotmail.com

G.-H. Dong (ed.), *Ambient Air Pollution and Health Impact in China*, Advances in Experimental Medicine and Biology 1017, DOI 10.1007/978-981-10-5657-4_12

throughout life by direct toxicity and via epigenetic mechanisms that mediate gene/environmental interactions. Beyond respiratory and cardiovascular diseases, air pollution also has adverse impacts on the development of impaired cognition, type 2 diabetes, cancers, and skin aging and even acts as a risk factor for obesity and neurodevelopment. It is urgent time that China should rapidly develop and deploy low-carbon technologies, reduce coal consumption, reform the electricity market, and implement carbon pricing policies [7]. However, presently, all too often policies relating to air pollution are directed toward acute episodes in summer or winter in the absence of a long-term strategy. The clear recognition that the damaging effects of air pollutant toxicants start at conception and extend across the whole of life demands a long-term set of solutions involving industry, central government, and local government.

The case for the devastating effects of air pollution on human health has been further emphasized by the severe pollution presenting in China, where coal and biomass burning contribute to the problem, and at the same time accompanied with a quick-increasing use of motor vehicles [7]. Furthermore, with the increase of aging populations, health risks in China will increase even if pollution levels remain constant, for older people are more susceptible to air pollution than younger people because of health problems such as myocardial infarction and stroke. Even with no changes in air pollution in China, the attributable deaths per capita will increase 20–30% during the next 15 years [9]. Affecting by population growth, the increase in deaths would be even greater even if China experiences no further changes in the levels of air pollution.

As the director general of the World Health Organization (WHO), Dr. Margaret Chan said that the harm caused by air pollution and exposure to hazardous chemicals was a new epidemic, overtaking major infectious diseases like AIDS, tuberculosis, and malaria [2]. So what is to be done? Just as the states reported by the Royal College of Physicians (RCP) and the Royal College of Physicians and Child Health (RCPCH): “everyone has some responsibility for reducing air pollutions; real change will only occur when everyone accepts this responsibility, and makes a concerted effort.” This includes the whole China, national and local governments, business and industry, as well as individuals in society at large. The following recommendations for future research were identified in this chapter:

12.1 The Synergistic Effect of Indoor Air Pollution and Climate Change

What we should pay attention is that ambient air pollution is not an isolated incident, which has been confirmed the synergistic effect with other environmental factors, such as climate change and indoor air pollution [6]. We spend 90% of our time indoors, and yet this is rarely taken into account when evaluating the health burden of ambient air pollution. It is well known that outdoor pollutants penetrate the home, schools, and workplaces; further, current trends to make buildings energy efficient by sealing them increase the accumulation of pollutants from furnishings, household products, and cooking [14]. In China, the burning of biomass for heating and cooking is a particular problem for women and children. Evidence indicated that particulate matter indoors may contain a different range of chemical from that encountered outsides [5, 10], including these from air fresheners [11] and microbial products [5, 8]. The combined exposures to outdoor and indoor pollutants create an even greater health hazard than outdoor pollution alone [3].

Furthermore, the adverse effects of air pollutants are further enhanced by the influence of climate change, with atmospheric conditions increasing accumulation of pollutants and formation of ozone and secondary particles, and it is forecasted that the situation will further worsen with climate change [9]. Many strategies that decrease air pollution are also beneficial in slowing down climate change. China is a vast geographical territory with contrasting climatic characteristics that influence the components and concentrations of air pollutants. Exploration of heterogeneous adverse effects of air pollution would benefit healthcare providers in individual regions. Evaluation of health responses to air pollution in northern versus southern China, and in Han Chinese versus other ethnic groups, would unravel the mechanisms and predictors of differential adverse effects in different geographic regions and ethnic groups [7]. Data examining dose–response relationships at high pollutant levels would be valuable for policy makers to tailor environmental protection regulations in individual regions [7].

12.2 Develop New Technologies to Improve Accurate Assessment of Air Pollution Exposure

Furthermore, in order to provide much more precise measurement of exposure levels and timely forecast the haze affairs, some new technologies, such as new smart monitoring, should be in priority developed to improve ambient air pollution monitoring, which can provide much more accurate and wider-ranging pollution monitoring programs so that we can tract population-level exposure to air pollution. In addition to fixed site monitoring, there have been considerable advances in small mobile devices that can be carried by individuals to monitor personal air pollutant exposure such as that developed by CleanSpace® that detects ambient levels of carbon monoxide or handheld particle monitors such as the Lighthouse Handheld 3016-IAQ monitor. The widespread use of such devices will serve to empower the public regarding their personal air pollution exposures.

12.3 Well-Designed Cohort Study of Sensitive Populations

Compared with the cohort studies conducted in general residents, well-designed studies should also be conducted in some sensitive populations. Children, older people, and people with chronic health problems are among the most vulnerable to air pollution. Public services must take account of this disproportionate harm through local tools, such as planning policies for housing and schools, equalities impact assessments, and joint strategic needs assessments. At an individual level, healthcare professionals should help vulnerable patients protect themselves from the worst effects of air pollution. The physician and other healthcare professionals have an important role to play in explaining air pollution and how to deal with it to patients.

12.4 Multi-omics Technologies in the Underlying Mechanisms Study

Much remains unclear regarding which pollutants account for most of the health effects and specific signaling pathways responsible for eliciting exacerbation of adverse effects. To fully appreciate the risks of air pollution to health, further research is urgently required on how pollutant mixtures impact on the human health. Beyond lung and cardiovascular diseases, research should accommodate systemic effects such as obesity, diabetes, and changes linked to dementia and cancer, as well as effects on the developing fetus and early childhood. The advent of multi-omics technologies including genomics, transcriptomics, proteomics, metabolomics, and microbiomics to analyze the complex multilevel interactions of pollution with humans is creating an exciting new science of exposomics in which exposures can be directly linked to disease causal pathways.

12.5 Benefits Evaluation of Improvement of Air Quality

The Olympic Blue and APEC Blue have ever confirmed the benefits of some drastic interventions [12, 13, 15, 16]. However, few studies have examined how improved air quality reduces exacerbation of chronic diseases on a long-term basis in China [7]. Reassuringly, implementation of air quality-control policies has led to reduced nitrogen dioxide and fine particulate mass ($PM_{2.5}$) levels and improved lung function development in boys and girls with and without asthma [4]. A longitudinal study involving 4,602 children (age range, 5–18 years) from 3 cohorts conducted during the 1993–2001, 1996–2004, and 2003–2012 years in 8 Southern California communities reported that decreases in ambient pollution levels were associated with statistically significant decreases in bronchitic symptoms in children [1]. Although the study design does not establish causality, the findings support potential benefit of air pollution reduction on asthma control. So, more study about the benefits assessment of long-term drastic interventions is needed.

In conclusion, China is becoming a suitable study site, providing an ideal opportunity to evaluate the effects of environmental pollution including air pollution on human health. Air pollution contributes to the pathogenesis and exacerbation of chronic diseases. Drastic measures—including industrial upgrading, quality improvement of vehicles, petrol, cooking fuel, and stoves, environmental health policy implementation, and surveillance—will translate into improved outcomes for patients with chronic diseases. The ambition and capacity of successful management of air pollution in China might serve as an example for developing countries where health impacts of air pollution are as serious as in China.

Acknowledgments This study was funded by Grants from the National Natural Science Foundation of China (No.81472936, 81673127, and 81673128), the National Key Research and Development Program of China (2016YFC0207000), the Guangdong Province Natural Science Foundation (2014A050503027), and the Fundamental Research Funds for the Central Universities (No. 16ykzd02, 17ykpy16).

References

1. Berhane K, Chang CC, McConnell R, Gauderman WJ, Avol E, Rapapport E, Urman R, Lurmann F, Gilliland F. Association of changes in air quality with bronchitic symptoms in children in California, 1993–2012. JAMA. 2016;315(14):1491–501.
2. CCAC Secretariat. WHO director general and UNEP executive director: "healthy environment important for human health". Paris: Climate and Clean Air Coalition; 2016. www.ccacoalition.org/en/news/who-director-general-and-unep-executive-director-healthyenvironment-important-human-health/ [Accessed 11 June 2017].
3. Farmer SA, Nelin TD, Falvo MJ, Wold LE. Ambient and household air pollution: complex triggers of disease. Am J Physiol Heart Circ Physiol. 2014;307:H467–76.
4. Gauderman WJ, Urman R, Avol E, Berhane K, McConnell R, Rappaport E, Chang R, Lurmann F, Gilliland F. Association of improved air quality with lung development in children. N Engl J Med. 2015;372(10):905–13.
5. Ghosh B, Lal H, Srivastava A. Review of bioaerosols in indoor environment with special reference to sampling, analysis and control mechanisms. Environ Int. 2015;85:254–72.
6. Gordon SB, Bruce NG, Grigg J, Hibberd PL, Kurmi OP, Lam KB, et al. Respiratory risks from household air pollution in low and middle income countries. Lancet Respir Med. 2014;2:823–60.
7. Guan W, Zheng X, Chung KF, Zhong NS. Impact of air pollution on the burden of chronic respiratory diseases in China: time for urgent action. Lancet. 2016;388:1939–51.
8. Happo MS, Sippula O, Jalava PI. Role of microbial and chemical composition in toxicological properties of indoor and outdoor air particulate matter. Part Fibre Toxicol. 2014;11:60.
9. Holgate ST. 'Every breath we take: the lifelong impact of air pollution' – a call for action. Clin Med (Lond). 2017;17(1):8–12.
10. Kabir E, Kim KH. An investigation on hazardous and odorous pollutant emission during cooking activities. J Hazard Mater. 2011;188:443–54.
11. Kim S, Hong SH, Bong CK, Cho MH. Characterization of air freshener emission: the potential health effects. J Toxicol Sci. 2015;40:535–50.

12. Lin H, Liu T, Fang F, Xiao J, Zeng W, Li X, Guo L, Tian L, Schootman M, Stamatakis KA, Qian Z, Ma W. Mortality benefits of vigorous air quality improvement interventions during the periods of APEC blue and parade blue in Beijing, China. Environ Pollut. 2017;220(Pt A):222–7.
13. Liu H, Liu C, Xie Z, Li Y, Huang X, Wang S, Xu J, Xie P. A paradox for air pollution controlling in China revealed by "APEC blue" and "parade blue". Sci Rep. 2016;6:34408.
14. Mortimer K, Ndamala CB, Naunje AW, Malava J, Katundu C, Weston W, Havens D, Pope D, Bruce NG, Nyirenda M, Wang D, Crampin A, Grigg J, Balmes J, Gordon SB. A cleaner burning biomass-fuelledcookstove intervention to prevent pneumonia in children under 5 years old in rural Malawi (the cooking and pneumonia study): a cluster randomised controlled trial. Lancet. 2017;389(10065):167–75.
15. Schleicher N, Norra S, Chen Y, Chai F, Wang S. Efficiency of mitigation measures to reduce particulate air pollution–a case study during the Olympic Summer Games 2008 in Beijing, China. Sci Total Environ. 2012;427–428:146–58.
16. Wang G, Cheng S, Wei W, Yang X, Wang X, Jia J, Lang J, Lv Z. Characteristics and emission-reduction measures evaluation of $PM_{2.5}$ during the two major events: APEC and parade. Sci Total Environ. 2017;595:81–92.

Index

G.-H. Dong (ed.), *Ambient Air Pollution and Health Impact in China*, Advances in Experimental Medicine and Biology 1017, DOI 10.1007/978-981-10-5657-4

MIX
Papier aus verantwortungsvollen Quellen
Paper from responsible sources
FSC® C105338

If you have any concerns about our products,
you can contact us on
ProductSafety@springernature.com

In case Publisher is established outside the EU,
the EU authorized representative is:
Springer Nature Customer Service Center GmbH
Europaplatz 3, 69115 Heidelberg, Germany

Printed by Libri Plureos GmbH
in Hamburg, Germany